17th Russian Youth Conference on Physics of Semiconductors and Nanostructures, Opto- and Nanoelectronics (RYCPS 2015)

Journal of Physics: Conference Series Volume 690

St. Petersburg, Russia
23 – 27 November 2015

Editors:

R.A. Suris
L.E. Vorobjev
D.A. Firsov
V.A. Shalygin

ISBN: 978-1-5108-2235-1
ISSN: 1742-6588

Printed from e-media with permission by:

Curran Associates, Inc.
57 Morehouse Lane
Red Hook, NY 12571

Some format issues inherent in the e-media version may also appear in this print version.

Printed by Curran Associates, Inc. (2016)

Institute of Physics
Dirac House, Temple Back
Bristol BS1 6BE UK

Phone: 44 1 17 929 7481
Fax: 44 1 17 920 0979

techtracking@iop.org

Additional copies of this publication are available from:

Curran Associates, Inc.
57 Morehouse Lane
Red Hook, NY 12571 USA
Phone: 845-758-0400
Fax: 845-758-2633
Email: curran@proceedings.com
Web: www.proceedings.com

Table of contents

Volume 690

17th Russian Youth Conference on Physics of Semiconductors and Nanostructures, Opto- and Nanoelectronics (RYCPS 2015)
23–27 November 2015, St. Petersburg, Russia

Accepted papers received: 1 February 2016
Published online: 26 February 2016
Preface

Papers

Bulk properties of semiconductors

Structure growth, surface and interfaces

Heterostructures, superlattices and quantum wells

Quantum dots, quantum wires and other low-dimensional systems

Novel materials

Journal of Physics: Conference Series **690** (2016) 011001 doi:10.1088/1742-6596/690/1/011001

Editorial Preface

The 17th Russian Young Conference on Physics of Semiconductors and Nanostructures, Opto- and Nanoelectronics was held on November 23 – 27 at Peter the Great St. Petersburg Polytechnic University.

The program of the Conference included semiconductor technology, heterostructures with quantum wells and quantum dots, opto- and nanoelectronic devices and new materials. A large number of participants with more than 200 attendees from many regions of Russia provided a perfect platform for the valuable discussions between students and experienced scientists.

Conference included two invited talks given by Doctor of Science S.A. Tarasenko ("Two-dimensional and three-dimensional topological insulators") and Doctor of Science M.V. Maximov ("Semiconductor lasers with low beam divergence"). Students, graduate and postgraduate students presented their results on plenary and poster sessions. The total number of accepted papers published in Russian (the official conference language) was 135. Here we publish 43 of them in English.

Like previous years, the participants were involved in the competition for the best report. Certificates and cash prizes were awarded to a number of participants for the presentations selected by the Program Committee. One special E.F. Gross Prize was given for the best presentation in semiconductor optics. Works with potential applications were recommended for participation in the following competition for support from the Russian Foundation for Assistance to Small Innovative Enterprises in Science and Technology.

The conference was supported by the Russian Foundation for Basic Research and the innovation company "ATC – Semiconductor Devices", St. Petersburg.

The official conference website is http://www.semicond.spbstu.ru/conf2015-eng.html

Editors

R.A. Suris
L.E. Vorobjev
D.A. Firsov
V.A. Shalygin

Journal of Physics: Conference Series **690** (2016) 011001 doi:10.1088/1742-6596/690/1/011001

Winner of the E.F. Gross Prize for the best presentation in semiconductor optics

Inna Kalituho

a student at Saint Petersburg Electrotechnical University 'LETI'

Presentation:

"*The proximity effect in ferromagnet-semiconductor hybrids*"

Winners of the competition for the best report

First Prize

Konstantin Denisov

a postgraduate student at Ioffe Institute
(Saint Petersburg)

Presentation:

"*Abnormal Hall effect in the carrier scattering by magnetic skyrmion*"

First Prize

Mikhail Fadeev

a student at Lobachevsky State University of Nizhniy Novgorod

Presentation:

"*The long-wavelength photoluminescence and stimulated emission in structures based on solid solutions HgCdTe*"

Peer review statement

All papers published in this volume of *Journal of Physics: Conference Series* have been peer reviewed through processes administered by the proceedings Editors. Reviews were conducted by expert referees to the professional and scientific standards expected of a proceedings journal published by IOP Publishing.

Estimation of doping density in low doped n-InAs by electrolyte-based capacitance-voltage measurements in the deep depletion mode

D Frolov and V Zubkov

Department of micro- and nanoelectronics, St. Petersburg Electrotechnical University "LETI", Prof. Popov str. 5, St. Petersburg 197376, Russia

E-mail: dsfrolov@etu.ru

Abstract. Doping density profiling in InAs is difficult due to lack of good quality Schottky contacts and the narrow bandgap nature of the material. The electrolyte can be used to form a Schottky-like contact but the inversion layer formation does not allow obtaining a dopant concentration in low doped InAs. To overcome this issue, a pulse CV technique has been implemented to drive a sample into the deep depletion mode which makes it possible to use the depletion approximation to calculate a carrier concentration. The measured capacitance-voltage characteristics were compared with the simulation and showed good agreement with it.

1. Introduction

The capacitance-voltage profiling technique requires a junction in which a space-charge region can be created [1]. Schottky contacts are traditionally used for this purpose, but due to formation of an accumulation layer at the surface, the junction between metal and n-InAs has a negative barrier height [2, 3]. The Schottky contact in InAs can only be formed in undoped or p-type structures [4]. As an alternative to the Schottky contact some other structures, like a MOS capacitor [5], a p-n barrier junction [6] or an electrolyte-semiconductor interface [7], can be used to make a space-charge region in InAs.

Choosing the electrolyte to form a Schottky-like junction has some advantages over other methods since it does not require complicated technological processes to create contacts and can be used together with the electrochemical etching of InAs. These advantages of electrolyte are implemented in the electrochemical capacitance-voltage technique (ECV) [8].

Although ECV is used for characterization of vast number of semiconductors, its application to InAs faced some difficulties [7, 9]. InAs is a narrow band gap material and its conduction band edge is located below the electrochemical potential of the electrolyte. Because of that, the accumulation layer is formed at InAs surface in thermal equilibrium, whereas in most of other materials the depletion occurs. That makes the CV behavior in InAs similar to one in metal oxide semiconductor (MOS) structures, with a Helmholtz layer at the electrolyte-semiconductor interface acting as an isolator. Additionally, in the standard CV method, the capacitance is measured with a small voltage step and a long integration time comparing to minority carriers response time. This slow variation of bias voltage leads to the formation of an inversion layer in InAs that makes the doping concentration extraction very difficult.

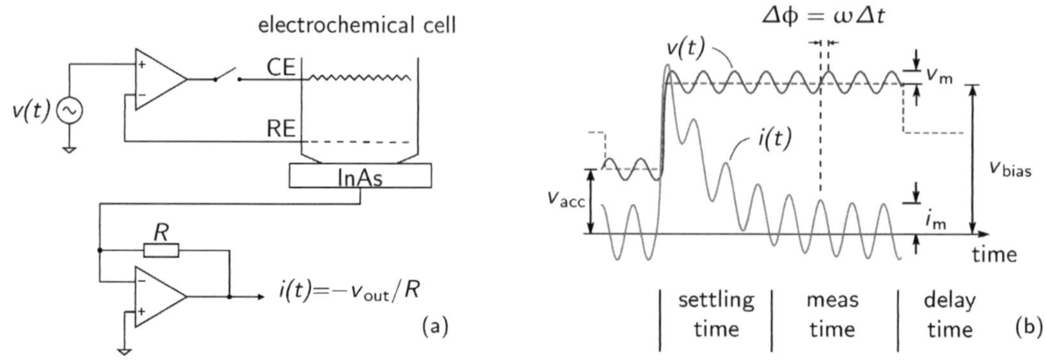

Figure 1. (a) Simplified circuit diagram of experimental setup for pulse CV measurements and (b) schematic representation of applied voltage and measured current.

Similarly to MOS structures, when a voltage pulse of a duration shorter than the thermal generation time constant is applied to the electrolyte-semiconductor interface, no inversion layer can be formed. This fact is used in certain CV techniques to drive a sample into the deep depletion mode. The deep depletion can be accomplished by a linear-sweep technique [10], a pulse CV technique [11] or by applying the fast ramp voltage [12]. In present work, a pulse CV technique has been used since it allows more precise control of the timing between measurement cycles.

2. Measurement technique

To implement pulse CV measurements the following instrumentation was assembled (figure 1(a)). The differential capacitance was measured using a standard technique by superposing a small ac voltage (of small amplitude) on the bias voltage and measuring the ac current through the sample [13]. An arbitrary waveform generator (PXI-5412) was used to generate both dc bias and ac test signal. The generator was connected to the electrochemical cell through an operational amplifier in the potentiostat configuration to control the voltage difference between the sample and the reference electrode (RE).

Because of the small energy bandgap of InAs, there is only a very small bias range to make capacitance-voltage measurements. Leaving the electrochemical cell connected to the circuit one can get uncontrolled formation of an anodic layer. To prevent the surface of the sample from oxidation during delay time between voltage steps, the output of an amplifier was connected to the counter electrode (CE) through the reed relay. The relay was controlled by PXI-4132 SMU (a source-measure unit) which was also used to generate the trigger signal to synchronize the start of the waveform generator with relay actuation.

The second operational amplifier in transimpedance configuration was used to make the direct conversion of the current into a voltage signal. The reference voltage and the voltage signal at the output of transimpedance amplifier were captured by a digital oscilloscope. The schematic representation of applied voltages and the measured current shown in figure 1(b).

To obtain the capacitance, the real and imaginary parts of the impedance have been calculated first, see equation (1).

$$Z = \frac{v_m}{i_m} e^{j\Delta\phi} \tag{1}$$

The impedance magnitude was calculated from the amplitude ratio of the voltage applied to the cell (v_m) and the displacement current (i_m) through it. The phase angle of complex impedance

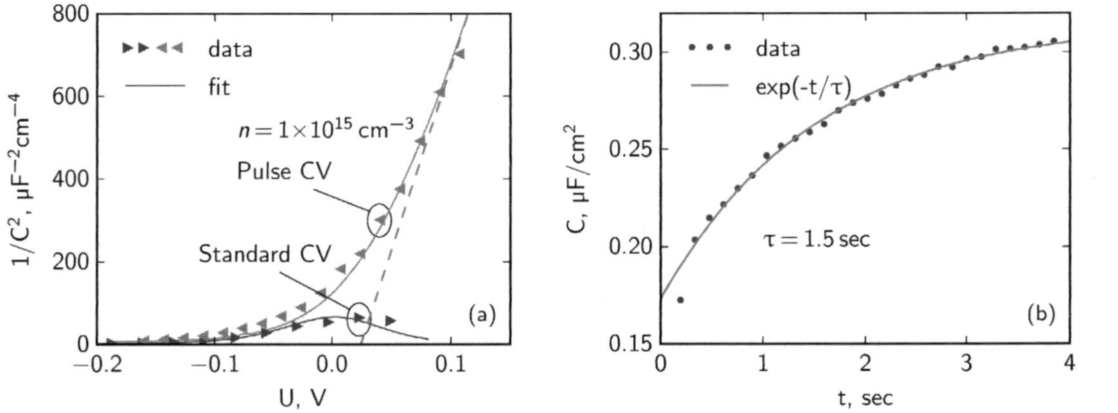

Figure 2. (a) Mott-Schottky plots for InAs (dots – experimental data, solid line – simulation, dashed line – linear approximation in the deep depletion mode) and (b) typical transient capacitance after bias step from accumulation to depletion.

was determined from the phase difference $\Delta\phi$ between ac test signal and ac current.

A parallel equivalent circuit was used for capacitance calculation due to the fact that InAs has a large surface conductivity which acts as a competitive path of current. The admittance of the parallel equivalent circuit may be written as

$$Y = 1/Z = G + j\omega C_p \tag{2}$$

where G is conductance and C_p is capacitance.

The described experimental setup was successfully tested with various combinations of resistors and capacitors connected in place of electrochemical cell at a range of frequencies up to 1 MHz.

3. Results and discussion

The epitaxial InAs structures were grown by HVPE on n^+-InAs substrates. Doping densities were studied in epi layer and in substrates. The pulse capacitance-voltage measurements were performed with the experimental setup described above. For standard measurements a LCR meter Agilent E4980A have been connected directly to the electrochemical cell. The aqueous solution of 0.1 M NaOH was used as an electrolyte. Contact area between the electrolyte and the sample was 0.1 cm^2. All CV measurements were performed at room temperature.

The model used for capacitance-voltage simulation was described in details in our previous work [9]. CV characteristics were calculated from the numerical solution of the nonlinear Poisson equation with modified Tomas-Fermi approximation. Deep depletion capacitance was calculated by setting the minority carrier concentration to zero. CV plots were calculated for doping concentration $N_D = 1 \times 10^{15}\,\mathrm{cm}^{-3}$.

The capacitance-voltage characteristics for InAs measured in standard and pulse modes are compared in figure 2(a). Both characteristics have been measured with the frequency of 300 Hz and amplitude (v_m) of 30 mV. The frequency was selected to minimize the dissipation factor, but to be high enough to have at least 10 cycles during the measurement time. In pulse CV at every measurement point the dc voltage was pulsed from an accumulation voltage (v_{acc}) to the voltage at which the data are to be taken (v_{bias}), as shown in figure 1(b). The settling time before the start of main measurements was about 10 msec. The delay time between each measuring cycle was ranged from 100 msec to 1 sec.

In the standard CV mode the inversion starts at positive biases, which results in an increase of capacitance and a drop on $C^{-2} - V$ curve. In contrast, the good linearity of Mott-Schottky plot can be seen in pulse CV at voltages above +0.05 V. Using the Mott-Schottky equation in this region we calculated the corresponding free carrier concentration to be of $1 \times 10^{15}\,\mathrm{cm}^{-3}$ that is in good agreement with the expected doping concentration. It should be noted that depending on the delay time between each measurement, the bias range, in which $C^{-2} - V$ follows the linear behavior is changing. It can be related to the growth and dissolution of anodic layer at the surface of InAs.

Figure 2(b) shows the transient capacitance measured with the bias step from accumulation to depletion. The experimental data were fitted by an exponential function. The time constant was calculated to be 1.5 sec which is much longer than the selected settling time.

4. Conclusions

At lower doping levels in n-InAs the standard (equilibrium) CV does not provide the correct value of dopant concentration. This happens for the reason that the inversion starts too early due to a small band gap of indium arsenide. To prevent the inversion layer formation, a pulse CV technique has been used. Here the fast bias pulse drives a sample into the deep depletion mode. The concentration obtained from the linear approximation of $C^{-2} - V$ in the deep depletion mode is in good agreement with the known doping concentration in low doped InAs. The proposed technique can be used for characterization of InAs with doping levels from $1 \times 10^{15}\,\mathrm{cm}^{-3}$ to $1 \times 10^{20}\,\mathrm{cm}^{-3}$. At doping levels below $1 \times 10^{15}\,\mathrm{cm}^{-3}$ the intrinsic carrier concentration will dominate at room temperature. At very high doping levels the parasitic conductivity through the surface will decrease the accuracy of capacitance measurement results.

References

[1] McGuire G E (ed) 1989 *Characterization of semiconductor materials: principles and methods* (*Materials science and process technology series* vol 1) (N.J.: Noyes Publications)
[2] Allen R E, Humphreys T J, Dow J D and Sankey O F 1984 *J. Vac. Sci. Technol. B* **2** 449–452
[3] Olsson L Ö, Andersson C B M, Håkansson M C, Kanski J, Ilver L and Karlsson U O 1996 *Phys. Rev. Lett.* **76** 3626–3629
[4] Walpole J N and Nill K W 1971 *J. Appl. Phys.* **42** 5609–5617
[5] Astromskas G, Storm K, Karlström O, Caroff P, Borgström M and Wernersson L E 2010 *J. Appl. Phys.* **108** 054306–5
[6] Hood A, Hoffman D, Wei Y, Fuchs F and Razeghi M 2006 *Appl. Phys. Lett.* **88** 052112
[7] Gopal V, Chen E H, Kvam E P and Woodall J M 2000 *J. Electron. Mater.* **29** 1333–1339
[8] Blood P 1986 *Semicond. Sci. Technol.* **1** 7–27
[9] Frolov D, Yakovlev G and Zubkov V 2015 *J. Phys. Conf. Ser.* **643** 012086
[10] Pierret R F 1972 *Electron Devices, IEEE Trans.* **19** 869–873
[11] Nicollian E H and Goetzberger A 1967 *Bell Syst. Tech. J.* **46** 1055–1133
[12] Nicollian E H and Brews J R 2002 *MOS (Metal Oxide Semiconductor) Physics and Technology* (NJ: Wiley)
[13] Meyer N I and Guldbrandsen T 1963 *Proc. IEEE* **51** 1631–1637

Structural, electrical and magnetic study of manganites $Pr_{0.6}Sr_{0.4}MnO_3$ thin films

D S Neznakhin[1], Yu E Samoshkina[2], M S Molokeev[2] and S V Semenov[2]

[1]Ural Federal University, Ekaterinburg 620000, Russia
[2]L.V. Kirensky Institute of Physics SB RAS, 660036 Krasnoyarsk, Russia

E-mail: d.s.neznakhin@urfu.ru

Abstract. Thin polycrystalline $Pr_{0.6}Sr_{0.4}MnO_3$ films were grown on the Y stabilized zirconium oxide substrates by magnetron sputtering using RF power and off-axis sputtering scheme with double cathodes. Only one polycrystalline phase with structural parameters consistent with that for the corresponding bulk sample was revealed in the films. Electric resistivity dependence on temperature demonstrates the shape characteristic for the substances with the Mott transition. The difference between magnetization temperature curves measured in the zero field cooling and field cooling modes was revealed. Magnetization field dependences were presented by the hysteresis loops changing their form with temperature.

I. Introduction

Substituted manganites with the perovskite structure and general formulae $R_{1-x}A_xMnO_3$ (where R is a trivalent lanthanide and A is a divalent alkaline earth metal) has attracted great attention in the past few decades [1-6]. The parent compounds $RMnO_3$ are known to be the Mott insulator with the antiferromagnetic ground state (AFM). At the partial replacement of the lanthanide ions with the divalent alkaline earth, a great variety of magnetic and electric phases appears in dependence on the divalent ion type and its concentration. In particular, for the region of x meanings from ~0.2 to ~0.5, the metallic ferromagnetic phase is usually realized.

Historically, the $La_{1-x}Sr_xMnO_3$ compounds were the main objects of investigation among manganites doped with Sr. As concerns Pr-Sr manganites, only a few works devoted to the magnetic and transport properties of the single- and polycrystalline bulk samples are available in literature [7-13]. Recently, we have investigated, for the first time, the visible magnetic circular dichroism (MCD) in $Pr_{1-x}Sr_xMnO_3$ thin films and have revealed remarkably different temperature dependences of the MCD maxima intensity at different wave lengths in the case of $Pr_{0.6}Sr_{0.4}MnO_3$ [14]. In this connection, a detailed investigation of electric and magnetic properties for the same samples seems to be of primary importance.

The aim of the present paper is a study of the temperature and field dependences of the electric resistivity and magnetization of the $Pr_{0.6}Sr_{0.4}MnO_3$ polycrystalline films in wide intervals of temperature and external magnetic field.

2. Experimental setup and samples

Films of different thicknesses (20, 30, 50, 80, and 130 nm) were prepared by the dc magnetron sputtering technique analogously to that presented in [15]. The single-phased $Pr_{0.6}Sr_{0.4}MnO_3$ targets were obtained by solid-state synthesis using stoichiometric Pr_2O_3, SrO, and MnO_2 powders as the starting materials. The residual gas pressure in the vacuum chamber was 3×10^{-6} Torr. The operating total pressure of the Ar and O_2 gas mixture (with a ratio of 4:1) was 3×10^{-3} Torr. The films were grown onto yttrium stabilized zirconium (YSZ) oxide (311) substrates which temperature during the sputtering was 750°C.

The crystal structure and phase purity of the samples were examined by the powder X-ray diffraction (XRD) with CuKα radiation. The electric conductivity dependence on temperature and external magnetic field was investigated using the standard four-point probe technique in the temperature interval 78–300 K. Magnetic measurements were carried out with a Quantum Design MPMS-XL7 EC superconducting-quantum interference device (SQUID) magnetometer in the temperature range of 5–300 K and in magnetic fields up to 20 kOe applied parallel and perpendicular to the films plane.

3. Results and discussion

XRD pattern for the synthesized films shows large peaks corresponding to the YSZ substrate and series of relatively narrow peaks with small intensities corresponding to only one polycrystalline phase $Pr_{0.6}Sr_{0.4}MnO_3$ without pronounced texture (figure 1). The crystal structure was refined within the orthorhombic Pnma space group with the lattice parameters and the average crystallites size presented in Table 1. Structural parameters of the studied films are consistent with the structural data for the bulk polycrystalline $Pr_{0.6}Sr_{0.4}MnO_3$ samples presented, for example, in Ref. [16] and also shown in Table 1.

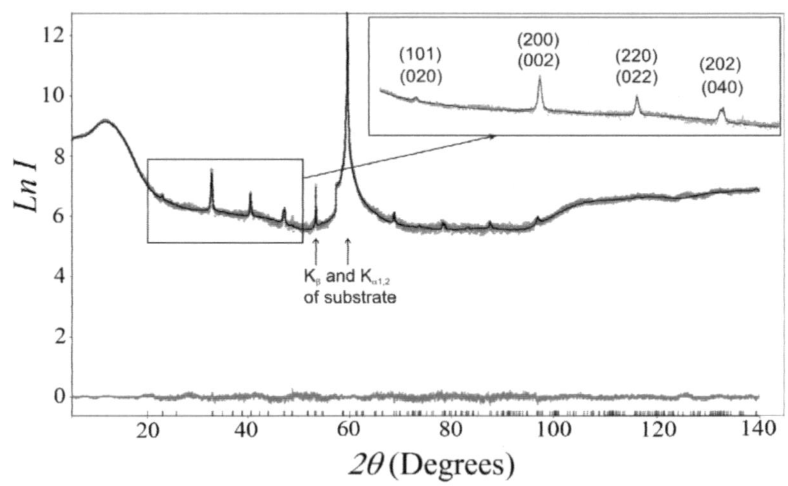

Figure 1. Room temperature XRD pattern for the $Pr_{0.6}Sr_{0.4}MnO_3$ film of 130 nm in thickness. The reflections from the substrate are indicated by arrows.

The temperature dependences of electric resistivity, *R(T)*, measured in the temperature interval 78–300 K both without magnetic field and under the magnetic field of 15 kOe are shown in figure 2 for the thickest film (130 nm). For the films of other thicknesses the pictures are similar. The *R(T)* curves show a single semiconductor-metal (called usually as insulator-metallic) transition at T_R, which is associated, in the case of manganites, with the competition between the double exchange and the super exchange interactions [16]. The resistivity value is seen to decrease in magnetic field, and T_R shifts to higher temperatures. Just the same *R(T)* and $T_R(H)$ behavior was observed in Ref. [16] for the polycrystalline bulk $Pr_{0.6}Sr_{0.4}MnO_3$ samples with the only difference: the T_R value in our case is distinctly lower. The magnetic field effect on the *R* and T_R behavior was explained in Ref. [17] by the fact that the applied magnetic field delocalizes the charge careers suppressing the resistivity causing the local ordering of the electron spins. Due to the ordering, ferromagnetic metallic state may suppress the higher temperature paramagnetic state resulting in an increase of T_R value under the influence of magnetic field. Usually, T_R coincides with T_C – temperature of the magnetic ordering, while in our case and also in Ref. [16] T_R is essentially lower as compared to T_C. It is possible to assume that the contribution of the inter-grain spin-polarized tunneling across the grain boundaries causes the T_R value lowering. The thin film geometry may affect the *R* value and temperature behavior too. This question needs further study.

Table 1. Structural parameters of the $Pr_{0.6}Sr_{0.4}MnO_3$ film of 130 nm in thickness.

Parameters	$Pr_{0.6}Sr_{0.4}MnO_3$ Our data	$Pr_{0.6}Sr_{0.4}MnO_3$ Ref. [16]
a (Å)	5.4465(7)	5.4805
b (Å)	7.721(1)	7.6726
c (Å)	5.422(1)	5.4401
V (Å^3)	228.00(6)	228.75
Average crystallites size (nm)	45 (7) along *b* axis 62 (3) along *a* and *c* axes	10.95

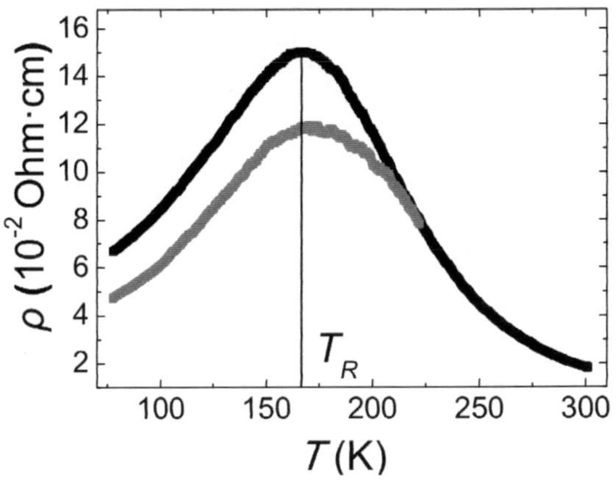

Figure 2. Temperature dependences of the resistivity in zero (black curve) and 15 kOe (red curve) magnetic field for the $Pr_{0.6}Sr_{0.4}MnO_3$ film of 130 nm in thickness.

The magnetization temperature dependences, *M(T)*, are shown in figure3 for the film with a thickness of 80 nm. The dependences were obtained for field cooling (FC) and zero field cooling (ZFC) modes at different values and orientations of the external magnetic field. In the FC procedure, the magnetization magnitude was measured at a cooling and heating of samples at the same magnetic field.

Consider first the FC process: for all *H* values and orientations, smooth curves are observed with an increase in the magnetization with lowering temperature. Just below T_C, the increase is not as sharp as in the case of bulk polycrystalline samples [9, 10]. Besides, the features at 100K and the temperature hysteresis observed in [10] are not revealed here. Note that such features were also not observed for stoichiometric $Pr_{0.6}Sr_{0.4}MnO_3$ polycrystals by some other authors [13]. The T_C value is approximately equal to 250 K, that is notably lower than the T_C value for the bulk polycrystalline samples (310 K) [10].

The FC and ZFC curves diverge from each other at some temperature T_{irr}, and then the ZFC curve runs below the FC curve demonstrating a broad maximum at a certain temperature T_m. At that, the ZFC magnetization diminishes with lowering temperature, sometimes, almost to zero. Such a behavior is typical for heterogeneous magnetic samples such as ensembles of superparamagnetic particles, or spin glasses, or polycrystalline samples with the random distribution of the easy magnetization axes of the crystallites. The temperatures T_{irr} and T_m as well as the ZFC maximum width depend on the magnitude and orientation of the external magnetic field. As it was mentioned above, the films investigated consist of randomly oriented nanocrystals. At the cooling of the films in a magnetic field, the magnetization direction in each crystallite is determined by the competition between the crystallographic anisotropy energy and the Zeeman energy.

For sufficiently large values of the external magnetic field, the crystallite magnetic moments arrange along the field and the FC curve follows the spontaneous magnetization of the material. At the cooling in the absence of the magnetic field, the magnetic moments align along the easy axis of each crystallite, and the sample does not have a total magnetic moment, or the moment is less as compared to the FC mode. Upon applying the field in the heating process, the magnetic moments of the crystallites start to adapt to the field direction what leads to an increase in the total magnetization of the films. However, the spontaneous magnetization of the material begins to decrease with the temperature increase. Thus, competition between these two processes leads to the appearance of a maximum in the ZFC curve.

Though such a picture explains qualitatively the ZFC magnetization temperature behavior, other mechanisms could be attracted to understand this phenomenon, in particular, realization of the spin-glass-like state in ZFC process. In this case, T_m corresponds to the temperature of the magnetic spins freezing, T_f. According to the spin glass theory, the T_f dependence on the value of the magnetic field applied during measurements should be described by a power law $T_f = a + bH^n$. For the Ising spin system the exponent $n = 2/3$. The analysis of the T_m dependence on the external magnetic field has shown the regularity (1) implementation in our case with $n = 2/3$. Nevertheless, further investigation is necessary to find an adequate mechanism describing the ZFC magnetization temperature dependence.

The typical magnetization curves *M(H)* obtained for two co-perpendicular orientations of an external magnetic field relative to the films plane at $T = 5$ K are shown in figure 4 (left). These are the symmetric hysteresis loops with the close values of the coercive fields H_C for the both orientations but with different values of the magnetic saturation field H_s and of different shape. At the perpendicular orientation of the magnetic field, the H_s value is noticeably higher.

The sharp H_C decrease and strong change of the loops shape with the temperature decrease from 90 to 5 K are the remarkable features of the films investigated. At $T = 90$ K, hysteresis, practically, absents; the magnetization curve consists, approximately, of two linear parts with the slope changing at H value close to H_C observed at 5 K; magnetic saturation is not reached even in the highest

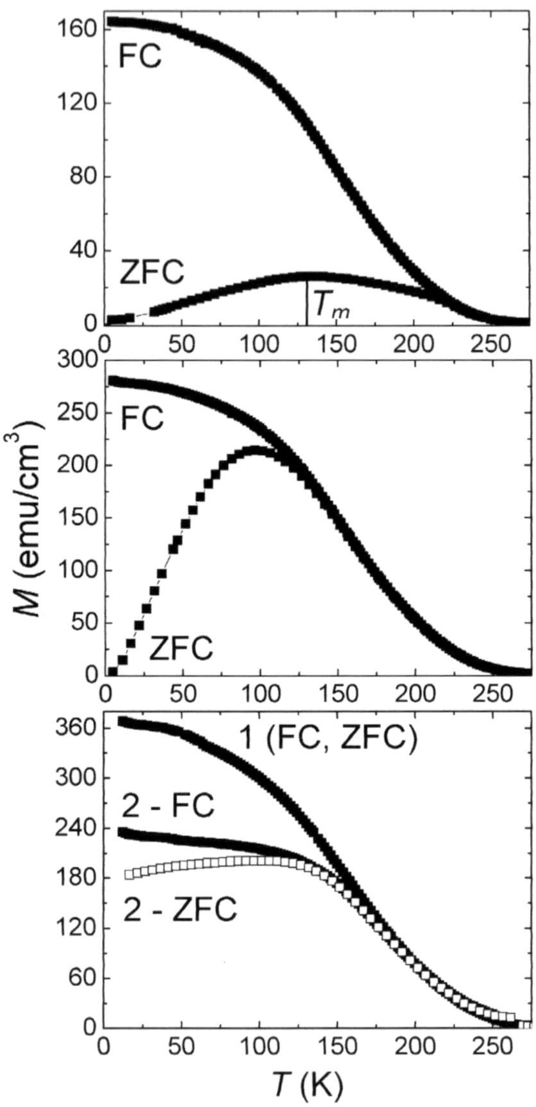

Figure 3. FC and ZFC magnetization temperature dependences for the $Pr_{0.6}Sr_{0.4}MnO_3$ film of 80 nm in thickness, recorded at $H = 100$ Oe directed in the film plane (top), $H = 500$ Oe directed in the film plane (in the middle), and $H = 3$ kOe directed in the film plane (curve 1) and perpendicular to the film plane (curves 2) – bottom.

magnetic field used being of 20 kOe (figure 4 (right)). The absence of magnetic saturation suggests a possible contribution of the Pr spins to the magnetic properties of the samples.

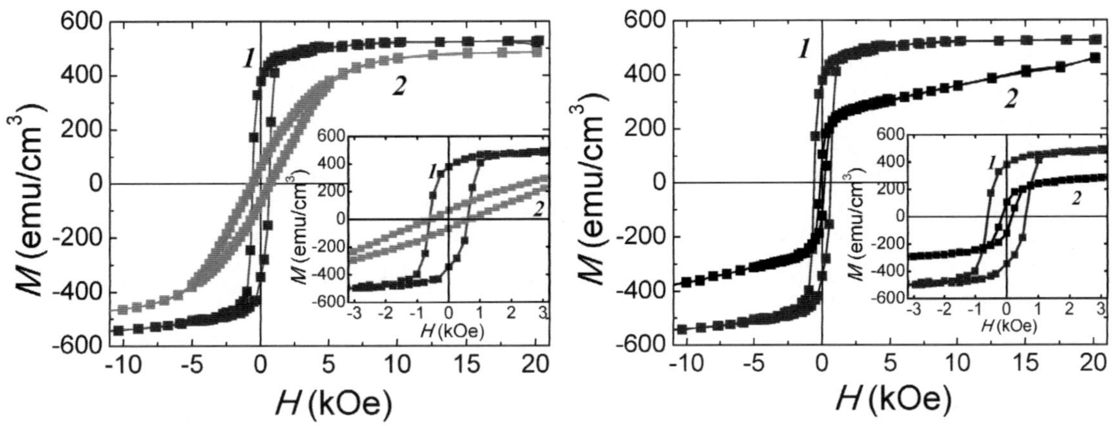

Figure 4. Hysteresis loops of the $Pr_{0.6}Sr_{0.4}MnO_3$ film with thickness of 80 nm: left – T = 5 K, with field H parallel (curve 1) and perpendicular (curve 2) to the sample plane; right – H parallel to the sample plane, T = 5 K (curve 1) and T = 90 K (curve 2). The inserts show the same dependences in the region of $H = \pm 3$ kOe.

The change of the magnetization curves shape with the temperature, in particular, the sharp H_C decrease when the temperature increases and the hysteresis loop appearance in the perpendicular field can be explained by the effect of the crystallographic anisotropy of the samples. It is known that the manganites crystallographic anisotropy increases approximately by an order of magnitude when the temperature decreases from T_C to liquid helium temperature [18]. The absence of magnetic saturation at temperatures exceeding ~25 K can be associated with the Pr ions effect on the manganite magnetic properties. Indeed, several authors considered a possible coupling between Pr *4f* and Mn *3d* spins when studying magnetic and transport properties of the praseodymium manganite doped with Pb or Ca [19, 20]. In particular, they found that magnetic ordering in similar compounds was mediated through the indirect exchange between *4f* and *5d* orbitals which, in turn, was in the direct exchange couplings with the *3d* Mn orbitals.

4. Conclusions

We have investigated the structure, resistivity and magnetic properties of $Pr_{0.6}Sr_{0.4}MnO_3$ polycrystalline films in detail. The resistivity temperature dependence is characteristic for the substances possessing the Mott metal-insulator transition. The magnetization temperature behavior is explained by two mechanisms: an effect of the crystallographic anisotropy and the presence of the spin glass state in the samples arising in the ZFC process. The character of the magnetization field dependences changes essentially with the temperature increase. The hysteresis, practically, disappears at higher temperatures that is associated with a decrease of the crystallographic anisotropy. The magnetic saturation is not reached even in the highest magnetic field used being of 20 kOe that ascribed to the Pr spins contribution to the total magnetization of the films.

Acknowledgments
The work was supported partly by RFBR, grant №14-02-01211, Grant of President of Russian Federation №NSh-2886.2014.2, and by The Ministry of Education and Science of the Russian Federation, project №2582.

References

[1] Ramirez AP 1997 *J. Phys.: Condens. Matter.* **9** 8171
[2] Haghiri-Gosnet A-M, Renard J-P 2003 *J. Phys. D: Appl. Phys.* **36** R127
[3] Coey J M D, Viret M, von Molnar S 1999 *Adv. in Phys*. **48** 167
[4] Chmaissem, Dabrowski B, Kolesnik S, Mais J, Jorgensen J D, Short S 2003 *Phys. Rev.B* **67** 094431
[5] Volkov N V 2012 *Physics-Uspekhi* **55** 250
[6] Yu-Kuai L, Yue-Wei Y, Xiao-Guang L 2013 *Chin. Phys. B.* **22** 087502
[7] Boujelben W, Cheikh-Rouhou A, Eellouze M, Joubert J C 2000 *J. Phase Transitions* **71** 127
[8] Rama N, Sankaranarayanan V, Rao M S 2006 *J. Appl. Phys*. **99** Q315
[9] Rößler S, Harikrishnan S N, Rößler U K, Kumar C M N, Suja E, Wirth S 2011 *Phys. Rev.B.* **84** 184422
[10] Repaka D V M, Tripathi T S, Aparnadevi M, Mahendiran R 2012 *J. Appl. Phys.* **112** 123915
[11] Thaljaoui R, Pekala K, Pekala M, Boujelben W, Szydlowska, Fagnard J-F, Vanderbemden P, Cheikhrouhou A 2013 *J. Alloys Comp.* **580** 137
[12] Kimura T, Ishihara S, Shintani H, Arima T, Takahashi K T, Ishizaka K, and Tokura Y 2003 *Phys. Rev. B* **68** 060403(R)
[13] Elleuch F, Triki M, Bekri M, Dhahri E, Hlil E K 2015 *J.Alloys Comp.* **620** 249
[14] Edelman I, Greben'kova Yu, Sokolov A, Molokeev M, Aleksandrovskiy A, Chichkov V, Andreev N, Mukovskii Y 2014 *AIP Advances* **4** 057125
[15] Mukovskii Y M, Shmatok A V, 1999*J. Magn. Magn. Matter* **196-197** 136
[16] Elleuch F, Triki M, Becri M, Dhahri E, and Hlil E K 2015 *J. All. Comp.* **620** 249
[17] Vencataiah G, Reddy P V 2005 *J. Magn. Magn.Mater*. **285** 343
[18] Steenbeck K, Hiergeist R, 1999 *Appl. Phys. Lett.* **75** 1778
[19] Dho J, Kim W S, Chi E O, Hur N H, Park, Ri H-C 2003 *Sol. St. Commun.* **125** 143
[20] Padmanabhan B, Suja E, Bhat H L, Rößler S, Dörr K, Müller K H 2006 *J. Magn. Magn.Matter* **307** 288

Photoluminescent tomography of ZnSe bulk crystals

A A Gladilin[1], V P Kalinushkin[1], O V Uvarov[1], E M Gavrischuk[2] and N A Timofeeva[2]

[1]A. M. Prokhorov General Physics Institute of the Russian Academy of Sciences, 38 Vavilov Street, Moscow, 119991, Russia
[2]G. G. Devyatykh Institute of Chemistry of High-Purity Substances of the Russian Academy of Sciences, 49 Troponina Street, Nizhny Novgorod, 603950, Russia

E-mail: agladilin@kapella.gpi.ru

Abstract. High potential of two-photon confocal microscopy for photoluminescent tomography of bulk semiconductor crystals are demonstrated in this work using ZnSe crystals as an example. The first results of investigation of ZnSe bulk crystals, both undoped and doped by 3*d* metals, are presented. Appearance of luminescence quenching centres in specific parts of crystals and grain recombination activity are demonstrated by the proposed method; these phenomena cannot be detected by other luminescence-based methods.

1. Introduction

Investigation of photoluminescence (PL) distribution in bulk semiconductors is of great interest for science and practical applications. Usually, the luminescence is caused by recombination of nonequilibrium carrier which is excited by radiation with quantum energy exceeding the bandgap width of the studied semiconductor. This radiation is absorbed by a thin near-surface layer with the depth of about a few micrometers in which the studied photoluminescence is generated; the obtained results are then attributed to the whole crystal bulk. That approach does not seem to be very reliable and the experimental procedure does not seem to be always correct. It has been shown previously [1–3] that results obtained using surface and bulk photoexcitations are often different. This especially relates to such problems as exploration of inhomogeneous materials produced by thermal diffusion of a dopant from the crystal surface. Investigation of bulk defects—such as grain boundaries in polycrystalline materials—and their properties which are usually different from the properties of the rest part of crystal is another challenging task.

This paper presents a method of photoluminescent tomography as a powerful tool for defect visualization and characterization in bulk semiconductor crystals.

2. Experimental technique

In the present investigation we used a Carl Zeiss LSM 710 NLO confocal microscope with a laser tunable in the wavelength range from 0.71 to 1.05 μm for two-photon excitation (the pulse duration was 150 fs, their frequency was 80 MHz). Average peak power was varied in the range from 0.1 to 1 W depending on the detection condition and was optimized to ensure the highest resolution of luminescent imaging of a sample. Photoluminescent (PL) spectra were registered in the range from 425 to 725 nm with the spectral resolutions of 10 nm. This instrument enabled the study of wide-bandgap semiconductor crystals, such as ZnSe with the bandgap $E_g = 2.71$ eV, using two-photon excitation and observing interband (460–480 nm) and impurity (480–715 nm) luminescence spectra. All measurements were carried out at room temperature.

The 3D PL imaging procedure was as follows. The exciting laser emission focused in a chosen local volume scanned an area in the focal plain. The emitted PL was detected by a photomultiplier tube. A dichotic mirror and interference filters were used to cut off the excitation light from the detector. PL spectra were obtained at each point of scanning. As a result, we scanned maps of PL spectra; each map was recorded from a volume close to a rectangular parallelepiped with a height of a few micrometers and a base of 1 × 1 mm^2. In such a way, we sequentially scanned studied crystals changing the depth of the mapped layer that allowed us to compose volumetric (3D) images at different PL wavelengths as well as so-called tomographic models (5D models in the X–Y–Z–λ–I_{PL} space, where X, Y and Z are the spatial coordinates, λ is the PL wavelength and I_{PL} is the PL intensity). Acquisition of a tomographic model with the dimensions up to 1 × 1 × 1 mm^3 and the spatial resolution of 25 μm took several minutes. Planar (2D) maps $I_{PL}(X, Y)$ for any value of λ and any depth in a crystal (Z) could be then extracted from the tomographic models.

We used samples of single crystalline and polycrystalline ZnSe and ZnSe:Fe to present the PL tomography. Polycrystals were prepared by CVD. Single crystals were obtained by vapor-phase free growth process on the ZnSe (111) substrate. Iron or chrome doping of ZnSe crystals was performed by the thermal diffusion from the iron (chrome) film deposited on a surface by the thermal deposition process. The concentration of Fe (Cr) was up to 5×10^{19} cm^{-3} near the surface and exponentially decreased with Z. Since iron (chrome) concentration decreased rapidly in the depth of a crystal, we succeeded to observe significant variations of characteristics in different regions of crystals depending on the depth from the surface. In addition, the presence of grains in polycrystalline ZnSe allowed us to demonstrate capabilities of confocal microscopy for investigation of macroscopic defects in the bulk of semiconductor crystals using two-photon excitation of PL.

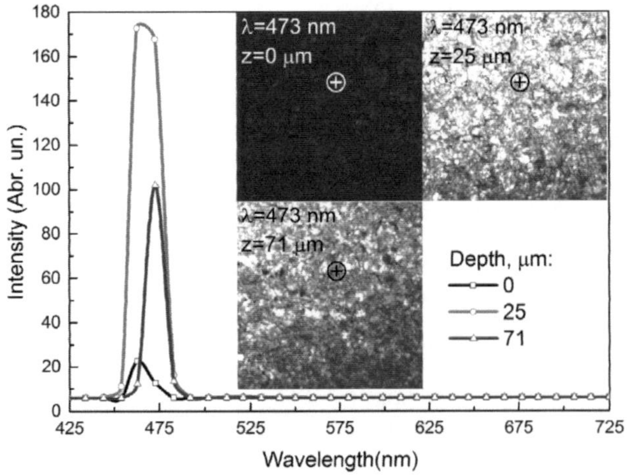

Figure 1. Photoluminescence spectra and 2D maps for the wavelength of 473 nm obtained from the depth of 0, 25 and 71 μm; points at which the spectra were recorded are shown by crosses; circles show areas over which the PL spectra are averaged.

3. Results and discussion

We start the presentation from the results concerning the original polycrystalline ZnSe. Figure 1 shows 2D maps of an undoped ZnSe crystal at the wavelength of 473 nm and different distances from the surface. The planar distribution of the PL intensity is seen to be strongly non-uniform. The near-surface layer ($Z = 0$) demonstrates a low average PL intensity. The corresponding spectrum contains only one band with a maximum at about 465 nm. The average PL intensity strongly but not monotonically depends on the depth. Initially it grows with Z reaching a maximum at $Z \approx 25$ μm; then it starts to decrease. The spectrum contains only one PL band at each depth. A maximum of this line moves to the long-wavelengths region with increase in Z. It "stops" at the wavelength of 475 nm at $Z \approx 50$ μm. PL maps of the sample are similar for all depths; they demonstrate a granular structure. Sizes of grains vary in the range of tens of micrometers. PL intensity inside grains varies by more than a decade for different regions with grain boundaries always showing a low PL intensity.

A high surface recombination rate at mechanically polished surface is the cause of low PL intensity in the near-surface layer ($Z = 0$). The observed PL band corresponds to the band-to-band recombination. Its shift to the long-wavelength region with growing Z is a result of high absorption of the short-wavelength part of the edge luminescence. As a consequence, only excitonic radiation is emitted from the depth of a crystal [4]. Absorption of this light is reduced but is nonzero; the intensity of the excitonic emission decreases for $Z > 30$ µm. A low PL intensity at grain boundaries can likely be explained by a high concentration of centers of non-radiative recombination. This is a commonly adopted explanation for all semiconductor materials. So, most of our results obtained for the undoped ZnSe crystals correspond to the available concepts supporting and "visualizing" them.

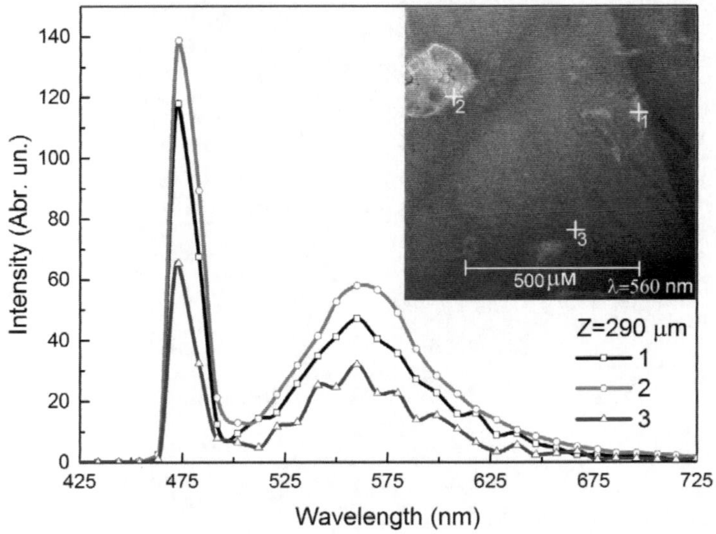

Figure 2. Photoluminescence spectra and a 2D map of a ZnSe:Fe crystal; the map was obtained at the wavelength of 560 nm from the depth of 290 µm, the spectra were obtained from different points indicated by numbered crosses in the PL map.

Let us dwell on the results relating to the ZnSe crystals doped by Fe. Characteristics of these crystals appeared to be very similar but they are strongly different from the results obtained for the as-grown samples. At first, PL spectra of the ZnSe:Fe crystal (figure 2) consisted of two bands: the excitonic one (475 nm) and a "broad" band in the range from 500 to 580 nm. The intensities of these bands were close and in some cases the intensity of the broad band exceeded the intensity of band-to-band PL. A characteristic feature of the broad band was its shape variation when PL was registered from different small (a few micrometers in size) regions of crystals (figure 2); note that this peculiarity was observed in both ZnSe:Fe and ZnSe:Cr crystals.

Secondly, distribution of the bulk PL intensity strongly differed in the doped and as-grown crystals. Figure 3 shows that intense PL, both excitonic and broad-band, is observed only at $Z > 100$ µm. Intensities of the PL bands are much less when closer to the surface. This result was common for this type of crystals; only the depth from which the PL intensity began to increase varied from 40 to 100 µm in different samples.

Interesting results were obtained for the features of the planar PL maps of ZnSe:Fe observed at different depths. Figure 4 shows 2D maps obtained at the wavelength of 473 nm (at the exitonic band) for one ZnSe:Fe sample which had an intense PL starting from $Z = 40$ µm. A PL contrast of grain boundaries is well recognized. The grain sizes are increased, as it takes place in annealed crystals. Up to the depth of about 120 µm, the contrast resembles that in the as-grown samples, i.e. the grain boundaries have lower band-to-band PL intensity than the grains. At deeper layers, the contrast decreases, disappears and then inverts at the depth of 260 µm: from this depth the grain boundaries have higher inter-band PL intensity than the grains. However, there remains a narrow domain at the middles of the grain boundaries with the decreased interband PL intensity (figure 4, $Z = 474$ μm, $\lambda = 473$ nm). The similar narrow domains at the middles of the grain boundaries are revealed close to the surface but in this case narrow zones of increased broad-band PL are observed inside the regions of the reduced interband emission (figure 4, $Z = 47$ μm, $\lambda = 541$ nm).

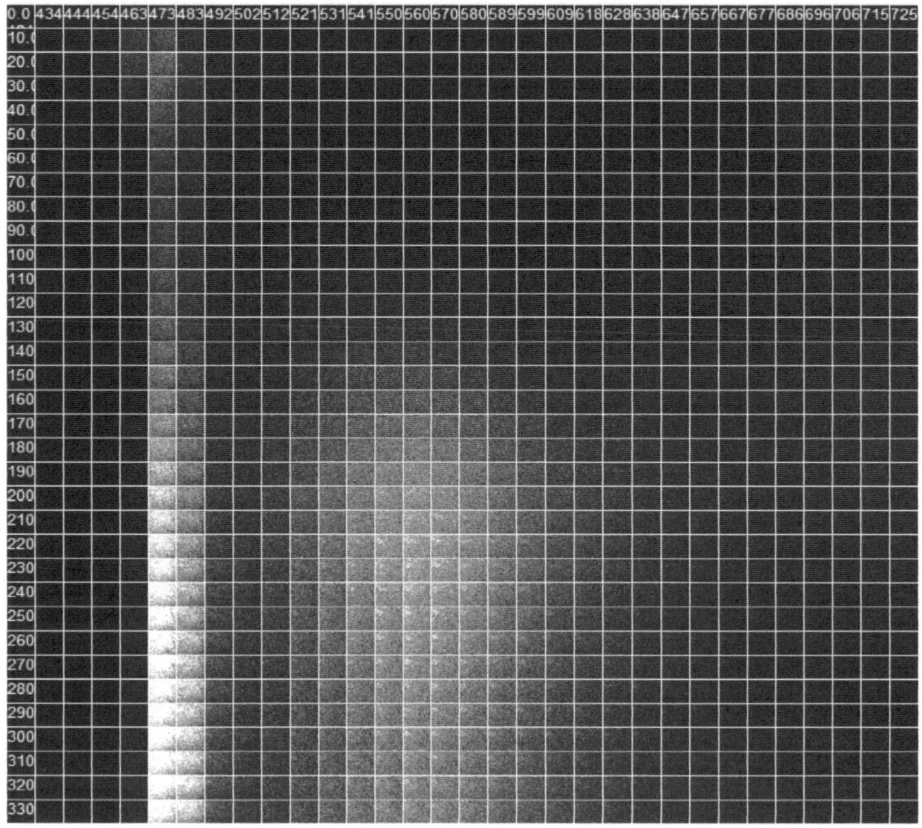

Figure 3. A tomography model of a ZnSe:Fe sample: each square represents a 2D map obtained at the wavelength indicated at the abscissa axis (nm) from the depth indicated at the ordinate axis (μm); the photoluminescence is seen to be emitted mainly from the bulk of the crystal.

Perhaps, it is associated with formation of a lot of centers during doping, which are competing recombination channels of non-equilibrium carrier. There are at least two ways of formation of these centers. Firstly, the formation of minor precipitates of the doping impurity is possible, which could be efficient centers of the non-radiative recombination. And secondly, atoms of the dissolved doping impurity may be the competing centers of the non-radiative recombination. In this case, they may act as non-radiative recombination centers or introduce PL bands, e.g., IR bands. PL appearing at different depths may be explained, e.g., by thickness variation of a surface layer removed by mechanical polishing after the dopant diffusion which was not precisely controlled. As a result, dopant concentrations at the surface and in the volume of different samples may vary.

The inverted recombination contrast at grain boundaries confirms the fact that composition of impurities and intrinsic defects in the bulk of crystals strongly depends on the depth. Probably, the mentioned above centers are contained in impurity atmospheres around grain boundary that explains the high contrast of the grain boundaries at $Z < 100$ μm. Concentration of these centers rapidly decreases with deepening in a crystal and the contrast is smoothed. The phenomena of high PL intensity at impurity atmospheres of grain boundaries in deep regions and high PL intensity of the broad band in the middles of impurity atmospheres of grain boundaries in near-surface regions call for further investigation.

In addition to the above mentioned centers, other centers formed in the process of doping were responsible for the appearance of the broad PL band. We observed this band in different iron-doped and chromium-doped crystals; consequently, the broad band is not connected with the dopant. This band likely refers to intrinsic defects, such as the interstitial zinc–vacancy zinc Frankel pair, and interstitial oxygen [4]. Note that a number of articles reported that these defects had many modifications and gave rise to numerous close PL lines [5]. Perhaps, this explains the observed variations of the shape of the broad band in the PL spectra obtained from different points in the crystals.

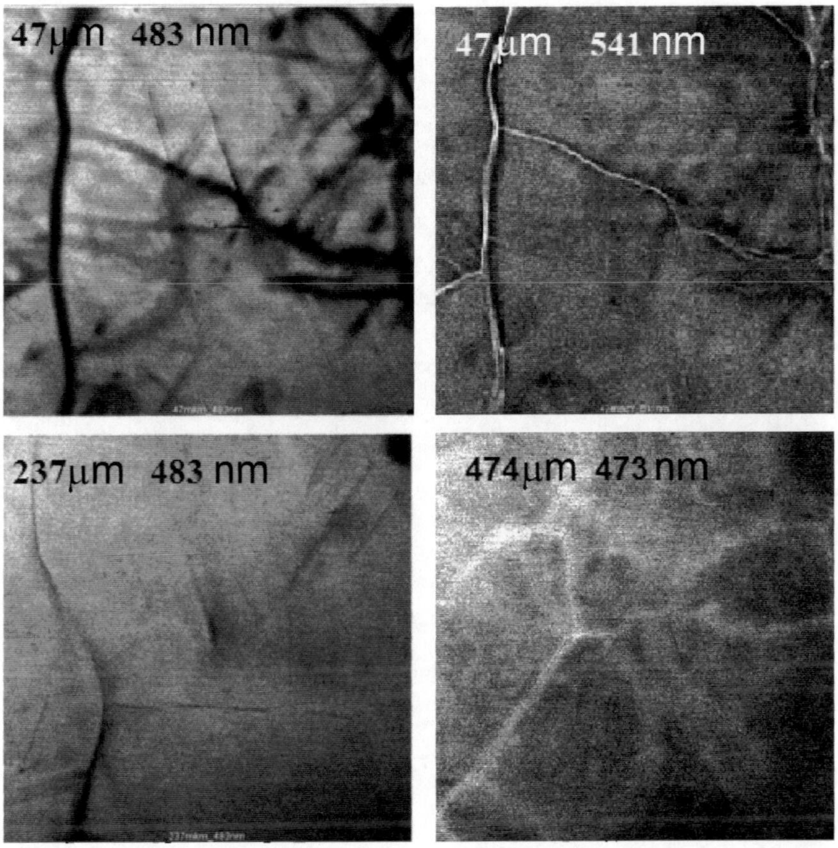

Figure 4. 2D maps of polycrystalline ZnSe:Fe obtained from different depths at three of wavelengths (the corresponding depths in micrometers and wavelengths in manometers are shown in each panel).

4. Conclusion

We have demonstrated that the PL tomography is a powerful tool for investigation of bulk semiconductors. This technique enables the express non-destructive 3D mapping of interband and defect or impurity related photoluminescence. Its capability could be increased by extension of detection range to near-IR and by utilization of an excitation source at longer wavelengths. If the detector spectral range is expanded to 1.5 μm, the industrially important materials, such as GaAs, InP, CdTe, could also be investigated.

Explorations of silicon and germanium are more difficult to conduct using this technique since the emissivity of Si and Ge is low. We suggest using capacitance methods or conductivity measurements at microwave frequencies for carrier detection in such researches.

Acknowledgments

We cordially thank Dr Victor Loschenov and Dr Anastasia Ryabova for their contribution to this research and beautiful micrographs obtained using the laser scanning microscope. We also thank Dr Vladimir Yuryev for fruitful discussions of this work and help in preparation of the paper. The Center of Collective Use of Scientific Equipment of GPI RAS supported this research by presenting admittance to its instrumentation. We acknowledge the support of this research. These studies were supported by by the Russian Science Foundation (Grant № 15-13-10028); the research was also funded through the Program of the Presidium of RAS "Fundamentals of the breakthrough dual-use technologies in the interests of the national security" (the project "A study of possibility of development of high-performanceIR lasers (4–5 μm) on the basis of iron doped ZnSe crystals with impact excitation of activator ions by hotelectrons")

References

[1] Yuryev V A, Kalinushkin V P, Zayats A V, Repeyev Y A and Fedoseyev V G 1996 *Notions and Perspectives of Nonlinear Optics* ed O Keller (Singapore: World Scientific) pp 665–70
Yuryev V A, Kalinushkin V P, Zayats A V, Repeyev Y A and Fedoseyev V G 1996 *Inst. Phys. Conf. Ser.* **149** 263–8
[2] Nitsuk Yu A, Vaksman Yu F and Yatsun V V 2012 *Semicond.* **46** 1265–9
[3] Nedeoglo D D and Simashkevich A V 1984 *Electical and Luminescent Properties of Zinc Selenide* (Chisinău: Ştiinţa)
[4] Rong F S, Barry W A, Donegan J F and Watkins G D 1996 *Phys. Rev.* B **54** 7779–88
[5] Morozova N K, Karetnikon I A, Blinov V V and Gavrischuk E M 2001 *Semicond.* **35** 24–32

Raman spectroscopy of monoatomic hydrogen at dislocations in silicon

N Vysotskii[1], A Loshachenko[1], E Borisov[1] and O Vyvenko[1]

[1]Saint-Petersburg State University, St. Petersburg, Uliyanovskaya 1, Russia

E-mail: a.loshachenko@spbu.ru

Abstract. A broad Raman peak at about 2000 cm^{-1} was found in hydrogenated silicon foils with a regular dislocation network (DN) produced by silicon wafer bonding. It was not observed in the reference foils without a dislocation network. The hydrogenation was performed from acid water solution at room temperature. The peak was ascribed to monoatomic hydrogen in silicon bond centered position that became stable in a close vicinity to the dislocation cores in opposite to the case of the regular lattice where its molecular form is energetically favorable.

1. Introduction

Hydrogen (H) when introduced in crystalline silicon interacts effectively with various lattice defects. Such interaction results in the formation of H-defect complexes that change electric properties of the defects giving rise in many cases to a reduction of their recombination activity. The latter is known as hydrogen passivation of recombination active defects and widely used now to improve the efficiency of multi-crystalline silicon (mc-Si) solar cells in photovoltaic industry.

The most powerful techniques to identify H configuration in silicon by its vibrational mode energy are infrared absorption and Raman spectroscopies. In regular Si lattice the configuration of interstitial hydrogen H which is stable at room temperature is the molecule H_2 centered in T–site giving rise to a Raman peak at about 3600 cm^{-1} [1,2]. According to the theoretical calculations, T-site H_2 configuration has significantly lower energy than the monoatomic one. The configuration of monoatomic H is in the center of Si-Si bonds (BC–site) that gives rise to the appearance of Si bond stretch mode in the vicinity of 2000 cm^{-1} (f.i. 1980 cm^{-1} for positive H^+ [3–6]). It is metastable and was observed only in proton irradiated silicon at liquid helium temperature and disappeared above 200 K.

Vibration modes of H in complexes with vacancies or impurities lye also in the region close to monoatomic H (1800-2200 cm^{-1} [3,7,8]) as well as with the silicon surface states and with grain boundaries of unknown structure in mc-Si (2000-2100 cm^{-1} [9,10]).

Previous experiments on H exo-diffusion from silicon and SIMS showed that hydrogen attracted by and accumulated at dislocations [11,12]. There are theoretical calculations of the stable configurations of H within a dislocation core [13,14] or at its vicinity [15]. In particular, it was predicted that in a close vicinity of a screw dislocation the H monoatomic configuration became more preferable that the molecular one. However, until now to the best of our knowledge there were no reports about successful measurements of vibration modes of hydrogen at dislocations.

In this work we investigated Raman spectroscopy of silicon foils prepared for transmission electron microscopy (TEM) investigations with a buried regular dislocation network (DN) that were hydrogenated by wet-chemical etching (WCE). The combination of a high DN density and the reduction of the background signal from the bulk allowed us to detect a Raman peak (at 2000 cm^{-1}, FWHM ~ 90 cm^{-1}) corresponding to the vibrational mode of monoatomic hydrogen both in p- and in n-type samples. The comparison of TEM images with the Raman signal spatial distribution revealed that the detected Raman peak is confined at DN and can be identified as a H-dislocation complex confirming the prediction of the theory [15].

2. Experimental

The samples were cut from hydrophilically bonded (100) Cz-Si doped with boron to about 10^{15} cm^{-3} or with phosphorus to 3×10^{14} cm^{-3}. DN was located at the depth of 160-180 nm from the surface. DN consisted of a square network of screw dislocations (~5-10 nm mesh) and rows of 60° dislocations (with an inter-dislocation distance of ~60 nm). The detailed information about the dislocation structure obtained with TEM, electrical levels and hydrogen out-drift processed was reported earlier and can be found elsewhere [16,17].

For Raman spectroscopy we used a thin plan view of TEM foils that enabled us to reduce drastically the background Raman signal. The foils were formed by conventional WCE (7:1 HNO_3:HF) that produces a beveled edge of the sample near the hole in their middle as well as the sample hydrogenation. The dislocated/reference foils were obtained by etching of back/front sides of the bonded wafers. The foils had a standard TEM diameter of 3 mm with a hole in the middle. After the TEM foils preparation, the foils were kept in the air for more than one month that was sufficient to diminish the Raman signal from hydrogen on the surface [18].

Raman spectroscopy was carried out by a Senterra unit with backscattering geometry at room temperature in the air using a 532 nm Nd laser for excitation as it is schematically shown in figure 1. One can easily recognize from figure 1 that there is a region of the foil at its internal edge where DN is absent. The lateral size of the dislocation-free region is as long as 20 micrometers that exceeds well the laser beam diameter of 6-7 um. That makes possible to acquire Raman spectra both from dislocation-free and from dislocation-rich regions of the sample. The exact position of the intersection of DN with the sample surface was obtained from TEM investigations. For this goal, a sequence of multiple TEM images with a sufficiently high magnification was obtained with a Zeiss-Libra 200 FE and they were then mated all together.

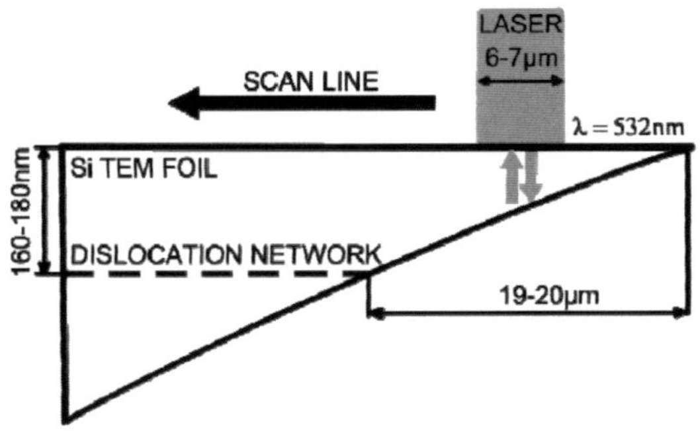

Figure 1. Schematic diagram of Raman spectroscopy measurements of a TEM foil with a dislocation network.

3. Results and discussion

The broad Raman peak in the vicinity of 2000 cm^{-1} with a full width at half maximum (FWHM) of about 90 cm^{-1} was observed both in p- and n- type foils in the region which contains DN (see in figure 2 (a)). As it was mentioned in Introduction, the position of the peak corresponds to the spectral range of the stretched mode of atomic hydrogen in BC-site. The peak was not observed in bulk samples, in dislocation-free reference foils prepared together with dislocated ones as well as under excitation of dislocation-free part of the foils as it is illustrated in figure 2 (b).

IOP Publishing
doi:10.1088/1742-6596/690/1/012004

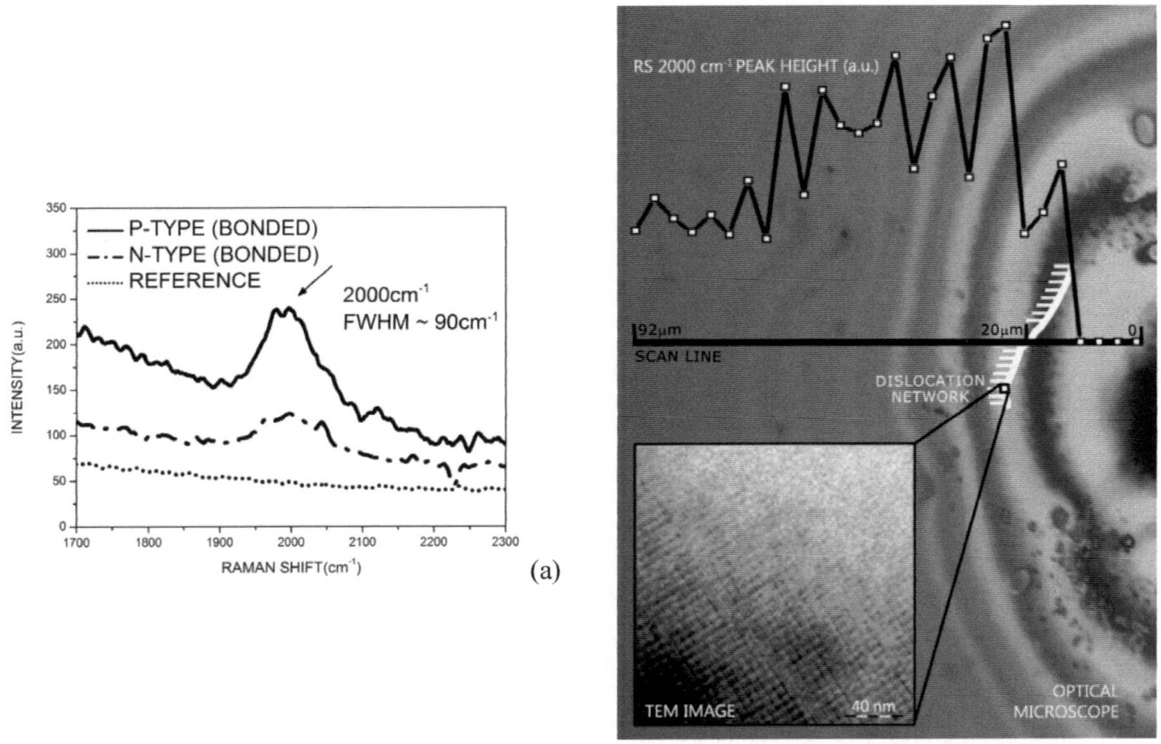

Figure 2. (a) Raman spectra for dislocated and reference Si samples. (b) Optical reflection micrograph of a n-Si foil with the position of DN defined by TEM (TEM mode) marked with a white comb. The profile of the 2000 cm^{-1} Raman peak (a black line with white square dots) was acquired along the straight horizontal black line.

One should firstly note that the dependences of the Raman peak intensity on the distance from the foil edge presented in figure 2b and figure 3 exhibit oscillations which are caused by the phenomenon of light interference in thin films and so will not be further discussed in details.

The intensity profiles of the 2000 cm^{-1} Raman peak presented in figure 2b and 3 can be divided into several parts. The first one is near the foil edge at the central hole where the Raman peak of 2000 cm^{-1} was absent while the signal of optical phonon mode of Si at 520 cm^{-1} [19] was already present (figure 3). The second part is characterized by a surge of the 2000 cm^{-1} peak intensity at the distance from the hole edge where DN intersects the foil surface. With moving further away from the foil edge, the intensity of the 2000 cm^{-1} Raman peak decreases but the intensity of the 520 cm^{-1} peak continues to increase with the distance. The increase of the silicon-lattice-related signal is obviously due to increase of the generation volume with the foil thickness whereas the decrease of the hydrogen-related signal is due to increase of the optical path length towards and from the DN plane that increases the light absorption. The theoretical curve of Raman signal profile calculated taking into account the excitation laser light absorption is down as the black dotted line in figure 3. The impact of the light absorption can be also noted from the examination of Raman profiles in figure 3 as a reduction of the interference oscillation amplitude at their most remote parts.

Figure 4 represents the results of thermal treatment of a p-type sample at a pressure of 10^{-5} Torr for one hour. The intensity of the hydrogen-related signal reduced upon annealing already at 150°C but it is still present even after the treatment at temperature as high as 500°C. This indicates the variety of binding strengths of hydrogen with the lattice. The thermal stability of the peak at 500°C is in accordance with previous data of exo-diffusion experiments that have shown not complete removal of hydrogen from dislocated silicon at this temperature [11].

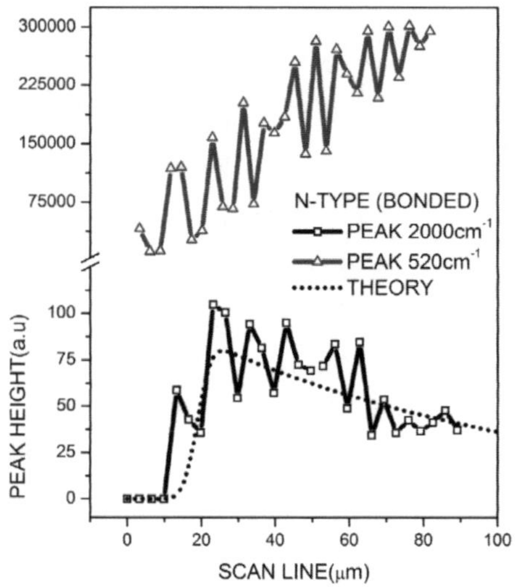

Figure 3. The intensity of the 2000 cm^{-1} peak as a function on the distance from the inner edge of the foil compared to the optical-phonon mode of Si 520 cm^{-1}. The dotted line shows the theoretical curve calculated with corrections for excitation laser beam width and light absorption.

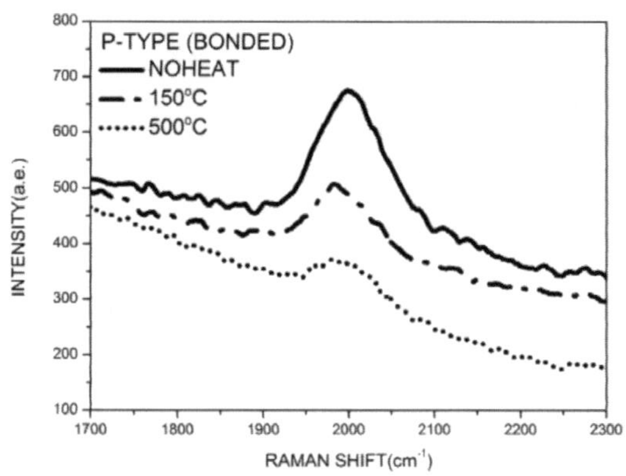

Figure 4. Raman spectra in a p-type bonded sample before and after annealing at different temperatures for one hour in vacuum (10^{-5} Torr).

4. Conclusion

In our work we provide first observation of a monoatomic hydrogen-related Raman peak at 2000 cm^{-1} at dislocations, using a TEM foil of bonded silicon wafer with well defined DN. We suggest that the origins of the Raman peak at 2000 cm^{-1} can be attributed to the stable BC-site Si-H configuration in the region close to the dislocation core. The observation of the stable signal corresponding to the silicon bond stretch mode of hydrogen in BC site confirms qualitatively recent results of theoretical calculations [14,15] predicting an increase of the binding energy of hydrogen in BC-Si atomic configuration in a close vicinity of the screw dislocation core due to lattice distortion. Temperature behavior of the peak height upon annealing and its broadness can be associated then with the variation of the binding energy for hydrogen in BC with the distance from the dislocation core.

5. Acknowledgements

We greatly acknowledge O. Konochuk for providing the samples. This work is supported by Russian Foundation for Basic Research (grant 14-02-31511-mol_a). The experimental results in this work were obtained using the equipment of Interdisciplinary Resource Center for Nanotechnology and Center for Optical and Laser Materials Research of St. Petersburg State University, Russia.

References

[1] Leitch A, Alex V and Weber J 1998 *Phys. Rev. Lett.* **81** 421–4
[2] Van De Walle C G and Goss J P 1998 *Phys Rev Lett.* **80** 2177–80
[3] Stein H J 1975 *J. Electron. Mater.* **4** 159–74
[4] Chang K J and Chadi D J 1989 *Phys. Rev. B* **40** 11644–53
[5] Budde M, Lüpke G, Parks Cheney C, Tolk N H and Feldman L C 2000 *Phys. Rev. Lett.* **85** 1452–5
[6] Van De Walle C G 1989 *Phys. Rev. B* **39** 10791–808
[7] Bergman K, Stavola M, Pearton S and Lopata J 1988 *Phys. Rev. B* **37** 2770–3
[8] Estreicher S K 1995 *Mater. Sci. Eng. R* **14** 319–412
[9] Chabal Y J 1991 *Phys. B Condens. Matter* **170** 447–56
[10] Kitahara K, Ishii T, Suzuki J, Bessyo T and Watanabe N 2011 *Int. J. Spectrosc.* **2011** 1–14
[11] Kveder V and Labusch R 1984 *Phys. Status Solidi a-Applied Res.* **84** 149-56
[12] Park Y, Lu J G and Rozgonyi G 2010 *Electron. Mater. Lett.* **6** 1–5
[13] Ewels C, Leoni S, Heggie M, Jemmer P, Hernandez E, Jones R and Briddon P 2000 *Phys. Rev. Lett.* **84** 690–3
[14] Matsubara M, Godet J and Pizzagalli L 2010 *Phys. Rev. B* **82** 34–7
[15] Matsubara M, Godet J and Pizzagalli L 2010 *J. Phys. Condens. Matter* **22** 035803
[16] Loshachenko A, Bondarenko A, Vyvenko O and Kononchuk O 2013 *Phys. Status Solidi (c)* **10** 36–9
[17] Loshachenko A S, Vyvenko O F and Kononchuk O 2013 *Solid State Phenom.* **205-206** 341–5
[18] Niwano M, Kageyama J I, Kurita K, Kinashi K, Takahashi I and Miyamoto N 1994 *J. Appl. Phys.* **76** 2157–63
[19] Russell J P 1965 *Appl. Phys. Lett.* **6** 223

Plasmon phonon modes and optical resonances in *n*-GaN

G A Melentev[1], D Yu Yaichnikov[1], V A Shalygin[1], M Ya Vinnichenko[1], L E Vorobjev[1], D A Firsov[1], L Riuttanen[2], S Suihkonen[2]

[1] Department of Physics of Semiconductors & Nanoelectronics, Peter the Great St. Petersburg Polytechnic University, Polytechnicheskaya 29, St.Petersburg 195251, Russia
[2] School of Electrical Engineering, Aalto University, Otakaari 1, Espoo 02150, Finland

E-mail: gamelen@spbstu.ru

Abstract. We study theoretically and experimentally transverse and longitudinal electromagnetic waves in *n*-GaN epitaxial layers. The studies are carried out on the epitaxial layers with various doping levels. Simulation of the reflectivity and absorptivity spectra is performed in a wide frequency range. Additionally, the radiation emission spectra are simulated for various temperatures of the GaN epitaxial layer. It is shown that the resonance peculiarities of the optical spectra are located closely to the frequencies of the coupled plasmon phonon modes. Experimental studies of the reflectivity spectra have been performed in the spectral range 2–20 THz. The experimental spectra are well fitted by the simulated ones and can be used for contactless determination of the electron concentration and mobility in GaN epitaxial layers. The experimental and theoretical results of the present work provide insights for the development of GaN-based devices aimed to absorb/emit terahertz or mid infrared radiation selectively.

1. Introduction

Gallium nitride is one of the most promising materials for applications in terahertz (THz) and mid infrared photonics. Various micro- and nanostructures based on GaN are used as THz detectors and emitters. The number of different physical effects used in these devices is constantly growing. In particular, one can mention the following recent investigations related to THz radiation emission in gallium nitride: THz radiation emission due to intracenter optical transitions in lightly doped *n*-GaN epitaxial layers [1], thermal radiation of hot electrons in the THz range in a single heterojunction AlGaN/GaN [2], THz emission due to plasmon excitation in GaN/AlGaN HEMT structures with submicron gate length [3], THz emission associated with surface plasmon polaritons in GaN-based microstructures [4]. It is feasible to obtain selective THz radiation emission/detection by means of bulk plasmon phonon polaritons in *n*-GaN epitaxial layers. To this end, in the present paper we investigate the optical properties of *n*-GaN/sapphire microstructures. In addition to transverse electromagnetic waves, longitudinal electromagnetic waves in *n*-GaN are also examined.

The plasmon–LO phonon interaction, i.e. an interaction between a longitudinal electromagnetic wave caused by free electron displacement relative to the positive background and longitudinal optical (LO) lattice vibrations, becomes significant when the plasma frequency $\omega_\text{P} = \left(\dfrac{4\pi e^2 N_e}{\varepsilon_\infty m_e}\right)^{1/2}$ is comparable with the LO phonon frequency ω_{LO} where e and m_e are the electron charge and the effective mass, respectively, N_e is the free electron concentration, ε_∞ is the high frequency permittivity. Firstly, we calculate the concentration dependence of the frequencies of the coupled plasmon–LO phonon modes which represent longitudinal electromagnetic waves. Secondly, we consider a few certain values of electron concentration corresponding to three frequency ranges, namely $\omega_P < \omega_{LO}$, $\omega_P \approx \omega_{LO}$, and $\omega_P > \omega_{LO}$ and simulate optical properties of the structures under investigation. Optical properties are determined by transverse electromagnetic waves (in the terms of quantum mechanics, bulk plasmon phonon polaritons) excited in the structures under optical excitation. We simulate spectra of the complex permittivity, refractive index and extinction coefficient for bulk *n*-GaN considering contributions of free electrons and lattice vibrations. Then we calculate reflectivity and absorptivity spectra of the *n*-GaN/sapphire structures taking into account interference effects. Finally, we simulate the radiation emission spectra for various temperatures of the GaN epitaxial layer.

Experimental part of our work is devoted to reflectivity spectra measurements for the structures with various doping levels of GaN epitaxial layer. The experimental spectra are compared with the simulated ones. In the last section of the paper we summarize the major results.

2. Theoretical calculations

2.1. Plasmon–LO phonon modes in n-GaN

For a doped polar semiconductor, the dielectric permittivity at the angular frequency ω is given by [5, 6]:

$$\varepsilon(\omega) = \varepsilon_\infty \left[1 + \frac{\omega_{\text{LO}}^2 - \omega_{\text{TO}}^2}{\omega_{\text{TO}}^2 - \omega^2 - i\omega\gamma} - \frac{\omega_{\text{P}}^2}{\omega^2 + i\omega\tau^{-1}} \right], \tag{1}$$

where ω_{TO} is the frequency of a transverse optical (TO) phonon which is connected with the LO phonon frequency by the Lyddane–Sachs–Teller relation: $\varepsilon_0/\varepsilon_\infty = \omega_{\text{LO}}^2/\omega_{\text{TO}}^2$, ε_0 is the static permittivity, γ is the damping constant for TO phonons, τ is the relaxation time determined by the mobility μ of electrons: $\tau = m_e\mu/e$.

Frequencies of the longitudinal electromagnetic waves which can propagate in a media can be found by solving the equation $\varepsilon(\omega) = 0$ [5] that gives two solutions for the doped polar semiconductor which are denoted below as ω_+^* and ω_-^*. Generally, ω_+^* and ω_-^* are complex quantities. The real part of these quantities ($\omega_+ \equiv \text{Re}\,\omega_+^*$ and $\omega_- \equiv \text{Re}\,\omega_-^*$) gives the frequencies of plasmon–LO phonon modes. The imaginary part gives their decrements. It should be noted that there is no dispersion for the longitudinal electromagnetic waves in the long-wavelength limit where the wavevector $q \to 0$. Usually plasmon–LO phonon modes are detected by means of Raman spectroscopy. In particular, plasmon–LO phonon modes in *n*-GaN were experimentally studied in [7–9]. However, optical detection of these modes by means of reflectivity spectra is also possible. For low doped *n*-GaN ($N_e < 2\cdot10^{17}$ cm^{-3}), such investigations were carried out in [10].

In the present subsection we simulate plasmon–LO phonon modes in *n*-GaN in a wide concentration range taking into account the damping constant for TO phonons (γ) and the relaxation time for free electrons (τ). In the next subsection we consider a correlation between the calculated frequencies of plasmon–LO phonon modes and peculiarities of the optical spectra (reflection and emission).

The concentration dependence of plasmon–LO phonon modes is presented in figure 1. To provide needed data on the dependence of τ upon N_e, we used the experimental dependence $\mu = \mu(N_e)$ obtained earlier on the basis of Hall measurements in *n*-GaN with various doping levels [11].

The simulation results based on these data are shown in figure 1 by green curves. The high frequency plasmon phonon mode ω_+ at low concentrations is close to the LO phonon frequency ω_{LO} and at high concentrations is close to the plasma frequency ω_P. The low frequency plasmon phonon mode ω_- at high concentrations is close to the TO phonon frequency ω_{TO}. All three these limit values are the same as ones given by a simplified theory which does not take into account phonon damping and electron relaxation processes (that is equivalent to the case of $\gamma = 0$ and $\tau = \infty$ in equation 1). But in the case of $\tau \neq \infty$, the general theory (green lines in figure 1) gives a significantly different result for ω_- for low concentrations in comparison with the simplified theory (black lines in figure 1). While the simplified theory [6] gives $\omega_- = \omega_P\sqrt{\frac{\varepsilon_\infty}{\varepsilon_0}}$, the general theory considered here gives much less values. Moreover, in accordance with our calculations, the low frequency plasmon phonon mode in *n*-GaN is completely overborne by electron scattering processes at concentrations $N_e < 1.8 \cdot 10^{17}$ cm^{-3}. It should be emphasized that both plasmon–LO phonon modes ω_+ and ω_- can be located rather far from ω_P and ω_{LO} especially in the case of $\omega_P \approx \omega_{LO}$.

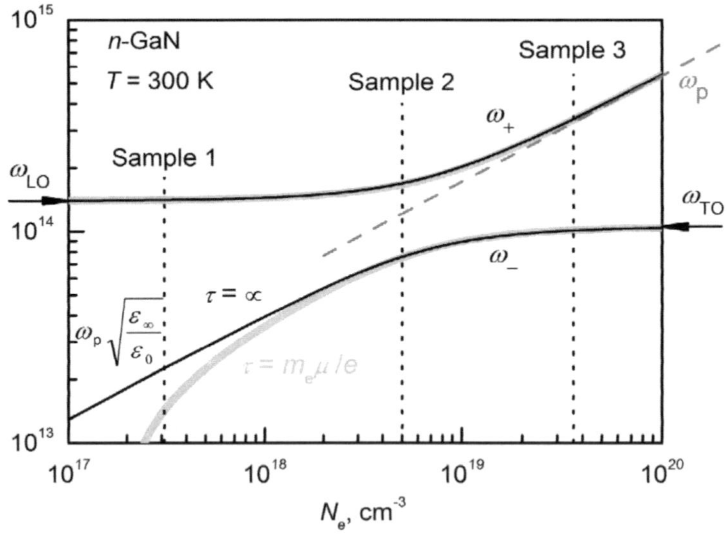

Figure 1. Concentration dependence of plasmon–LO phonon modes in *n*-GaN. Green lines: calculation taking into account the damping constant for TO phonons and the relaxation time for free electrons. Black solid lines: calculation for the collisionless electron plasma ($\tau = \infty$). Dotted lines denote electron concentrations in three samples under study.

2.2. Reflectivity and absorptivity spectra of n-GaN epitaxial layers

We simulate reflectivity and absorptivity spectra for n-GaN epitaxilal layers grown on a sapphire substrate. We consider three values of electron concentration, namely, $N_e = 3.1 \cdot 10^{17}$, $5 \cdot 10^{18}$ and $3.6 \cdot 10^{19}$ cm^{-3} (named as sample 1, sample 2 and sample 3) corresponding to three frequency ranges $\omega_P < \omega_{LO}$, $\omega_P \approx \omega_{LO}$, and $\omega_P > \omega_{LO}$, respectively.

Firstly, with the help of the Fresnel equation:

$$(n+ik)^2 = \varepsilon(\omega), \quad (2)$$

we simulated spectra of the refractive index n and the extinction coefficient k for bulk n-GaN. In the calculations we used the following values of n-GaN parameters: $\varepsilon_\infty = 5.4$, $\varepsilon_0 = 9.5$, $\hbar\omega_{TO} = 69.3$ meV, $m_e = 0.2\, m_0$ [12, 13], $\gamma = 2.4{\cdot}10^{11}$ s^{-1}, $\mu = 122$ cm^2/V·s.

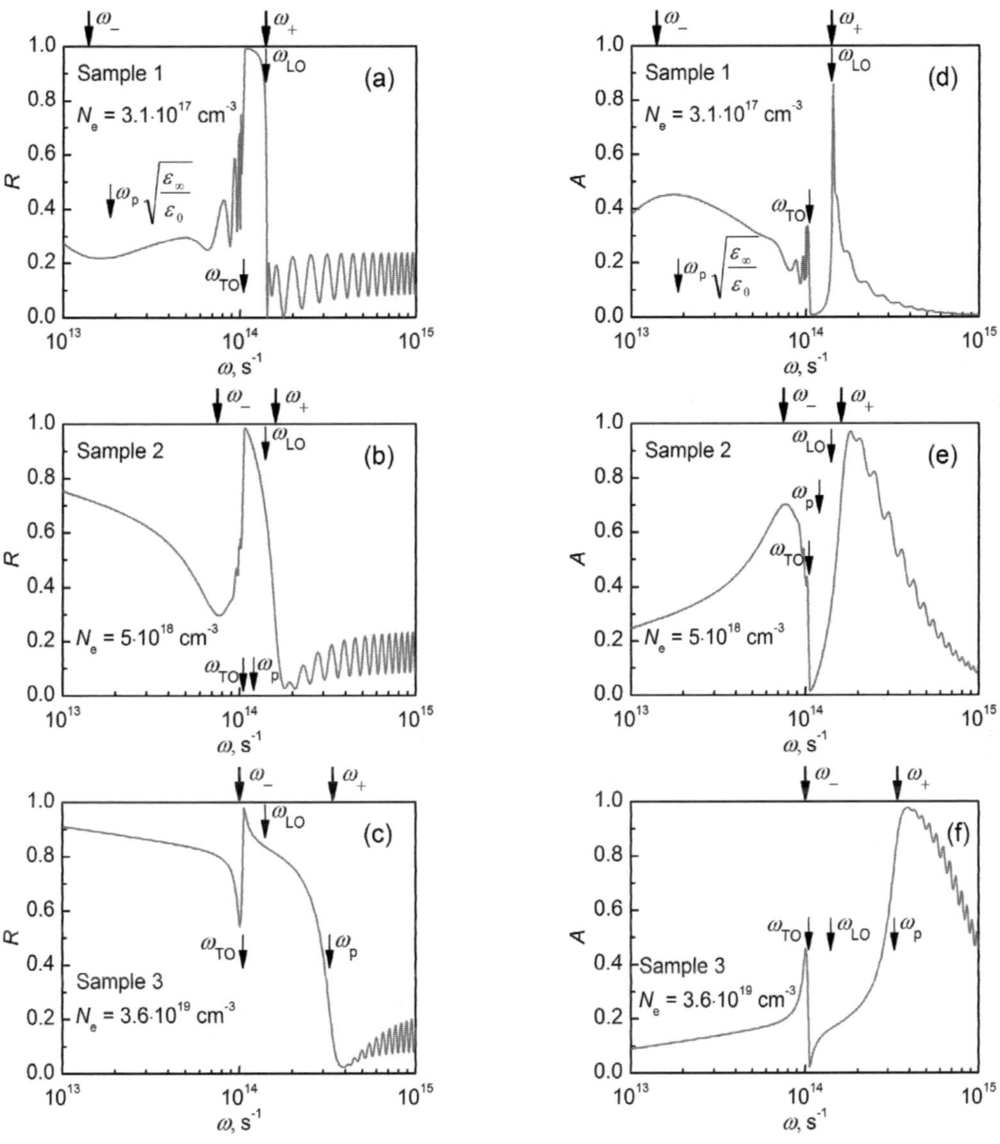

Figure 2. Reflectivity (a), (b), (c) and absorptivity (d), (e), (f) spectra of n-GaN/sapphire structures under normal incidence of radiation (calculation).

Then we calculated the reflectivity (R) and transmittivity (T) spectra for different thicknesses of the epitaxial layer taking into account interference effects. Calculations were based on the analytical formulas from [14]. The value of the refractive index of the sapphire substrate was put as 2.9 [15]. In figure 2 we present the simulation results on the reflectivity of the GaN epilayer with the thickness of 6.2 μm under normal incidence of radiation.

One can see, that for the low-doped sample 1 (figure 2 (a)) the calculated reflectivity spectrum has a usual shape in the *Reststrahlen band* (in the spectral range between ω_{TO} and ω_{LO}), and the plasmon phonon mode ω_+ practically coincides with ω_{LO}. The reflectivity R has a deep minimum at the frequency $\omega \approx \omega_+ \approx \omega_{LO}$. In the case of $\omega_P \approx \omega_{LO}$ (sample 2, figure 2 (b)), the reflectivity spectrum changes significantly and one can see two reflectivity minimums at the frequencies of plasmon phonon modes ω_- and ω_+. Finally, in the case of the heavily doped sample 3 (figure 2 (c), the superposition of the Reststrahlen band and plasma reflection band takes place at frequencies $\omega \leq \omega_{LO}$ and one can see a very unusual sharp dip in the reflectivity spectrum at $\omega \approx \omega_- \approx \omega_{TO}$ and the second reflectivity minimum at $\omega \approx \omega_+ \approx \omega_P$. Thus, analysis of the calculated reflectivity spectra shows that the reflectivity minimums are always very close to the frequencies of the plasmon phonon mode ω_- and ω_+. There is one exception only: in the low-doped sample with low electron mobility the reflectivity minimum $\omega \approx \omega_-$ is inexpressive.

Using the data on the calculated transmittivity of the GaN/sapphire structures under normal incidence of radiation, we then determined absorptivity spectra with the help of the equation $A = 1 - T - R$. The results of simulation for all three samples are presented in figure 2 (d), (e), (f). Analysis of the calculated absorptivity spectra shows that they are characterized by the intensive absorption lines at the frequencies of the plasmon phonon mode ω_- and ω_+. These absorption lines are especially well pronounced in cases of $\omega_P \approx \omega_{LO}$ and $\omega_P > \omega_{LO}$ (figure 2 (e) and (f)). In the latter case, the absorption line at $\omega \approx \omega_- \approx \omega_{TO}$ represents a very sharp resonance.

It should be noted that in addition to abovementioned peculiarities, the reflectivity and absorptivity oscillate according to the frequency in the spectral ranges where the epitaxial layer is transparent enough. These oscillations are caused by the interference of electromagnetic waves in the epitaxial layer.

3. Experimental results and their discussion

Wurtzite gallium nitride epitaxial layers doped with silicon were grown by MOVPE on a *c*-plane sapphire substrate covered by a 2 μm undoped GaN buffer layer. Two structures were grown. In one structure *n*-GaN epitaxial layer is characterized by the following parameters: thickness d = 10 μm, electron concentration $N_e = 5\cdot 10^{18}$ cm^{-3} and mobility μ = 189 cm^2/V·s (sample 2). The other structure contains a heavily doped *n*-GaN epilayer: d = 6.2 μm, $N_e = 3.6\cdot 10^{19}$ cm^{-3}, μ = 122 cm^2/V·s (sample 3). The concentration and mobility were determined by means of Hall and conductivity measurements at room temperature. It is known that at such doping levels GaN demonstrates formation of an impurity band [11] and free carriers are not frozen even at liquid helium temperatures.

The reflectivity spectra were investigated at room temperature using a Fourier spectrometer and a pyroelectric detector in the spectral range 8–80 meV (2–20 THz). A globar and a mercury lamp were used as sources of radiation. The incidence angle θ was about 11°, reflectivity measurements were carried out with an angular aperture of $\Delta\theta$ = 16°. An aluminum mirror was used as the reflectivity reference.

Experimental reflectivity spectra demonstrate the same peculiarities as were predicted by theoretical simulation (figure 2 (b), (c)). Because it is difficult to determine the exact value of the reference mirror reflectivity, we normalized the experimental spectra to the local maximum of reflectivity in the vicinity of ω_{TO}. The normalized experimental reflectivity spectra are presented in figure 3 by black lines. The theoretical spectra were normalized by the same manner and were fitted to the experimental ones. Three quantities, namely N_e, μ and γ, were used as fitting parameters. The theoretical reflectivity spectra at the optimum values of the fitting parameters (N_e', μ' and γ') are shown in figure 3 by red lines. There is a good agreement between the experimental and theoretical spectra for both samples. In particular, both spectra demonstrate a local reflectivity minimum in the close vicinity of the frequency of plasmon polariton mode $\hbar\omega_{-}$ (figure 3 (a), (b)).

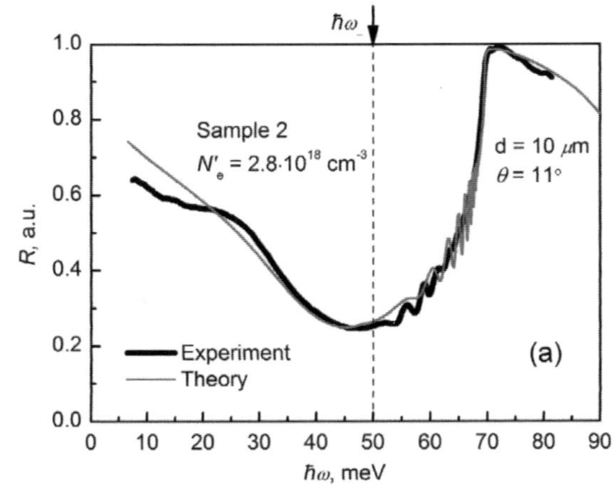

Figure 3. Normalized reflectivity spectra of the samples 2 and 3: experiment (black curves) and theoretical fit (red curves). Dotted lines denote the spectral position of the low frequency plasmon phonon mode $\hbar\omega_{-}$.

It should be noted that the values of the electron concentration and mobility obtained from the reflectivity spectra ($N_e' = 2.8\cdot10^{18}$ cm^{-3} and $\mu' = 180$ cm^2/V·s for the sample 2, $N_e' = 1.9\cdot10^{19}$ cm^{-3} and $\mu' = 179$ cm^2/V·s) differ slightly from the values determined by Hall measurements. The latter are related to the average values across the whole GaN layer, but the former characterize only a subsurface region of the layer. The value of the damping constant for TO phonons (γ') obtained from the reflectivity spectra is $7.5\cdot10^{11}$ s^{-1}. Thus, study of the reflectivity spectra is an effective method of contactless determination of electron concentration, mobility and TO phonon damping constant of *n*-GaN epitaxial layers.

4. Expected emission spectra

In the section 3 it was demonstrated that the reflectivity spectra considered in section 2 fit well the experimental results on radiation reflection for the samples 2 and 3. It is reasonable to suppose that the simulated absorptivity spectra for these samples (see figure 2 (e), (f)) will describe well the radiation absorption in the samples. In accordance with Kirchhoff's law, for a body at thermal equilibrium the emissivity is equal to the absorptivity. That is why one can describe blackbody-like emission of the considered samples by means of Planck theory. The spectral density of radiation emission from an *n*-GaN epitaxial layer in the direction perpendicular to the sample surface in the limits of solid angle $\Omega << 2\pi$ is given by:

$$D(\omega,T) = \frac{\hbar\omega^3}{4\pi^3 c^2\left[\exp(\frac{\hbar\omega}{k_B T}) - 1\right]}\cdot A(\omega)\cdot\Omega, \qquad (3)$$

where T is the temperature of the *n*-GaN epitaxial layer, k_B is the Boltzmann constant, c is the light velocity.

Figure 4 demonstrates the simulation results on the spectral dependencies of D for the samples 2 and 3 at different temperatures T. In every sample at high enough temperature one can see two selective emission bands in the mid infrared range in the vicinity of plasmon phonon modes ω_- and ω_+. The emission band in the vicinity of ω_- is much narrower than the one in the vicinity of ω_+. The increase of electron concentration results in a "blue" shift of both emission bands. There is an opportunity to modulate the epilayer temperature by means of electric current, for instance, applying periodical pulses of electric voltage. This allows to develop emitters of modulated mid infrared radiation.

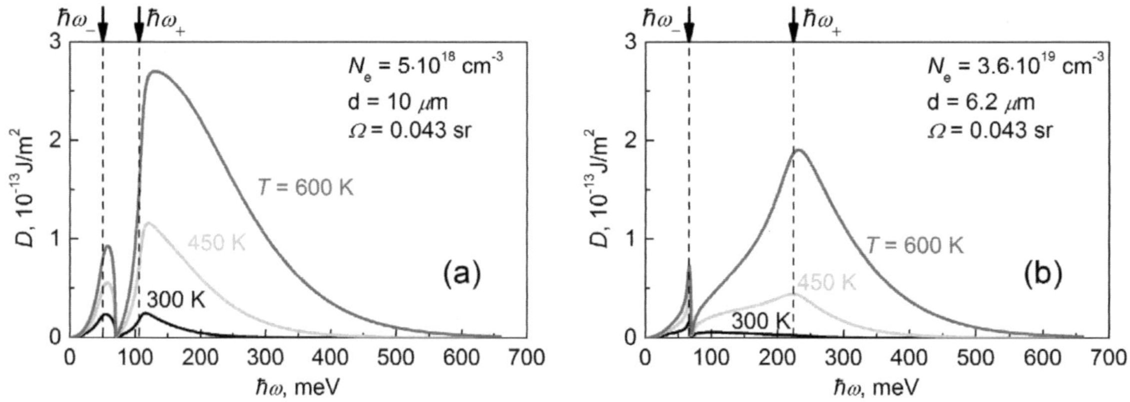

Figure 4. Dependence of the spectral density of radiation emission on photon energy. Theoretical simulation for the samples 2 (a) and 3 (b) at different temperatures. The solid angle $\Omega = 0.043$ sr corresponds to an angular aperture $\Delta\theta = 16°$.

5. Conclusions

Transverse and longitudinal electromagnetic waves in *n*-GaN epitaxial layers have been studied theoretically and experimentally. Epitaxial layers with electron concentration in the range 10^{17} – 10^{19} cm^{-3} have been examined. Simulation of the reflectivity and absorptivity spectra has been performed in the terahertz and mid infrared spectral ranges (the angular frequency range of 10^{13} – 10^{15} s^{-1}). Additionally, the radiation emission spectra have been simulated for various temperatures of the GaN epitaxial layer. It has been shown that the resonance peculiarities of the optical spectra are located closely to the frequencies of the coupled plasmon phonon modes. Experimental studies of the reflectivity spectra have been performed in the spectral range 8–80 meV (2–20 THz). The experimental spectra are well fitted by the simulated ones and can be used for contactless determination of the electron concentration, mobility and TO phonon damping constant in GaN epitaxal layers. The experimental and theoretical results of the present work provide insights for the development of GaN-based devices aimed to absorb/emit terahertz or mid infrared radiation selectively. In particular, electrically controlled emitters of modulated mid infrared radiation can be developed.

Acknowledgements

This work was supported by the Russian Foundation for Basic Research (grant 16-32-60085-mol-a-dk) and the Ministry of Education and Science of the Russian Federation (state assignment).

References

[1] Shalygin V A *et al* 2009 *J. Appl. Phys.* **106** 123523
[2] Shalygin V A, Vorobjev L E, Firsov D A, Sofronov A N, Melentyev G A, Lundin W V, Nikolaev A E, Sakharov A V and Tsatsulnikov A F 2011 *J. Appl. Phys.* **109** 073108
[3] El Fatimy A *et al* 2010 *J. Appl. Phys.* **107** 024504
[4] Melentyev G A, Shalygin V A, Moldavskaya M D, Panevin V Yu, Vorobjev L E, Firsov D A,

Nykänen H, Riuttanen L, Svensk O and Suihkonen S 2015 *J. Phys.: Conf. Series* **586** 012005
[5] Yu P Y and Cardona M 2010 *Fundamentals of Semiconductors* 4th ed (Berlin–Heidelberg: Springer)
[6] Ukhanov Yu I 1977 *Optical Properties of Semiconductors* in Russian (Moscow: Nauka)
[7] Kozawa T, Kachi T, Kano H, Taga Y, Hashimoto M, Koide N and Manabe K 1994 *J. Appl. Phys.* **75** 1098
[8] Perlin P, Camassel J, Knap W, Taliercio T, Chervin J C, Suski T, Grzegory I and Porowski S 1995 *Appl. Phys. Lett.* **67** 2524
[9] Harima H, Sakashita H and Nakashima S 1998 *Mater. Sci. Forum* **264-268** 1363
[10] Kuroda N, Kitayama T, Nishi Y, Saiki K, Yokoi H, Watanabe J, Cho M, Egawa T and Ishikawa H 2006 *Jpn. J. Appl. Phys.* **45** 646
[11] Agekyan V F *et al* 2015 *Phys. Solid State* **57** 787
[12] Barker A S and Ilegems M 1973 *Phys. Rev.* B **7** 743
[13] Davydov V Yu, Kitaev Yu E, Goncharuk I N, Smirnov A N, Graul J, Semchinova O, Uffmann D, Smirnov M B, Mirgorodsky A P and Evarestov R A 1998 *Phys. Rev.* B **58** 12899
[14] Born M and Wolf E 2003 *Principles of Optics: Electromagnetic Theory of Propagation, Interference and Diffraction of Light* 7th ed (Cambridge: Cambridge University Press) pp 735–789
[15] Voronkova E M, Grechushnikov B N, Distler G I and Petrov I P 1965 *Optical materials for infrared technology* in Russian (Moscow: Nauka)

Influence of indium doping on the electrical properties of $Ge_2Sb_2Te_5$ thin films for nonvolatile phase change memory devices

P I Lazarenko[1], A A Sherchenkov[1], S A Kozyukhin[2], A V Babich[1], H P Nguen[3], S P Timoshenkov[1], D G Gromov[1], A O Yakubov[1] and D Y Terekhov[1]

[1] National Research University of Electronic Technology, 124498 Zelenograd, Russia
[2] Kurnakov Institute of General and Inorganic Chemistry of the Russian Academy of Sciences, 119991 Moscow, Russia
[3] Industrial University of Ho Chi Minh City, Ho Chi Minh City, Viet Nam

E-mail: aka.jum@gmail.com

Abstract. In this article the influence of different amounts of In (0, 0.5, 1 and 3 wt. %) on the thermal and electrical properties of $Ge_2Sb_2Te_5$ thin films for nonvolatile phase change memory devices is investigated. Crystallization temperature, resistivity, width of the optical band gap, Urbach energy, activation energy and type of conductivity are estimated for all investigated compounds. Storage and data processing times of the PCM cells on the basis of investigated materials were calculated. Nonmonotonic concentration dependences of properties were observed.

1. Introduction

Memory devices based on the flash technology are the most widespread today. However, flash technology has a number of drawbacks [1]. In this regard, development of new types of memory is quite essential. Promising memory types at this moment are magneto-resistive (MRAM), ferroelectric (FRAM), resistive (RRAM) and phase change (PCM) memories [1–3].

In this paper we investigated the materials for phase change memory application. Principle of operation of the PCM is based on the rapid reversible phase transformation between amorphous and crystalline states, which takes place in PCM materials under low-energy external influences. The material of $Ge_2Sb_2Te_5$ (GST225) is considered to be the most promising for use in PCM devices. It should be noted that some applications (for example in cosmic units) impose specific requirements on the properties of PCM materials. So, we must be able to control the properties of PCM materials. However, maintaining control of the properties is a tough problem due to the insensitivity of most chalcogenide glassy semiconductors to the doping because of Fermi level pinning.

One of the possible effective ways for the modification of the properties of materials is using substitutional doping [4,5]. Indium is in the same period of the Periodic table as antimony and has a close atomic radius (0.159 nm for Sb and 0.166 nm for In). By replacing antimony atoms with indium, the binding energy decreases from 277 kJ/mol for the Sb–Te bond to 218 kJ/mol for that of In–Te [6]. In this case, it is a good reason to believe that indium can substitute antimony. It was shown [7] that incorporation of indium into GST225 shifts peaks of Raman spectra to lower frequencies that can be caused by the incorporation of indium atoms into both germanium telluride and antimony telluride sublattices. Indium is also a good choice because recently it was reported that the introduction of indium into GST225 increased an optical contrast and changed the crystallization speed [8].

However, the influence of indium doping on the electrical properties of GST225 is not well understood. Thus, in this study we investigated the influence of indium doping on the electrical properties of $Ge_2Sb_2Te_5$ thin films, which are critical for the operation of PCM devices.

2. Experimental setup and samples

The initial $Ge_2Sb_2Te_5$ alloys doped with different amounts of In [0 wt. % (alloy #1), 0.5 wt. % (alloy #2), 1 wt. % (alloy #3) and 3 wt. % (alloy #4)] were synthesized with using of quenching technique [9]. The materials (99.99% purity) were sealed in evacuated ($5\cdot10^{-3}$ Pa) quartz ampoules then step by step heated to 800 °C in a rocking furnace to ensure homogeneous melt. The melt was quenched to the air.

Thin films were prepared by thermal evaporation of the synthesized GST225 in vacuum. Residual pressure in the chamber was 10^{-3} Pa. The substrate temperature during evaporation did not exceed 50°C.

The films thicknesses were determined by atomic force microscopy (AFM, NT-MDT SolverPro). The structure of obtained thin films was determined by X-Ray diffraction (XRD, Rigaku D/MAX). According to XRD, as-deposited films have the amorphous structures.

X-Ray Microanalysis (CAMEBAX) was used to determine thin films compositions. Results of composition analysis are presented in table 1. Analysis revealed agreement between compositions of the initial bulk alloys and respective deposited thin films.

Table 1. Results of composition analysis of deposited thin films

Initial bulk alloy	Thin film composition according to X-Ray microanalysis (at. un.), accuracy ±5%				Indium content in thin films (wt.%, calculation)
	Ge	Sb	Te	In	
# 2	0.2236	0.2234	0.5483	0.0046	0.45
# 3	0.1984	0.2649	0.5280	0.0087	0.96
# 4	0.1910	0.2673	0.5064	0.0353	3.80

The planar structures containing Al electrodes with a fixed interelectrode distance and the investigated thin films deposited upon them were fabricated on oxidized c-Si substrates to study their electrical properties. Interelectrode distances were formed by the photolithography method and were equal to 7 mcm. Planar structures were mounted in the glass-to-metal package.

Measurements of current-voltage characteristics (I-V characteristics or CVC) were carried out in the isothermal regime at the temperature range from room temperature to 70 °C with a step of 10 °C and voltage range from 0.1 to 10 V with a step of 0.1 V. Temperature dependences of resistivity of planar samples were measured in the heat-up regime in the temperature range from room temperature to 300 °C with a step of 0.1 °C and a heating rate of 1 °C.

The set-up on the basis of a multimeter HP 34401A-01 and a power supply Agilent E3674 was used for the investigation of thermopower and determination of the type of conductivity for the investigated films. The samples containing Al electrodes and the investigated thin films deposited upon them were fabricated on pyroceramics substrates. The distances between electrodes were 5 mm, and temperature gradient was 5 °C.

Differential scanning calorimetry (DSC-50, Shimadzu) at 8 different heating rates from 1 to 90 deg/min in a nitrogen flow (20 ml/min) was used to examine the thermal properties, activation energies and times of crystallization of investigated materials. Thin films were scraped off from c-Si wafers and pressed into Al pans. The samples had masses of few milligrams. Empty Al pans were used as references.

3. Experimental results

Ohmic regions at low electric field strength ($E \leq 10^3$ V/cm) were observed for all investigated compounds by I-V measurements. This was a reason why the value of electrical strength of 10^3 V/cm was chosen for measurements of temperature dependences of resistivity (figure 1).

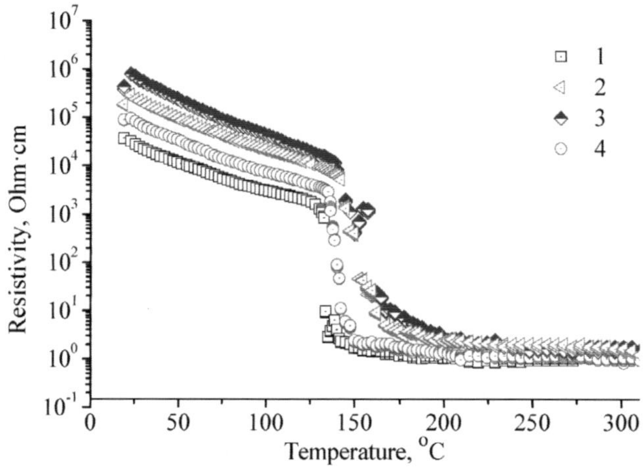

Figure 1. Temperature dependencies of resistivity for the thin films of GST225 with different contents of In: 1 – 0 wt. % In; 2 – 0.5 wt. % In; 3 – 1 wt. % In; 4 – 3 wt. % In.

Exponential temperature dependences in the range from room temperature to 120 °C were observed for all compounds. The temperature dependence of the conductivity of amorphous semiconductor typically shows an exponential behavior [10]. This allows to get the information on the activation energy of conductivity:

$$\sigma = \sigma_0 \, exp\left(-\frac{E_a}{kT}\right), \tag{1}$$

where σ is the conductivity, σ_0 is a constant, E_a is the activation energy of conductivity, k is the Boltzmann constant, and T is the absolute temperature.

Results of the activation energies calculations are presented in table 2. A width of optical band gap (E_g) and Urbach energy (E_0) were estimated by means of spectrophotometry and ellipsometry and are also shown in the table.

Table 2. Energy parameters for GST225 + In thin films

In, wt. %	E_g, eV	E_0, eV	E_a, eV
0	0.61	0.13	0.29
0.5	0.67	0.17	0.29
1	0.62	0.18	0.32
3	0.64	0.15	0.30

Results of thermopower measurements showed that holes are the majority carriers in the investigated thin films. According to the dual-channel model proposed by P. Nagels [11], p-type conductivity in chalcogenide glassy semiconductors is formed by the charge carrier transport via the delocalized states of the valence band, and localized states of the valence band tail. As the values of E_a

of the investigated thin films are close to the half of the optical band gap width, the main contribution in the conductivity is introduced by the transport of charge carriers via the delocalized states. It should be noted that the incorporation of indium into $Ge_2Sb_2Te_5$ leads to the increase of Urbach tails, which pointed out to the growth of the structure disorder.

Sharp drops in resistivity for all investigated materials were observed in the range from 120 to 170 °C. Taking into account literature data [12] and our results obtained from DSC (see figure 2) and XRD measurements, this sharp drops can be attributed to the phase transitions from amorphous to metastable fcc (NaCl type) crystalline state.

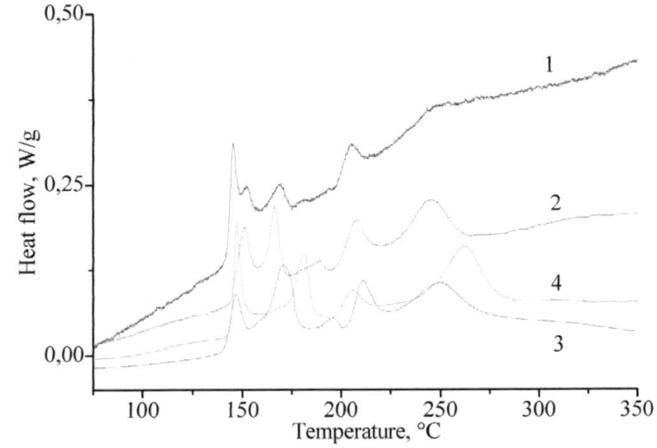

Figure 2. DSC scans of as-deposited GST225 thin films with different contents of In: 1 – 0 wt. % In; 2 – 0.5 wt. % In; 3 – 1 wt. % In; 4 – 3 wt. % In.

The onset crystallization temperatures determined from temperature dependencies of resistivity (T_{on} is the temperature of the beginning of sharp drop) for compounds with different In content (table 3) adequately correlate with those found by DSC. Discrepancies can be attributed to the difference in the heating rates, which were lower in the case of resistivity measurements.

Table 3. Energy parameters for GST225 + In thin films

In, wt. %	ρ vs T				DSC
	T_{on}, °C	ρ_{25}, Ohm·cm	ρ_{200}, Ohm·cm	ρ_{25}/ρ_{200}	T_{on}, °C
0	132	$1.6\cdot10^4$	1.6	$1\cdot10^4$	141
0.5	136	$4.8\cdot10^4$	1.9	$2.5\cdot10^4$	145
1	143	$2\cdot10^5$	1.7	$1.2\cdot10^5$	150
3	140	$6\cdot10^4$	2.9	$2.1\cdot10^4$	152

Incorporation of In leads to the increase of resistivity of amorphous films and of the difference in the resistivity values for amorphous and crystalline states (see table 3), which is important for the reliable work of PCM cells. Except for the thin film of alloy #4, the contrasts of resistivities are not very different from the undoped GST225. The symbols ρ_{25} and ρ_{200} in table 3 represent resistivities of the amorphous films at 25 °C and crystalline films at 200 °C, respectively.

Understanding the kinetics of phase transition allows predicting the behavior of materials under certain temperature conditions, which is important for the development and optimization of technology. In particular, the kinetics of crystallization determines the performance and stability of the phase change memory cell. Parameters determining crystallization kinetics of the investigated films, the so called kinetic triplet (effective activation energy E_{ac}, pre-exponential factor A, and reaction model) were estimated by using the previously proposed procedure [13] and are presented in table 4.

Table 4. Determined kinetic parameters of crystallization process for GST225 + In thin films

In, wt. %	E_{ac}, eV			ln$A$, c^{-1}			Model
	α=0.05	α=0.95	Average values	α=0.05	α=0.95	Average values	
0	2.08	1.42	1.75	55.50	41.89	48.61	2nd and 3rd order reaction models
0.5	1.93	1.64	1.87	50.73	33.43	42.78	
1	1.63	1.21	1.52	51.14	33.85	43.21	
3	1.76	1.58	1.80	46.13	45.89	49.72	

Determined kinetic triplets were used to predict the possible storage and data processing times of PCM cells on the basis of the investigated materials with crystallization at room temperature and at 400 °C, respectively. We used the procedure proposed in the work [13]. The calculated data on processing and storage times for PCM cells on the basis of investigated materials are presented in figure 3. As can be seen, the compositions with 0.5 and 3 wt. % In allow to increase sufficiently a storage time in comparison with the undoped material (figure 3). On the other hand, crystallization time at 400 °C decreases nearly by an order of magnitude for the thin film of alloy #3 in comparison with that of the undoped GST225.

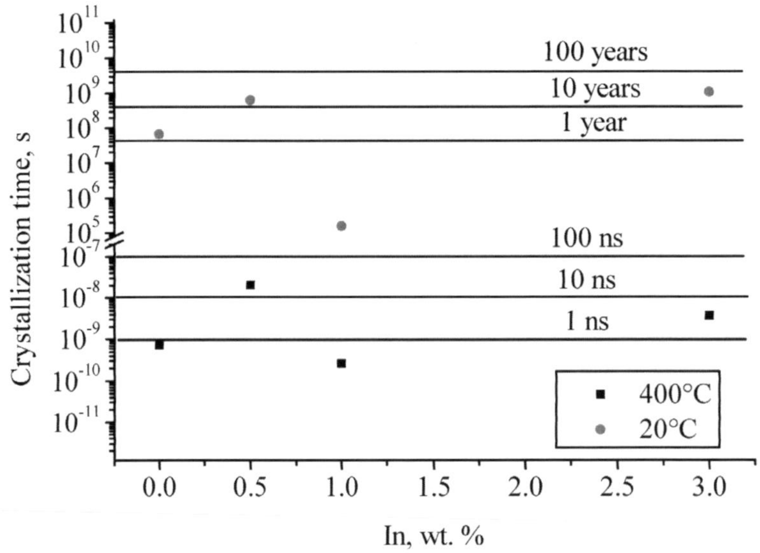

Figure 3. The dependencies of crystallization time on concentration of In in GST225 thin films. Calculations were carried out for two temperatures (20 and 400 °C).

It should be noted, that deviations from monotonous concentration dependences for thermal and electrical parameters were observed consisting in the appearance of extremes on these dependences.

This effect can be explained by using the percolation theory according to which the formation of an infinite cluster is accompanied by critical phenomena [5].

4. Conclusion

The influence of In doping on the properties of $Ge_2Sb_2Te_5$ thin films was investigated. Introduction of low concentrations of In (upon 3 wt. %) in $Ge_2Sb_2Te_5$ thin films is accompanied by the changes of crystallization temperature, resistivity, optical band gap width, Urbach energy, the activation energy of conductivity, storage and data processing times. Introduction of In to $Ge_2Sb_2Te_5$ allows to control the properties of materials for the phase change memory application, which is important for the optimization of PCM technology.

Acknowledgments

This work was supported by the Ministry of Education and Science of Russian Federation (FPP, project N 14.578.21.0085, project ID RFMEFI57814X0085).

References

[1] Krynder M H and Kim C S 2009 *IEEE Transactions On Magnetics.* **45** 3406
[2] Belov A N, Pyatilova O V and Vorobiev M I 2014 *Advances in Nanoparticles* **3** 1
[3] Solnyshkin A V, Kislova I L, Silibin M V and Kiselev D A 2014 *Ferroelectrics* **469** 144
[4] Wang K, Steamer C, Wamwangi D, Ziegler S and Wuttig M 2007 *Microsyst Technol.* **13** 203
[5] Kozyukhin S, Sherchenkov A, Babich A, Lazarenko P, Nguyen H P and Prikhodko O 2014 *Canadian Journal of Physics* **92** 684
[6] Lide D R; Haynes W N *CRC Handbook Chemistry and Physics 90th ed*, 2009 ed D R Lide (Boca Raton, CRC Press)
[7] Kozyukhin S, Veres M, Nguyen H P, Ingram A and Kudoyarova V 2013 *Physics Procedia* **44** 82
[8] Wang K, Steimer C, Wamwangi D, Ziegler S and Wuttig M 2005 *Appl. Phys. A* **80** 1611
[9] Kozyukhin S, Sherchenkov A, Gorshkova E, Kudoyarova V and Vargunin A 2010 *Phys.Status Solidi C* **7** 848
[10] Mott N. and Davis E 1979 *Electronic Processes in Non-Crystalline Materials* (Oxford, Oxford Univ.)
[11] Nagels P 1985 *Topic in Applied Physics* **36** 113
[12] Friedrich I, Weidenhof V, Njoroge W, Franz P and Wuttig M 2000 *Journal Of Applied Physics* **87** 4130
[13] Sherchenkov A, Kozyukhin S and Babich A 2014 *J Therm Anal Calorim.* **117** 1509

Journal of Physics: Conference Series **690** (2016) 012007 doi:10.1088/1742-6596/690/1/012007

On peculiarities of the electron transport in doped manganites

M V Fetisova, O A Martynova, and V E Gasumyants

Peter the Great St. Petersburg Polytechnic University, 29 Polytechnicheskaya str., St. Petersburg, 195251 Russia

E-mail: vgasum@yandex.ru

Abstract. A systematic experimental study of the temperature dependences of the resistivity and thermopower, S, in hole- and electron-doped manganites of the $Re\text{MnO}_3$ (Re = La, Ce) systems doped by strontium in a wide range of content was performed. The main specific features of the thermopower behavior at temperatures both below and above the temperature of the magneto-ordered transition were revealed and discussed. The $S(T)$ dependences were observed to be quite complicated in both temperature ranges. The $S(T = 300\text{ K})$ values are negative for the $Ce_{1-x}Sr_xMnO_3$ samples, while for the $La_{1-x}Sr_xMnO_3$ ones they are positive for slightly doped samples (at $x < 0.2$), but negative for a high doping level ($x > 0.333$). The thermopower results are qualitatively discussed with respect to peculiarities of the conduction process.

1. Introduction

Although manganites have been known since 1950s, their intensive researches began after 1993 due to discovering the effect of the colossal magnetoresistance in these materials [1–3]. The specific feature of these materials influencing strongly their physical properties is a very complicated magnetic phase diagram [4]. As a result, when varying temperature and doping level different types of phase transitions can occur and manganite samples exhibit complex magnetic properties [5, 6]. Under varying sample composition the temperature of the magneto-ordered transition, T_c, can change in a wide range and the magneto-ordered state can be either ferromagnetic or antiferromagnetic of various types [4-6]. In spite of extensive discussions and a huge number of publications devoted to the study of different manganite properties, the nature of the colossal magnetoresistance effect remains unclear. However, unusual properties of these materials go well beyond this phenomenon. In particular, questions on the mechanism of the conduction process in different temperature ranges (both below and above T_c) and approaches allowing one to describe the experimentally observed behavior of the transport coefficients are still undetermined and call for further investigations of the electron transport peculiarities in manganites of different compositions.

Such a situation is related to the fact that the presence of strong correlation effects is characteristic of manganites. This affects all the experimentally observed properties of these materials and makes it impossible to use the classical models based on the single-electron approximation for their interpretation and analysis. For this reason, a lot of models were proposed to describe unusual properties of manganites (see, for example, reviews [5, 6]), but at present none of them can be considered as generally accepted. In particular, some proposed approaches have aimed to explain experimental results on a character of the temperature dependences of the resistivity and thermopower and their modifications under doping influence. One of the purposes of such studies was to clear up the conductivity type in manganites. It is well-known, that the charge-carrier type is usually determined from the experimental investigations of either the Hall coefficient or thermopower. However, these coefficients in manganites can often have different signs in different temperature ranges [7, 8]. Besides, contrary to the conductivity and magnetoresistance effect, the peculiarities of the behavior of other transport coefficients (in particular, the thermopower) have not been studied in details and most of the experimental results are related to a limited range of doping.

For the above reasons, this paper is devoted to a systematic experimental study of the specific features of the electron transport in manganites of different systems (both hole- and electron-doped) when changing the doping level in a wide range.

2. Samples and Experimental Details

The ceramic samples of $La_{1-x}Sr_xMnO_3$ ($x = 0.15$, 0.2, 0.28, 0.333, 0.38, 0.5, 0.666) and $Ce_{1-x}Sr_xMnO_3$ ($x = 0.5$, 0.67) were used for investigations. Such compositions were chosen to trace the modification of the electron transport in a wide range of doping and also to compare the effect of strontium influence in cases of different manganite systems. All the samples were prepared by the standard solid-state processing technique from the high-purity oxides mixed in the required proportions. X-ray diffraction analysis has shown them to be almost of single phase with amount of foreign impurities not exceeding 1-2%. The sample homogeneity was controlled by measuring the local values of the thermopower in various points on the sample surface at room temperature.

The resistance temperature dependences, $R(T)$, were measured by the standard four-probe low-frequency ac ($f = 20$ Hz) method. The thermopower, S, was measured by a differential method relative to copper electrodes at the temperature difference between the two ends of a sample of about 2K and then calculated by correcting for the absolute thermopower of copper. Both transport coefficients were measured in the temperature range of $T = 4.2$–300 K.

3. Experimental Results and Discussion

The $R(T)$ dependences are typical for ceramic manganite samples. As an example, figure 1 shows the results obtained for two samples of the $La_{1-x}Sr_xMnO_3$ system and $Ce_{1-x}Sr_xMnO_3$ samples. The T_c value for the $La_{1-x}Sr_xMnO_3$ sample with $x = 0.15$ is about 270 K and at high temperatures ($T > T_c$) resistance increases exponentially with decreasing temperature, reaches a maximum at $T \approx T_c$ after that decreases sharply at temperatures down to 140 K. At lower temperatures the resistance increases again, that is most likely associated with an influence of grain boundaries in a ceramic sample having a high resistance. For $La_{1-x}Sr_xMnO_3$ samples with a higher strontium content, the T_c value exceeds 300 K in agreement with the data of other studies [9, 10]. For this reason, a sharp drop of the resistance becomes not observable and its value increases with decreasing temperature throughout the measured temperature range. Thus, most of the experimental results obtained for the $La_{1-x}Sr_xMnO_3$ system are related to temperatures below T_c, giving us a possibility to use these samples for revealing specific features of the thermopower behavior in the magneto-ordered state.

The $R(T)$ dependences for $Ce_{1-x}Sr_xMnO_3$ samples are mainly characterized by the same peculiarities. In this case the magneto-ordered transition appears at much lower temperatures and T_c value decreases slightly with increasing strontium content (from 105 K for the sample with $x = 0.5$ to 100 K for the one with $x = 0.67$, see figure 1). Thus, in cases of $Ce_{1-x}Sr_xMnO_3$ samples we can obtain the results characterizing their properties at temperature both below and above T_c. Note that the observed T_c values correspond well to published results for manganites of the Ce-based system [11]. As for the temperature dependence of the resistance, in the $Ce_{0.33}Sr_{0.67}MnO_3$ sample a characteristic peak on the $R(T)$ dependence is clearly seen near $T \approx T_c$; at lower temperatures the resistance decreases. The values of resistance for $Ce_{0.5}Sr_{0.5}MnO_3$ are much higher (by about two orders of magnitude), besides the transition to the magneto-ordered state manifests itself significantly weaker and a drop in the resistance value at $T < T_c$ is weak enough when comparing to the sample with $x = 0.67$. These facts are probably the results of a grain boundaries influence. Nevertheless, for both $Ce_{1-x}Sr_xMnO_3$ samples the $R(T)$ dependences at $T > T_c$ are qualitatively similar to each other.

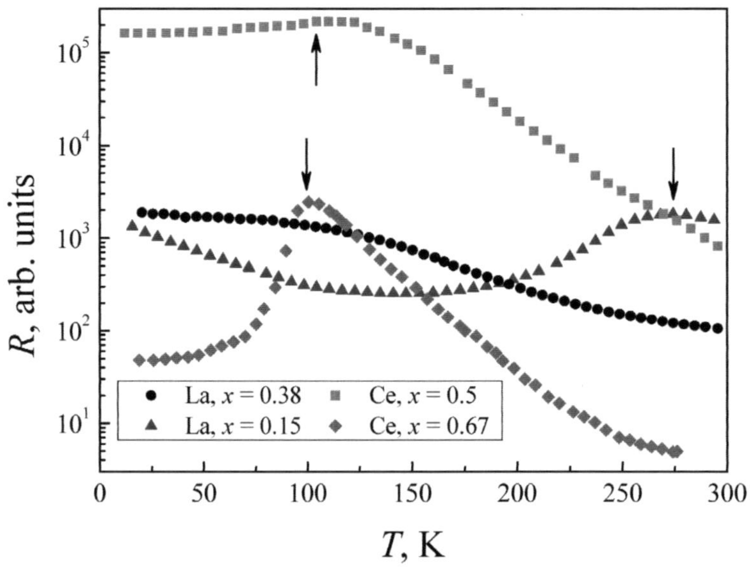

Figure 1. Resistance vs temperature for $La_{1-x}Sr_xMnO_3$ (La, x = 0.15, 0.38) and $Ce_{1-x}Sr_xMnO_3$ (Ce, x = 0.5, 0.67) samples. Arrows indicate the temperatures of the magneto-ordered transition.

The main aim of our work was to study the thermopower behavior. This coefficient is well-known to be much less sensitive to an influence of grain boundaries and other defects compared to the resistance. As a result, the thermopower study in ceramic samples allows one to obtain the information on the electron properties of a material with a given composition eliminating (or at least minimizing) the effect of sample macrostructure.

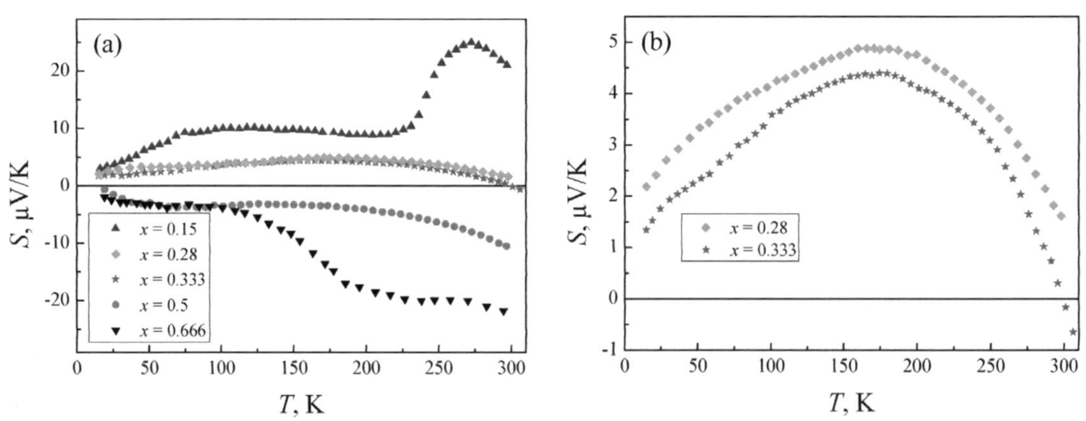

Figure 2. Temperature dependences of the thermopower for $La_{1-x}Sr_xMnO_3$ samples.

The $S(T)$ dependence for $La_{1-x}Sr_xMnO_3$ samples are presented in figure 2 (a). One can see that in a wide rage of doping there is a systematic variation of the thermopower with increasing lanthanum content. For the sample with $x = 0.15$ the thermopower is positive and for temperatures near T_c the $S(T)$ dependence corresponds qualitatively to the temperature dependence of resistance: at $T > T_c$ the thermopower increases slightly up to 24 µV/K and then falls sharply down to 9 µV/K at $T = 210$ K. In the low-temperature region (at $T < T_c$, i.e. in the magneto-ordered state) the thermopower shows a slight increase (up to 10.3 µV/K), demonstrates a broad maximum at $T = 115$ K, and then falls down to 2.8 µV/K at $T = 14$ K. For other samples with $x \leq 0.5$ having a high T_c the thermopower at $T < 300$ K does not demonstrate abrupt changes (like the case of the $R(T)$ dependences), but a character of its temperature dependences is completely similar to that observed for the sample with $x = 0.15$ at $T < T_c$. The only feature which can additionally be noted is that in case of $x = 0.5$ a slight decrease of the negative thermopower values is observed at $T < 90$ K.

With increasing strontium content, the S ($T = 300$ K) value decreases noticeably, becomes very close to zero for intermediate doping ($x = 0.28$-0.333) and negative for higher strontium contents (see figure 3). The experimental results obtained for two samples (with $x = 0.28$, 0.333) demonstrating very small thermopower values are additionally shown in figure 2 (b). It is seen that even in case of such small S values the $S(T)$ dependences are very similar in shape. It should be noted that the thermopower for the $La_{0.667}Sr_{0.333}MnO_3$ sample shows the sign change at a temperature near $T = 300$ K. One can propose that the analogous property is characteristic of the $La_{0.72}Sr_{0.28}MnO_3$ sample, but at higher temperatures.

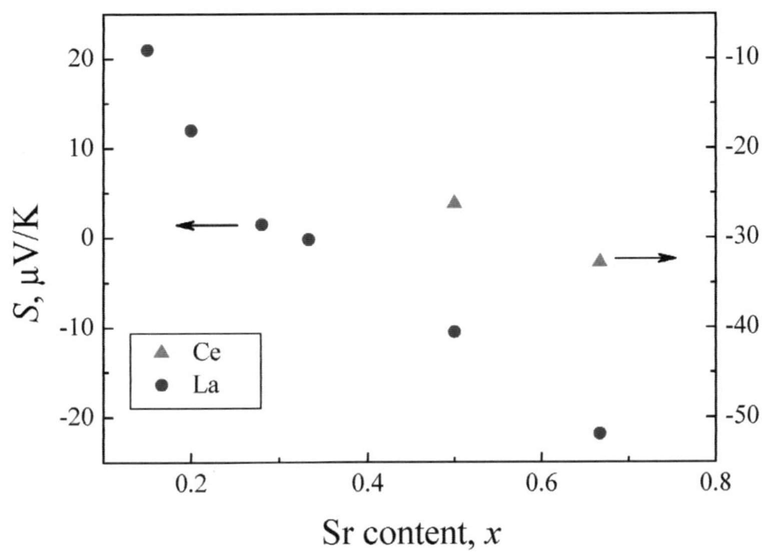

Figure 3. Thermopower value at $T = 300$ K vs strontium content for $La_{1-x}Sr_xMnO_3$ (La) and $Ce_{1-x}Sr_xMnO_3$ (Ce) samples.

Thus, for all samples with $x \leq 0.5$ the main peculiarities of the $S(T)$ curves at $T < T_c$ remain qualitatively unchanged, namely, the thermopower in the magneto-ordered state changes slightly with decreasing temperature and demonstrates a maximum whose position depends on the strontium content. However, the $S(T)$ dependence obtained for the $La_{0.334}Sr_{0.666}MnO_3$ sample is characterized by the presence of additional features. In this case the thermopower, being negative in the whole measured temperature range, changes slightly at high temperatures ($T > 190$ K), falls (in absolute value) sharply enough as temperature decreases from 190 K to 115 K, and changes slightly again at lower temperatures. Such distinctive features of the thermopower behavior are most probably related to the fact that the type of the magneto-ordered state in the $La_{1-x}Sr_xMnO_3$ system changes with increasing strontium content. The samples with $x = 0.28$-0.5 are of ferromagnetic phase in the whole measured temperature range, while the $La_{0.334}Sr_{0.666}MnO_3$ sample is paramagnetic at high temperatures and antiferromagnetic at low temperatures [9, 12]. Besides, in this sample a transition from the tetragonal to the orthorhombic structure occurs with decreasing temperature [9, 12]. These peculiarities result in the observed modification of the thermopower temperature dependence.

The results on the $S(T)$ dependences obtained by us for samples of the $La_{1-x}Sr_xMnO_3$ system are in qualitative agreement with the published data for hole-doped manganites of different systems. For example, a decrease in the thermopower value with increasing sodium content in $La_{1-x}Na_xMnO_3$ was observed in [13], a change in the thermopower sign in the magneto-ordered state was observed in $Nd_{1-x}(Sr,Pb)_xMnO_3$ [14] and $La_{0.6}Eu_{0.07}Sr_{0.33}MnO_3$ [15], the experimental $S(T)$ dependences for $La_{0.67-x}R_xSr_{0.33}MnO_3$ (R = Eu, Gd) [7] look to be similar to those obtained by us for the $La_{1-x}Sr_xMnO_3$ samples with $x = 0.28$, 0.333. This suggests that the specific features of the thermopower behavior in the magneto-ordered state including both a character of the $S(T)$ dependences and the thermopower variation under doping observed for samples of the $La_{1-x}Sr_xMnO_3$ system are typical for hole-doped manganites as a whole. Besides, these results clearly indicate that to describe the thermopower in the magneto-ordered state one should involve a two-band model and take into account a contribution of both hole and electrons in the conduction process.

The $S(T)$ dependences for $Ce_{1-x}Sr_xMnO_3$ samples are shown in figure 4. The thermopower at $T = 300$ K is negative, increasing strontium content leads to a decrease in the $S(T = 300$ K) value like the case of the $La_{1-x}Sr_xMnO_3$ system (see figure 3). As temperature decreases, the absolute value of S changes at first slightly, but in the temperature range from 190 K to 115 K for the sample with $x = 0.5$ and from 250 K to 150 K for the one with $x = 0.67$ falls sharply down to very small values. The latter points to an existence of two types of charge carriers (electrons and holes) at $T > T_c$, the relationship between whose concentrations changes strongly with decreasing temperature. Note that analogous results (a negative thermopower at T = 120-350 K and a sharp decrease in its absolute value at temperatures from 200 K to 120 K) was obtained for the $Sr_{0.9}Ce_{0.1}MnO_3$ sample [16].

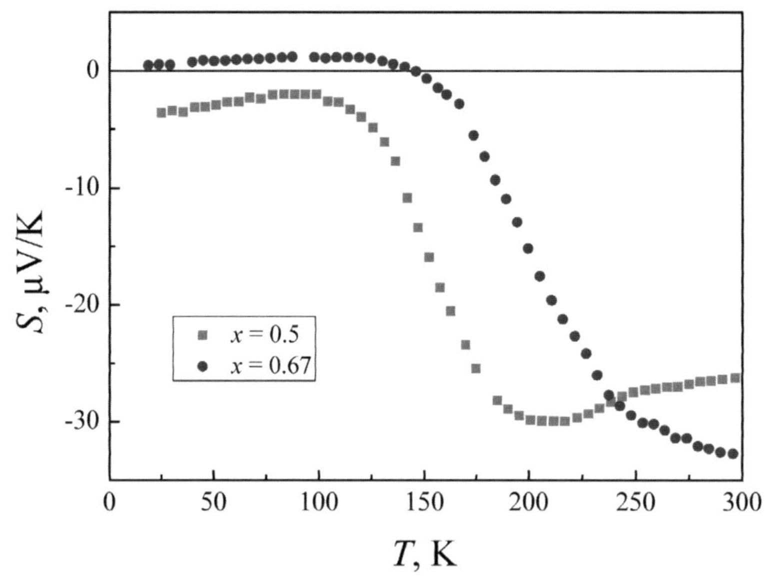

Figure 4. Temperature dependences of the thermopower for $Ce_{1-x}Sr_xMnO_3$ samples

It should be noted that despite a minor change in the T_c value in the studied $Ce_{1-x}Sr_xMnO_3$ samples (being about 5 K) obtained from the resistance data (see figure 1), the temperature range of a sharp thermopower decrease shifts upwards by about 50 K with increasing x. This clearly shows that strontium doping influences strongly the concentrations of free electrons and holes or their ratio at temperatures above T_c.

At low temperatures (below T_c) the thermopower for both samples changes insignificantly, while specific features of its behavior depend on the strontium content in $Ce_{1-x}Sr_xMnO_3$. For the sample with $x = 0.5$ a slight increase in the negative thermopower values is observed in the range from T_c to 20 K, while for the one with $x = 0.67$ the thermopower becomes positive at $T \approx 145$ K and decreases weakly at further temperature decrease. The observed peculiarities of the thermopower behavior give evidence of a strong modification of the conduction process at $T \approx T_c$.

4. Conclusions

In summary, this paper presents the results of the systematic study of the thermopower behaviour in manganites of the $La_{1-x}Sr_xMnO_3$ and $Ce_{1-x}Sr_xMnO_3$ systems. Increasing strontium content was observed to result in a decrease of the thermopower value for both investigated systems. In $Ce_{1-x}Sr_xMnO_3$ samples the thermopower at room temperature is negative, while in $La_{1-x}Sr_xMnO_3$ it is positive for a low doping level, very close to zero for intermediate doping ($x = 0.28$–0.333), and negative for a higher doping level. The peculiarities of the temperature dependences of the thermopower in different temperature ranges (both below and above the temperature of the magneto-ordered transition) characteristic of hole- and electron-doped manganites are revealed and discussed. The observed specific features of these dependences indicate that both below and above T_c two types of charge carriers (electrons and holes) give contributions to the conduction process, so that an interpretation of the electron transport in manganites requires applying two-band models.

Acknowledgments

The study was supported by the grant of the St. Petersburg government for young scientists.

References

[1] Chanara K, Ohno T, Kasai M , Kozono Y 1993 *Appl. Phys. Lett.* **63** 1990
[2] von Helmolt R, Wecker J, Holzapfel B, Schultz L, Samwer K 1993 *Phys. Rev. Lett.* **71** 2331
[3] Jin S, Tiefel T H, McCormack M, Fastnacht R A, Ramesh R, Chen L H 1994 *Science* **264** 413
[4] Dunaevkii S M 2004 *Phys. Solid State* **46** 193
[5] Nagaev E L 2001 *Phys. Reports.* **346** 387
[6] Skryabin Yu N, Izyumov Yu A 2001 *Physics – Uspekhi* **44** 109
[7] Bebenin N G, Zainullina R I, Mashkautsan V V, Ustinov V V, Vasil'ev V G, Slobodin B V 2001 *Phys. Solid State* **43** 501
[8] Bebenin N G, Zainullina R I, Mashkautsan V V, Burkhanov A M, Ustinov V V, Vasil'ev V G, Slobodin B V 1998 *J Exp. Theor. Phys.* **86** 534
[9] Chmaissem O, Dabrowski B, Kolesnik S, Mais J, Jorgensen J D, Short S 2003 *Phys. Rev. B* **67** 094431
[10] Hemberger J, Krimmel A, Kurz T, Krug von Nidda H-A, Ivanov V Yu, Mukhin A A, Balbashov A M, Loidl A 2002 *Phys. Rev. B* **66** 094410
[11] Mandal P, Hassen A, Loidl A. 2004 *Phys Rev B* **69** 224418
[12] Szewczyk A, Gutowska M, Dabrowski B 2005 *Phys. Rev. B* **72** 224429
[13] Vergara J, Ortega-Hertogs R J, Madurga V, Sapina F, El-Fadli Z, Martınez E, Beltran A, Rao K V 1999 *Phys. Rev. B* **60** 1127
[14] Crespi V H, Lu L, Jia Y X, Khazeni K, Zettl A, Cohen M L 1996 *Phys. Rev. B* **53** 14303
[15] Zainullina R I, Bebenin N G, Mashkautsan V V, Burkhanov A M, Sukhorukov Yu P, Ustinov V V, Vasil'ev V G, Slobodin B V 2000 *Phys. Solid State* **42** 292
[16] Lu W J, Zhao B C, Ang R, Song W H, Sun Y P 2005 *Physica B* **367** 243

Journal of Physics: Conference Series **690** (2016) 012008 doi:10.1088/1742-6596/690/1/012008

Luminescence of a-screw dislocations in low-ohmic GaN

O S Medvedev[1, 2], O F Vyvenko[1] and A S Bondarenko[1]

[1]V.A. Fok Institute of Physics, St. Petersburg State University, 198504 St. Petersburg, Russia
[2]IRC Nanotechnology, Research Park,St. Petersburg State University, 198504 St. Petersburg, Russia

E-mail: o.s.medvedev@spbu.ru

Abstract. Straight segments of a-screw dislocations introduced by indentation or scratching of the basal (0001) or prismatic (01-10) surfaces of intentionally undoped low-ohmic GaN were found to radiate a doublet of narrow luminescent lines in the spectral region at about 3.1 - 3.2 eV. The components of the doublet were separated by about ~30 meV and accompanied with LO-phonon replicas. The intensity of the dislocation-related luminescence was significantly higher than that of free exciton in dislocation-free regions indicating a giant oscillator strength of dislocation-related luminescence optical transition that was proposed to occur between the electron states of the stacking fault and the hole states of the bounded partials in the core of the *dissociated* screw dislocation.

1.Intoduction

Gallium nitride is a promising direct-wide-gap semiconductor for high power and optical electronic applications. Nowadays, GaN films are grown mostly heteroepitaxially on diverse substrates with a large lattice mismatch. As the result, heteroepitaxial GaN contains a high density of dislocations. The impact of dislocations on the electronic properties of semiconductors may be defined by a variety of reasons such as the dislocation-induced elastic lattice strains, local electronic states or bands of their own cores, which in turn, can exhibit diverse atomic structure, as well as the point defects attracted to dislocations or introduced/changed during their motion. A few sub-band gap luminescent bands with the maxima at photon energies of about 3.2 eV [1] and 3.4 eV [2] have been attributed to as-grown dislocations while the bands at 2.9 eV [3], 3.12 eV [4], 3.18 eV [5,6], 3.35 eV [7] were found in GaN after plastic deformation but the interpretation of their origin is still under discussion.

Recently we reported [5,6] about the observation of a strong cathodoluminescence (CL) band with quantum energies around 3.1-3.2 eV due to a-screw dislocations in *low-ohmic* GaN samples. It was found that the dislocation related luminescence (DRL) band exhibited a doublet spectral structure and its intensity exceeded well one of the free exciton band. In this paper we present the results of further investigations of the DRL properties including the dependences of the energy position of DRL on the temperature and internal strains which appeared to be similar to those ones of free exciton. TheDRL origin is discussed and is assumed to be exciton bound by the 1D confined states of dissociated screw dislocations.

2. Experimental

We investigated 200 μm thick intentionally undoped GaN (0001) single crystals grown by the hydride vapor phase epitaxy technique with a density of grown-in dislocations of about 10^6 cm^{-2} and free electron concentration in range from 10^{16} cm^{-3} to 10^{18} cm^{-3} measured by Raman spectroscopy and capacitance-voltage techniques. Cathodoluminescent (CL) measurements were performed with a scanning electron microscope (SEM) Zeiss Supra 40 equipped with a Gatan MonoCL+ system at accelerating voltages of 3-30 kV with electron beam current in ranges from 0.2 nA to 10 nA in the temperature ranges from 70 K to 300 K.

Fresh dislocations were introduced at room temperature either by the indentation with a Vickers indentor of as-grown basal (0001) or polished (10-10) prismatic sample surface under the loads between of 0.1 and 5 N or by its scratching with a diamond scriber.

3. Results

Figure 1(a) represents CL spectra taken at two different positions of electron beam. CL spectrum of dislocation-free regions (figure 1 (a), dotted line) of all investigated samples consisted of the line of free exciton (FE) at 3.46 - 3.47 eV, and two low intense bands at 3.27 eV and 3.4 eV that were previously ascribed to the recombination of donor-acceptor pairs or the second phonon replica of FE and exciton bound on point defects, respectively [1]. Spectral measurements on the a-screw dislocation-rich regions revealed the appearance of a set of additional intense narrow lines at around 3.1-3.2 eV, (figure 1 (a), solid line) with the main doublet denoted as DL_L and DL_H accompanied with LO-phonon replicas. The integrated intensity of DRL lines was at least two times higher than one of FE. The full width at half maximum of both DRL lines was of about 16 meV at 70 K being close to that of FE while the energy separation between DL_L and DL_H components was of about 30 meV.

An examination of the CL image near the scratch on the (0001) surface in figure 1 (b) revealed straight lines stretched in all possible {12-10} directions. According to the crystallography of wurtzite structure and to the data of TEM [3,8], such lines are nothing else than a-screw dislocation with the Burgers vector of 1/3*a* [12-10]. Under the electron beam excitation these bright lines moved quickly and partly or completely disappeared at both room and liquid nitrogen temperatures. Note that in the CL image in figure 1 (b) the tiny bright dots with dark halos around are caused by optical scattering of outcoming CL near the pits grown-in threading dislocations (TD). The pits were clearly seen on the corresponding secondary electron image (not presented) and will not be discussed further.

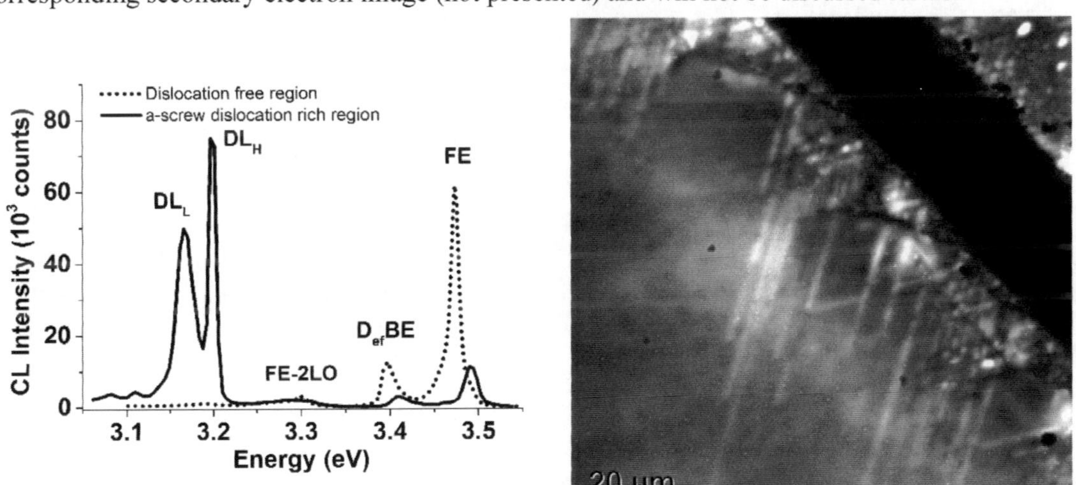

(a) (b)

Figure 1. (a) – CL spectra obtained in the spot mode from a screw dislocation rich region (solid line) and far away from the scratch (dot line).The Accelerating voltage was 10 kV, the electron beam current was 0.5 nA and the temperature was 70 K;
b) – Panchromatic CL images near the indentation prick on the (0001) surface.

An additional information about the origin of DRL was obtained from the spectral behavior of its components on mechanical strains and on temperature. Mechanical strains are inhomogeneous in the vicinity of the indentation prick or of the scratch. The strains are of a complicated character but in any case they decrease with the distance from the impact place that allows one to obtain the dependence of the DRL spectral position with respect to the position of FE. Figure 2 (a) represents the DRL spectra taken at three different distances from a scratch. Note that the intensity of DRL near the scratch is

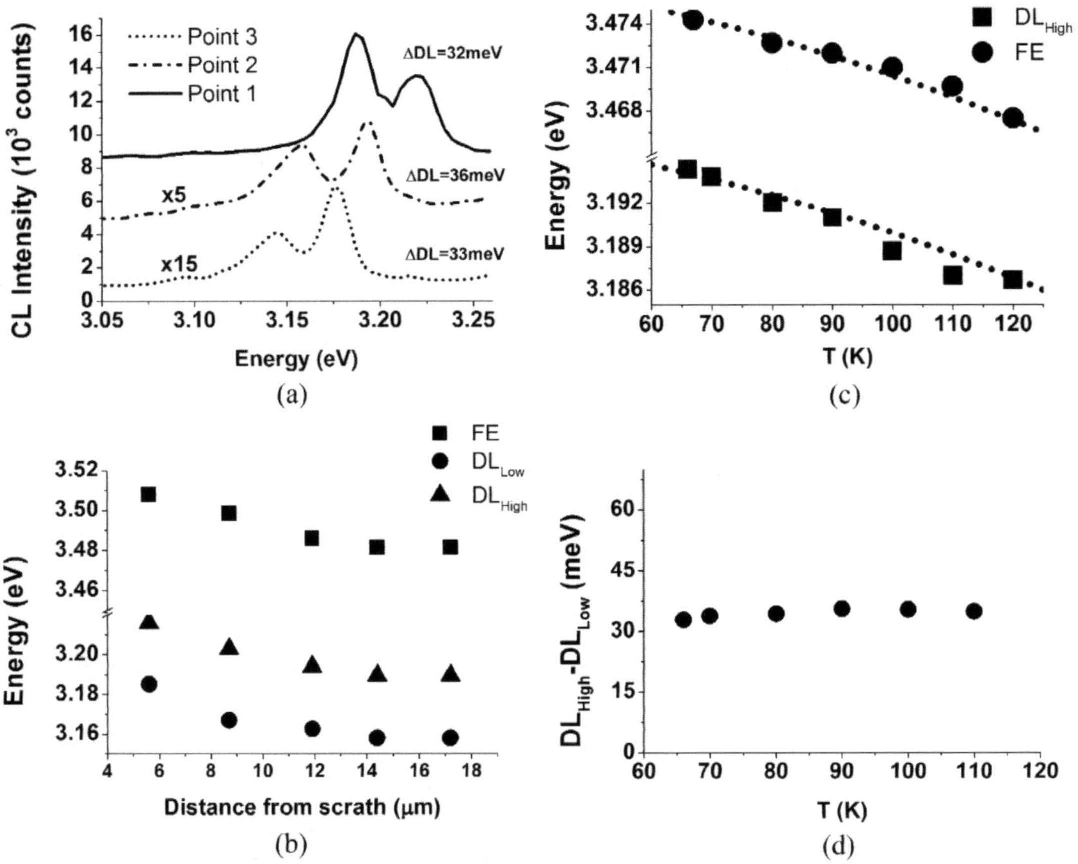

Figure 2. (a) – Spectra taken at different distances from a scratch (Point 1 –10 μm, Point 2 –20 μm, Point 3 –30 μm);(b) – Energy positions of FE and each DRL band at different distances from the scratch; (c) – Temperature dependences of FE and DL_H bands; (d)– Dependence of energy separation of DRL components on the temperature

much higher than far from it. This is due to a higher density of dislocations near the scratch. One can see from figure 2 (b) that both components of the DRL doublet exhibit a blue shift when the distance to the scratch decreases, i.e. upon the strain increase. The strain-induced spectral shift of every DRL line coincides quantitatively with the shift of FE (figure 2 (b)). In opposite to that, the energy separation between DL_L and DL_H peaks did not change upon the strain being of about 30 meV in all investigated samples and it was independent on the temperature (figure 2 (d)). The temperature dependence of the DRL spectral position in a range from 70 to 120 K also repeats the behavior of FE (igure 2 (c)) like the strain dependence. The shape of the temperature dependence for FE and DL_H bands can be well fitted with the Varshni approximation (dot lines in the figure 2 (c)).

4. Discussion

The origin of the luminescence localized along the dislocation lines with an emission energy which is significantly less than the semiconductor band gap can be associated either with the local electronic states of dislocations themselves or of point defects in a close vicinity to the dislocation cores.

The point defects localized at dislocations [4] seem unlikely to be an origin of DRL since in this case it is difficult to explain the high dislocation mobility observed in our experiments as well as the other DRL properties. Accordingly, the doublet-line DRL with an energy of about 0.3 eV below the GaN band gap is an intrinsic property of the screw dislocations in *low-ohmic* GaN. On the other hand, Albrecht et al [7] reported recently about a *single* line of DRL with the peak position at 3.35 eV also due to straight individual a-screw dislocations in *semi-insulating* GaN. That line is, thus, red shifted only by 0.14 eV with respect to the GaN band gap energy and vanishes above 100 K. These two observations imply that the DRL spectral position of the dislocations of the same type depends strongly on the sample doping level that, in turn, can be explained by their different core structures.

There are two main core structures of screw dislocations. It can be perfect or dissociated into two 30° partial dislocations bounded the stacking fault (SF) ribbon. The dislocations in semi-insulating GaN [7] were found to be perfect. The deformation potential theory (DFT) based on symmetry considerations [9] predicts that a perfect screw dislocation can produce only shallow bound hole states, confined within a thin cylinder of shear lattice strains around them, which lye less than 0.1 eV above the valance band of GaN. Very recent DFT calculations for a supercell of 400 atoms [7] revealed the conduction band bending due to shear strai ns which does exists in the analytical model [9]. The sum of the conduction and valence band bending around the perfect screw dislocation could be as high as 0.2 eV that was sufficient to explain the DRL red shift of 0.14 eV in *semi-insulating* GaN but is still too small to explain the red shift of 0.3-0.34 eV in the low-ohmic material. In this way, we came to the conclusion that the additional DRL red shift of about 0.15 eV is due to the splitting of the dislocation core.

We have no data about the dislocation core structure of the dislocations freshly introduced at room temperature in the low-ohmic samples. However, previously reported TEM investigations of grown-in a-screw dislocations in undoped GaN revealed that they were dissociated with the lateral width of I_2 SF of about 5 nm [10].

The energy level of the core states of 30° partials estimated theoretically was of about 1.1 eV [11] above the valence band and, thus, can not be responsible for the observed DRL. SFs in wurtzite (WZ) structure can be imagined as a thin layer of sphalerite phase with the band gap of 3.27 eV. They were proposed to be considered as quantum wells (QW) [12,13] with the conduction band offset of 0.27 eV at interface of type II heterostructure [13]. Luminescence of SFs in GaN was rather extensively investigated and a correlation between the CL energy positions and the SF atomic structure was established (see the recent review [14]). The only I_2 can be obtained by a simple glide in the basal plane and, thus, can be a part of a gliding dislocation. The energy positions reported for I_2 luminescent lines are in the same ranges of 3.32-3.36 eV [14] giving the energy position of the I_2SF level of about 0.1-0.15 eV below the conduction band. That value together with the band bending due to the deformation potential at SF bounding partials gives by an order of magnitude just the DRL red-shift in low-ohmic GaN. In addition, the properties of DRL are similar to FE (line width, high intensity, depending on the voltage and temperature) that allows to describe DRL as an exciton bound at the confined states of screw dislocations. The observed energy of the optical transition of DRL is the exciton binding energy, which is of the order of tens of meVs.

The observed SF ribbon width in the dislocation core of 5 nm is comparable with the exciton radius in GaN of about 3 nm, transforming an extended 2D SF into a 1D-like system. As it was shown both theoretically [15] and experimentally [16], the decrease of the QW lateral dimension gave rise to an increase of the exciton binding energy and its oscillator strength that might give an additional small red shift and explain the enhanced DRL yield as well as the doublet line structure as being due to the ground and first excited state of the 1D system.

The reason for the different core structures of screw dislocations in the semi-insulating and low-ohmic samples might be the impact of the Fermi level on the expansion of SFs as it was originally suggested to explain the "recombination enhanced dislocation glide" (REDG) in SiC [17,18]. It was proposed [18,19] that SF nucleation and expansion in SiC are driven by a reduction in the electronic energy of the system that occurs when conduction electrons become trapped within the SF-induced QWs. In our samples with a net donor density more than 10^{16} cm^{-3}, the Fermi level at room temperature is situated less than 0.15 eV below the bottom of the conduction band that makes it possible to fill the SF quantum confined states with electrons and, accordingly, stimulate SF nucleation [17]. An indirect confirmation of that assumption is our observation of enhanced dislocation mobility under the electron beam excitation that is known as the recombination enhanced dislocation glide and that was explained in SiC just with the model [17] mentioned above.

Conclusions

A-screw dislocations introduced by room temperature plastic deformation in low-ohmic GaN were shown to be efficient sources of intrinsic dislocation-related luminescence (DRL) with an emission energy of about 0.3 eV less than the band gap, which exhibit the following properties. The brightness of DRL exceeded FE intensity; DRL spectrum consists of doublet of narrow lines separated by 30 meV; the DRL line behavior upon mechanical strains and temperature is very similar to that of FE; the exciton bound at the confined states of a-screw dislocations is assumed to be the mechanism responsible for DRL.

It is suggested that the dislocation core structure in GaN might experience the changes upon the impact of the equilibrium and excess free electron concentration (the Fermi level position being either perfect or dissociated by the mechanism similar to that proposed for SiC [17–19].

Acknowledgments

Yuri Shreter and Vlad Voronenkov are greatly acknowledged for providing the samples and helpful discussions.

References:

[1] Reshchikov M A and Morkoç H 2005 *J. Appl. Phys.* **97** 061301
[2] Shreter Y G, Rebane Y T, Davis T J, Barnard J, Darbyshire M, Steeds J W, Perry W G, Bremser M D and Davis R F 1997 *Mater. Res. Soc.* **449** 683–8
[3] Albrecht M, Strunk H P, Weyher J L, Grzegory I, Porowski S and Wosinski T 2002 *J. Appl. Phys.* **92** 2000
[4] Huang J, Xu K, Fan Y M, Wang J F, Zhang J C and Ren G Q 2014 *Nanoscale Res. Lett.* **9**
[5] Medvedev O, Vyvenko O, Bondarenko A and Voronenkov V 2014 *12th international workshop on Beam Injection Assessment of Microstructures in Semiconductors* (Tsukuba) p MoA4
[6] Medvedev O S, Vyvenko O F and Bondarenko A S 2015 *Semiconductors* **49** 1181–6
[7] Albrecht M, Lymperakis L and Neugebauer O 2014 *Phys. Rev. B* **241201** 1–4
[8] Ratschinski I, Leipner H S, Heyroth F, Fränzel W, Moutanabbir O, Hammer R and Jurisch M 2011 *J. Phys. Conf. Ser.* **281** 012007
[9] Rebane Y T and Steeds J W 1993 *Phys. Rev. B* **48** 963–72
[10] Zakharov D, Liliental-Weber Z, Wagner B, Reitmeier Z, Preble E and Davis R 2005 *Phys. Rev. B* **71** 235334
[11] Savini G, Blumenau A T, Heggie M I and Öberg S 2007 *Phys. Status Solidi* **4** 2945–9
[12] Rebane Y T, Shreter Y G and Albrecht M 1997 *Phys. Status Solidi* **164** 141
[13] Stampfl C and Van de Walle C 1998 *Phys. Rev. B* **57** R15052–5
[14] Lähnemann J, Jahn U, Brandt O, Flissikowski T, Dogan P and Grahn H T 2014 *J. Phys. D. Appl. Phys.* **47** 423001
[15] Christol P, Lefebvre P and Mathieu H 1993 *J. Appl. Phys.* **74** 5626
[16] Zhu Q, Ganière J D, He Z B, Karlsson K F, Byszewski M, Pelucchi E, Rudra A and Kapon E 2010 *Phys. Rev. B* **82** 165315

[17] Galeckas A, Linnros J and Pirouz P 2006 *Phys. Rev. Lett.* **96** 1–4
[18] Caldwell J D, Stahlbush R E, Ancona M G, Glembocki O J and Hobart K D 2010 *J. Appl. Phys.* **108**
[19] Pirouz P 2013 *Phys. Status Solidi Appl. Mater. Sci.* **210** 181–6

Charge transport processes in proustite crystals Ag_3AsS_3

N Yu Guniya[1], R A Castro[1] and E B Shadrin[2]

[1]Herzen State Pedagogical University of Russia, 48 Moyka river Embankment, St. Petersburg 191186, Russia
[2]Ioffe Physical Technical Institute, Russian Academy of Sciences, Polytechnicheskaya 26, St. Petersburg 194021, Russia

E-mail: naala4kaka@mail.ru

Abstract. Structural characteristics and charge transfer processes in Ag_3AsS_3 crystals were studied by dielectric spectroscopy and scanning electron microscopy methods; the relaxation parameters of the system were evaluated.

1. Introduction

Proustite (light red silver ore) Ag_3AsS_3 single crystal is a ferroelectric with a melting point of 580 °C. This crystal system is characterized by high values of dielectric permittivity, advantageous for its application in nonlinear optical devices and electron acoustic systems [1]. Currently, proustite crystals are used for the development of infrared emission transducers. At the moment, there are practically no data on the characteristics of polarization processes in the high frequency range and the effects of such factors as illumination, introduction of admixtures of various metals, etc. on dielectric properties of this class of materials. The study of electric and dielectric properties is of particular importance both for specifying the electronic processes in different materials and for understanding the nature of their fundamental properties [2, 3]. In this regard, the objective of this work was to define the specific features of charge transfer mechanisms in proustite crystals Ag_3AsS_3 using dielectric spectroscopy method.

2. Experimental methodology

Measurements were taken with a "Concept 81" spectrometer, NOVOCONTROL Technologies GmbH & Co. The samples were single-crystal wafers of about 1 mm thick and 1 cm^2 in area, with an even surface perpendicular to the third-order axis. Measurements of dielectric spectra (components of the complex conductivity σ') in proustite crystal samples were taken in the temperature range T = 173 ... 473 K. Temperature precision was ±0.5°C. The measuring voltage applied to the sample was 1.0 V. Investigation of the film surface structures and the elemental composition of the samples was performed with a Carl Zeiss EVO 40 scanning electron microscope (SEM). The table 1 presents the data on the weight % and atomic % content of chemical elements in proustite crystal samples.

Table 1. Elemental composition of Ag_3AsS_3 samples.

Element	Nominal concentration	Mass fraction	Weight %	Weight % Sigma	Atomic fraction %
S	4.11	0.9846	18.53	0.21	41.76
As	2.64	0.9432	12.43	0.28	11.99
Ag	13.23	0.8513	69.04	0.31	46.25
Total			100		

3. Results and discussion

In the low frequency range, we observe a continuous increase in conductivity with an increase in frequency, i.e., the dispersion of type $\sigma \sim \omega^S T^L$ is detected, which, in general, is typical of solids, where transport processes are caused by the jumps of charge carriers (figure 1). This transfer mechanism was first proposed by Pollack [4] for chalcogenide glasses and later was applied to a wide class of materials by other authors [5, 6]. We can assume the existence of a hopping transfer mechanism for charge carriers in proustite crystals, in which silver ions make thermally activated jumps through the lattice. The correctness of this assumption stems from the more pronounced dispersion ε' in low frequency range, observed with increasing temperature. The changing pattern of dielectric permittivity response on the frequency change can result from the increasing pace of thermal excitation of silver ions. It should be noted that the ion hopping transport contributes to both components of the complex dielectric permittivity, since silver ions can oscillate in the lattice sites as dipoles and can move along the lattice sites thus creating quasi dipoles; they can also move across the lattice and perform hopping charge transport. The exponent s of the frequency dependence decreases with increasing temperature and takes values in the range (s = 0.45 ... 0.70), which is also indicative of a hopping mechanism of conductivity.

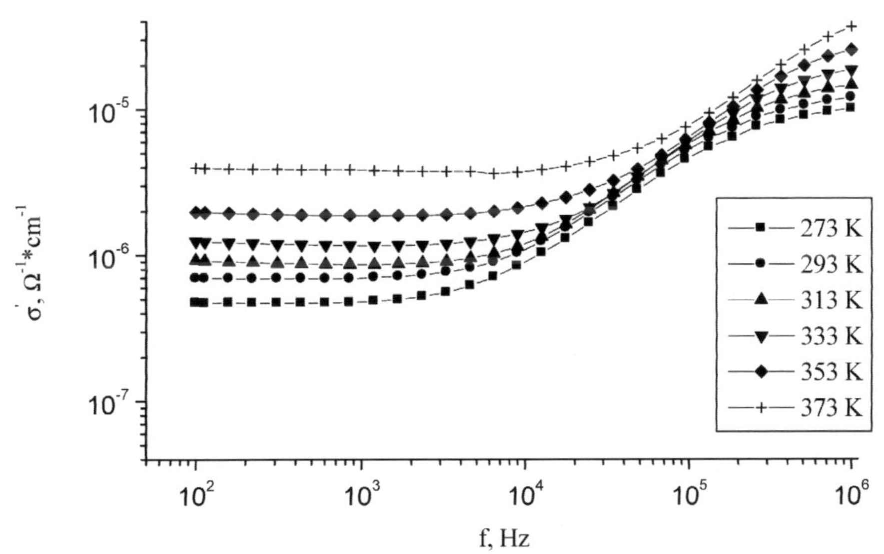

Figure 1. Frequency dependence of the real part of complex conductivity at different temperatures.

We have revealed an exponential temperature dependence of specific conductivity with two regions (figure 2) with activation energy $E_1 = 0.10 \pm 0.01$ eV and $E_2 = 0.30 \pm 0.01$ eV, respectively. It can be assumed that the charge transport is a thermally activated process; the activation energy on alternating current (E_a (ω)) for various frequencies is part of the well-known Arrhenius expression [7]:

$$\sigma_{ac}(\omega) = \sigma_0 \exp\left(\frac{E_a(\omega)}{k_B T}\right), \tag{1}$$

where σ_0 is a constant.

Changes in the temperature dependence pattern can be associated either with the presence of impurities and defects in the structure, or with the transition from the mixed to the ionic conductivity.

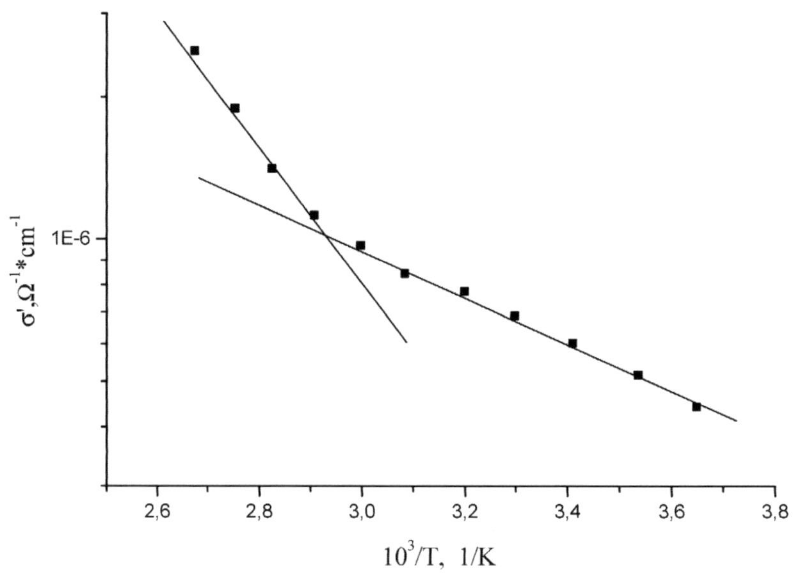

Figure 2. Temperature dependence of the specific conductivity of proustite crystal samples at frequency $f = 10^2$ Hz.

At higher frequencies, the specific conductivity depends weakly on the frequency over the entire temperature interval (figure 3). At low temperatures, a decrease in conductivity is even observed at the higher frequency range. This can be associated with the fact that the rapid change in the field polarity prevents the ordered motion of silver ions, which are, as noted earlier, primary charge carriers. As the temperature increases, there comes a point when carriers take on enough energy to perform jumps, which in combination with possible phase transitions leads to the changing nature of dispersion σ'.
The temperature dependence of σ' is presented in figure 4, which shows that the specific conductivity also increases exponentially with the absolute temperature. The value of activation energy E_a at $f = 10^8$ Hz was calculated and found to be 2.40 ± 0.01 eV.

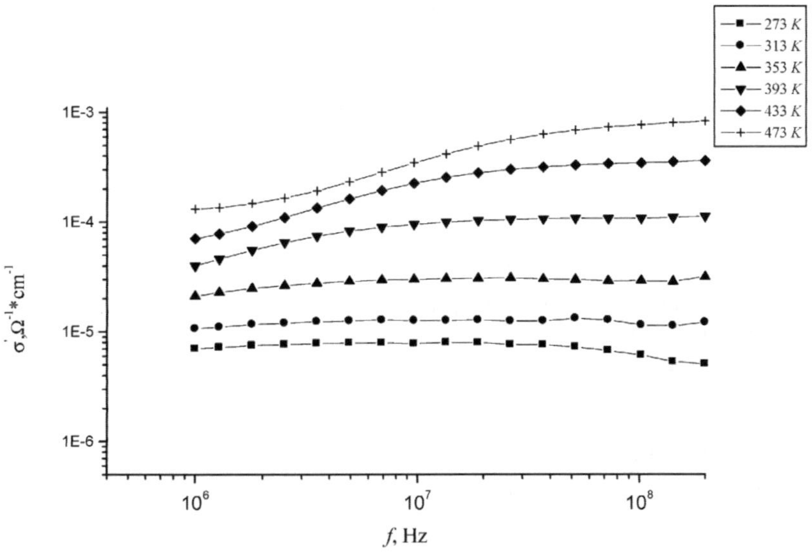

Figure 3. High-frequency dependence of the real part of complex conductivity at different temperatures.

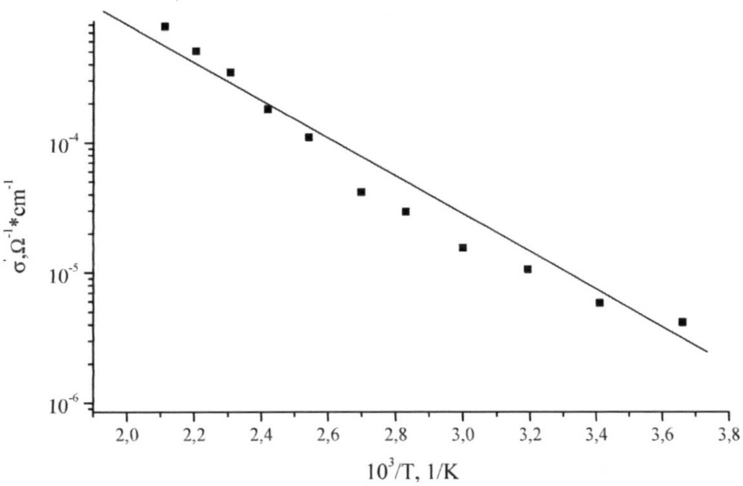

Figure 4. Temperature dependence of the real part of complex conductivity at frequency $f = 10^8$ Hz

4. Conclusions

A continuous increase in specific conductivity with increasing frequency has been observed in the low frequency range, i.e., the dispersion of type $\sigma \propto \omega^s T^L$ is detected, which in general is typical of solids, where transport processes are caused by hopping mechanism of charge transport. The exponent s of the frequency dependence decreases with increasing temperature and takes values in the range (s = 0.45 ... 0.70). An exponential temperature dependence of specific conductivity with two plots with activation energy $E_1 = 0.10 \pm 0.01$ eV and $E_2 = 0.30 \pm 0.01$ eV, respectively, has been revealed.

Changes in the temperature dependence pattern can be associated either with the presence of impurities and defects in the structure, or with the transition from the mixed to the ionic conductivity.

In the frequency range $f = 10^6 \ldots 10^9$ Hz, the specific conductivity is virtually independent of frequency; at low temperatures, a certain decrease in conductivity is observed with an increase in frequency, which can be associated with the fact that a rapid change in the field polarity prevents the ordered motion of silver ions.

References

[1] Rud' V Yu, Rud' Yu V and Terukov E I 2010 *Semiconductors* **44** 1025
[2] Castro R A, Bordovsky V A and Grabko G I 2010 *Technical Physics Letters* **36** 9
[3] Castro R A, Bordovsky V A and Grabko G I 2009 *Glass Physics and Chemistry* **35** 54
[4] Elliott S R 1987 *Advances in Physics* **36** 135
[5] Chen R H, Wang R J, Chen T M and Shern C S 2000 *Journal of Physics and Chemistry of Solids* **61** 519
[6] Murowski L and Barczynski R J 1995 *Journal of Non-Crystalline Solids* **185** 84
[7] Gupta V, Bamzai K K, Kotru P N and Wanklyn B M 2006 *Materials Science and Engineering* **130** 163

Investigation of spinodal decomposition of InGaAsP solid solutions grown by the MOCVD technique

A E Marichev[1,2], B V Pushnyi[1,3] and R V Levin[1,3]

[1] Ioffe Institute, St. Petersburg, Russia
[2] Saint-Petersburg State Electrotechnical University "LETI", St. Petersburg, Russia
[3] Submicron Heterostructures for Microelectronics Research and Engineering Center of the RAS, St. Petersburg, Russia

E-mail: segregate1@yandex.ru

Abstract. Technology for growing layers of InGaAsP solid solutions with *Eg* ~ 1.0 - 1.2 eV on InP substrates by the MOCVD technique in the spinodal decomposition region has been elaborated. Results of the investigation of solid solution layers by the methods of photoluminescence, reflectance anisotropy spectroscopy and fractal analysis are presented.

1.Introduction

Fabrication of high effective photodetectors for powerful conversion of laser radiation requires direct band gap materials with lattice perfection and high mobility of generated charge carriers. Such materials possess high photoluminescence (PL) intensity.

The most appropriate material for photodetectors of laser radiation with wavelength $\lambda = 1064$ nm (1,165 eV) is the InGaAsP solid solutions isoperiodic with InP, the band gap of which is in the range of 1.0 – 1.15 eV (300 K). However, it is known that these solid solutions grown at 600 – 627°C fall within the spinodal region [1], that complicates obtaining the photoactive layers with an optimal thickness of 1-2 μm.

2. Experimental details

First, the range of spinodal region for InGaAsP solid solutions isoperiodic with InP was determined. Calculations performed by K. Onabe [1] within the theory of quasi-regular solutions for InGaAsP ones show that there exists a wide region of spinodal decomposition in the system mentioned above at 600°C. Figure 1 shows that, the solid solution with lattice parameters similar to those of InP and the band gap in the range of 0.9 - 1.05 eV falls within the spinodal region.

Calculations performed according to the same model using thermodynamic parameters of Ilegems and Panish [2] give results significantly different from those of Onabe. According to the calculations based on the model of lattice delta-parameter, performed by De Kremu [3], Ipatova [4], and Stringfellow [5], the system of InGaAsP solid solutions at the same temperatures has wider regions of immiscibility than those obtained by calculations made within the model of quasi-regular solutions.

Thus, our calculations have confirmed that the required solid solutions fall within the spinodal region, and it is necessary to design a technology for fabrication of InGaAsP solid solutions layers with parameters (thickness and PL efficiency) appropriate for photodetectors.

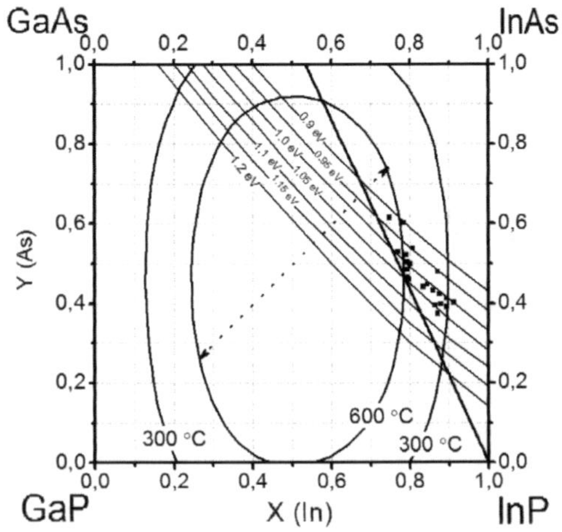

Figure 1. Diagram of compositions for InGaAsP solid solutions. Spinodal isotherms for 300°C and 600°C, isoenergetic lines for 0.9 - 1.2 eV are constructed using the model from [6] and the isoperiod with InP in the concentration square of GaInAsP solid solutions. Dots indicate compositions of specimens obtained in the present work.

All layers were grown in a MOCVD installation AIXTRON AIX-200 at a pressure of 100 mbar and a temperature of 600°C in a flow of purified hydrogen with a dew point of not worse than -100°C. The total hydrogen flow through the reactor was equal to 5 slpm.

Metal-organic compounds TEGa and TMIn were used as Ga and In sources, respectively, and hydrides Arsin(AsH3) and Phoshine(PH3) – as As and P sources. Samples doping was not carried out.

The structures were grown on InP(100) substrates with misorientation of 4° towards the (111)A plane (doped with Te, $n = (1 - 5)\times10^{18}$ cm^{-3}) on 1 μm thick InP buffer layers.

The layer material composition was determined by the X-ray spectral analysis, the layer thickness – by SEM and the surface quality – by atomic force microscopy. To estimate the material quality, the following methods were applied: PL, reflectance anisotropy spectroscopy and multifractal analysis of a surface.

3. Results of investigations

3.1. Reflectance anisotropy spectroscopy.

The following samples of InGaAsP solid solutions have been investigated (table 1):

Table 1. Characteristics of the tested samples.

Sample No	Thickness, nm	*Eg*, eV	$\Delta a/a$
I253	1000	1.0	0.015
I255	1000	1.0	$3.88*10^{-3}$
I297	50	1.0	0
I298	200	1.0	$<2.1*10^{-3}$
I299	500	1.0	$<2.1*10^{-3}$
I300	1000	1.0	$<2.1*10^{-3}$

Where $\Delta a/a$ – lattice mismatch between InP and InGaAsP solid solutions, *Eg* - the energy band gap.

Reflectance anisotropy (RA) spectra of InGaAsP layers are shown in figure 2, RA spectra of InP are multiplied by 10. Spectra were obtained in air conditions, the samples were covered by a thin layer of natural oxide. As one can see, all samples possess optical anisotropy which is significantly higher than in the case of InP layer. In all the spectra, derivative-like structures are clearly seen in the region of 2.7 - 3.2 eV, which includes energies of E_1 - $E_1 + \Delta_1$ optical transitions of InP and InGaAsP solid solutions, and also in the region of E_0' and E_2 optical transitions 4.5 - 5.2 eV. The highest amplitude of these signals is observed for the 200 nm thick layer. These signals can be associated with anisotropic electric field or anisotropic stress present in the samples, which results in appearing of derivative-like signals. The anisotropic stress gives the same signal and also the background (negative for [110] stress and positive for [1 - 10]). The additional anisotropic background also present in the spectra can be explained by presence of anisotropic roughness on the surface. Presence of peculiarities in the RA spectra of solid solutions at energies of optical transitions in In_xGaAsP (energies are marked by black arrows according to LB) confirms that the obtained layers have the required compositions. Besides the mentioned peculiarities, only ones at the energies of transitions in InP are seen. That means the absence of any other binary compositions in the obtained layer.

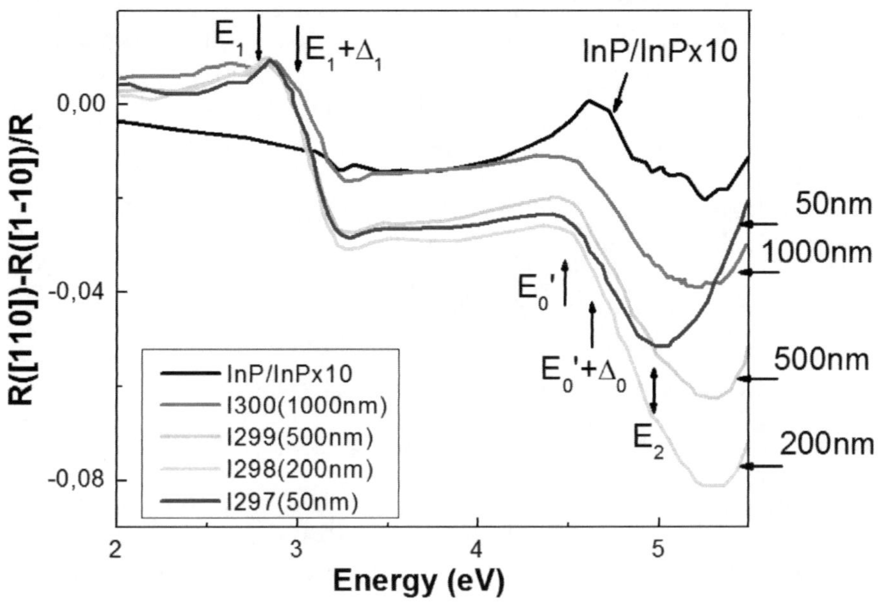

Figure 2. RA spectra for the samples I297 - I300 of InGaAsP layers grown on In(001) of different thicknesses

Figure 3 presents the RA spectra of 1 μ thick films (I253, I255 and I300). A substantial weakening of the I253 and I255 spectra amplitudes indicates the decrease of the anisotropic relief of the surface and of a stress in the films. Besides, additional peculiarities appeared in the spectral region of 2.5 - 3.3 eV. In figure 4, these spectra are compared with those of the (001) surfaces of the (001) oxidized surfaces of doped GaAs and InAs. Black arrows indicate peculiarities present in the I300 sample spectrum (2.89eV – InGaAsP and 3.28eV – InP). The peculiarities mentioned above are associated with the E_1 - $E_1 + \Delta_1$ transitions in the mentioned compounds. Thus, it can be said that GaAs and InAs binary compounds are present in the I253 and I255 films.

Thus, it is established that there is no spinodal in layers (I297-I300), and the layers have a uniform composition, but stresses inducing changes in the surface relief are observed. In the layers of 1000 nm thick I253 and I255, inclusions of different fractions are observed, i.e. decomposition of the solid solution is present.

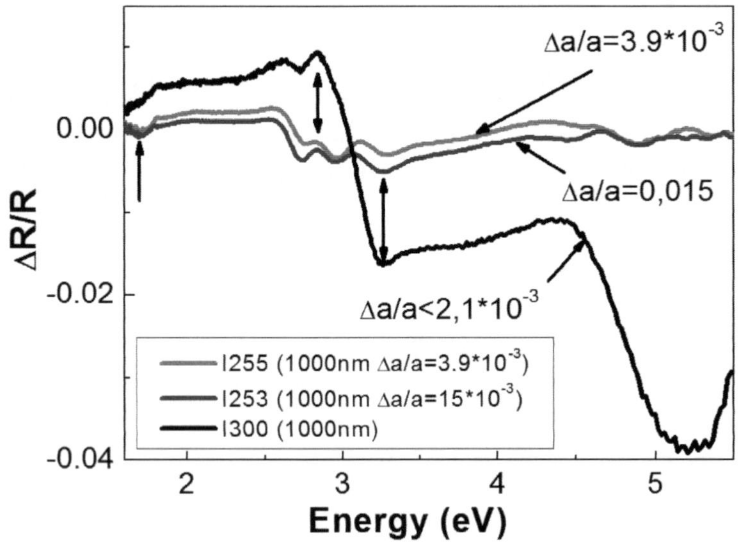

Figure 3. RA spectra of 1000 nm thick samples of InGaAsP layers grown on In(001)

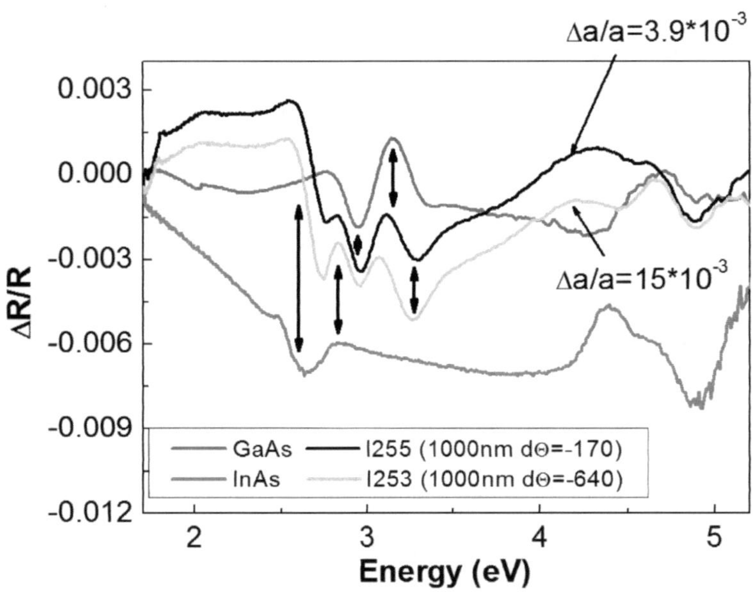

Figure 4. RA spectra of 1000 nm thick samples of InGaAsP layers grown on InP(001). Comparison with RA spectra of (001) naturally oxidized surface of GaAs and InAs samples with n ~ 2×10^{18} cm^{-3} doping.

3.2. Multifractal analysis.

According to the data of the multifractal analysis of the atomic force microscope images of samples surface, it is seen that a small roughness (Table 2) is formed on the surface of all layers. Probably, its appearance takes place after cooling, and appearing tension results in solid solution decomposition on the surface and a break of the top layer part. A high roughness (Table 3) is formed only for thick layers practically in the entire depth and partially in the buffer layer relieving stresses in the bulk, which improves the layer quality.

Table 2. Small roughness.

Sample No	d, nm	h, nm	L_o , μm	L_d , μm	σ, din·cm^{-2}
I300	1000	25	0.14	0.145	1.4x 10^{10}
I299	500	17	0.12	0.12	1x 10^{10}
I298	200	20	0.15	0.16	8x10^{9}
I297	50	21	0.15	0.15	1x 10^{10}

where σ - tension, h –roughness height, L_d – difference between roughnesses, L_o – width at the base of the roughness, d – sample thickness

Table 3. High roughness.

No	d, nm	h, nm	L_o , μm	L_d , μm	σ, din·cm^{-2}
I300	1000	45	0.85	0.9	1.4x 10^{9}
I299	500	31	1.2	1.3	3.6x 10^{8}
I298	200	absent			
I297	50	absent			

where σ - tension, h –roughness height, L_d – difference between roughnesses, L_o – width at the base of the roughness, d – sample thickness.

The tension is most effectively lowered in the sample (I300) by forming a high roughness. Figures 5 - 6 present pictures of typical surfaces obtained by an atomic force microscope.

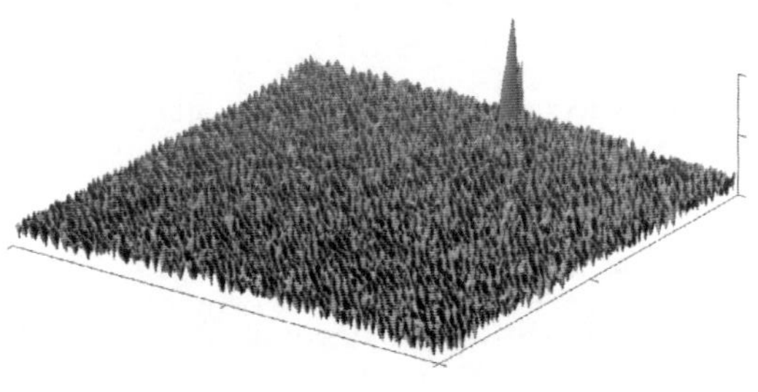

Figure 5. Pictures of the surface of a 50 nm thick sample

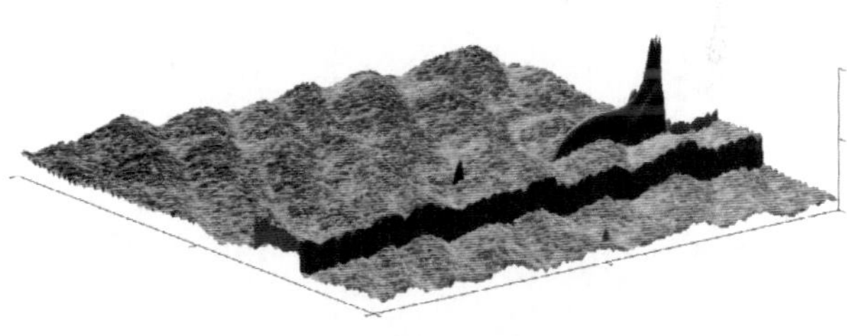

Figure 6. Pictures of the surface of a 1000 nm thick sample.

Data of the multifractal analysis do not contradict to the observations of the reflectance anisotropy spectroscopy: stresses grow with layer thickness but the solid solution decomposition is absent. At the layer thickness greater than 200 nm, the surface roughness becomes high, stresses are relieved and the solid solution decomposition is probable.

3.3. Photoluminescence. The PL method gives us information on the material quality. The higher the PL intensity, the more perfect is the material. Table 4 presents the results of the PL investigations of the grown samples.

Table 4. Results of the PL investigations of the grown samples.

Sample No	Thickness, nm	$\Delta a/a$	PL intensity (a. u.)	FWHM PL (meV)
I253	1000	0.015	694,3	20
I255	1000	3.88×10^{-3}	1875	10
I297	50	-	601,9	28
I298	200	$<2.1\times10^{-3}$	2722	20
I299	500	$<2.1\times10^{-3}$	$1{,}59\times10^{5}$	20
I313	(InGaAsP = 100 nm) 10 replications	-	$1{,}2\times10^{6}$	21
I314	(InGaAsP = 100 nm/InP 20 nm) 7 replications	-	$9{,}8\times10^{5}$	24

where - $\Delta a/a$ – lattice mismatch between InP and InGaAsP solid solutions, PL intensity – photoluminescence intensity for solid solutions, FWHM PL – the full width at a half maximum photoluminescence.

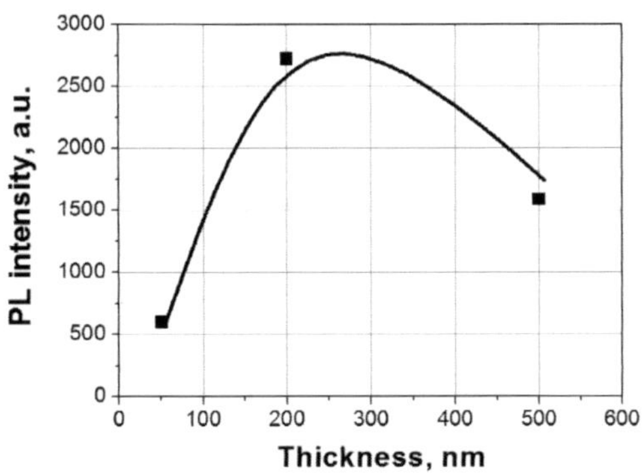

Figure 7. Dependence of the PL intensity on the thickness.

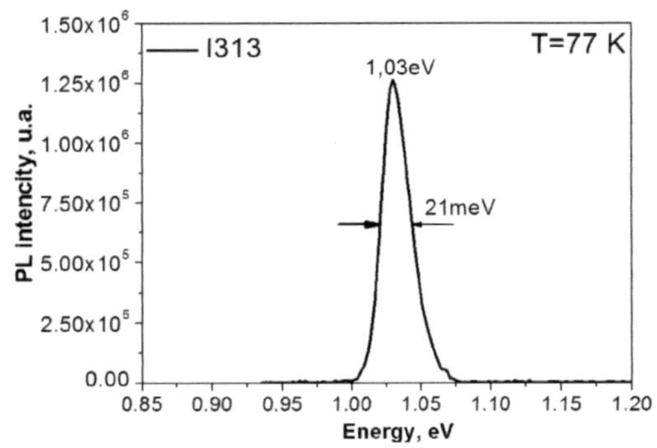

Figure 8. PL spectrum of the best sample I313($In_{0.8}Ga_{0.2}As_{0.46}P_{0.54}$).

Analysis of the PL spectra confirms the observations of the previous investigations – with growing layer thickness up to 100 - 200 nm (figure 7), the solid solution spinodal decomposition is not observed, the PL intensity rises, and the PL intensity decay takes place.

On the basis of all investigations, a structure with a layer thickness of 1000 nm composed of 10 100nm thick layers (I313) has been fabricated. The PL intensity of such a structure has increased by 3 orders of magnitude. Figure 8 presents the PL spectrum of this structure.

Acknowledgements
The authors thank N.M. Shmidt, A.B. Gordeyeva, P.N. Brunkow and S.I. Troshkov for their help in the research and discussion of the results.
This work has been supported by the Russian Science Foundation, grant No 14-29-00178.

References
[1] Onabe K 1982 *Jpn. J. Appl. Phys.* **21** 797
[2] Panish M B, Ilegems M 1972 *Paramon Press* **7** 39
[3] de Cremoux B 1982 *J. Physique* **43** 5

[4] Ipatova I P, Shchukin V A, Malyshkin V G, Maslov A Yu and Shchukin V A 1993 *Semiconductors* **27** 1062
[5] Stringfellow G B 1982 *J. Cryst. Growth* **58** 194
[6] Shim K, Rabitz H 1999 *Journal Of Applied Physics.* **85** 7705

Self-catalyzed GaAs and InAs nanowire growth (Monte Carlo simulation)

A G Suprunets[1,2], M A Vasilenko[1,2] and N L Shwartz[1,2]

[1] Rzhanov Institute of Semiconductor Physics, 630090 Novosibirsk, Russia
[2] Novosibirsk State Technical University, 630073 Novosibirsk, Russia

e-mail: nastysuprunets@yandex.ru

Abstract. A new lattice Monte Carlo model of $A^{III}B^{V}$ nanowire growth according to the vapor-liquid-solid mechanism is suggested. The peculiarities of this model and the procedure of energy parameters estimation for the InAs system are described. The characteristics of InAs and GaAs self-catalyzed nanowire growth are obtained using the suggested model. A comparison of InAs and GaAs self-catalyzed nanowire growth characteristics using the suggested Monte Carlo model is performed.

1. Introduction

Attention to nanowires (NWs) on the base of III-V materials can be attributed to their promising applications in opto- and nanoelectronics [1–3]. The first method of semiconductor NW formation is growth from catalyst drops according to the vapor-liquid-solid (VLS) mechanism [4]. For $A^{III}B^{V}$ semiconductors self-catalyzed NW growth is promising when one of the elements composing the semiconductor – metal (In or Ga) is used as a catalyst [5–7].

Theoretical works helped to explain some experimental results and revealed that self-catalyzed growth is more sensitive to the arsenic flux than the growth from golden droplets [8, 9]. Experimentally, it was demonstrated that InAs NWs had a narrower temperature range for growth in comparison with GaAs NWs [10]. Recently, it has been demonstrated that trials to extend the temperature range in a high temperature direction results in catalyst-free InAs nanowires growth [11]. There are only few works devoted to MBE growth of InAs NWs according to the VLS mechanism [6, 7]. Monte Carlo (MC) simulation can help in clearing up the details of nanostructure growth. Up to the present, only a number of articles have been devoted to MC simulation of nanowire growth according to the VLS mechanism [12, 13].The results of MC simulation of GaAs NW growth with gold and gallium drops as catalysts were presented in [13]. In the present work a modified variant of the lattice MC model of self-catalyzed NW growth is suggested. The procedure of energy parameters estimation for the InAs system is described. The comparison of InAs and GaAs self-catalyzed nanowire growth characteristics is fulfilled using this model.

2. Monte Carlo model

Simulation was carried out using the program package SilSim3D [14]. The present MC model is similar to that presented in [13]. However, instead of the 6-component system suggested in [13], now we consider the 5-component system: As_2, As, Ga (In), Ga_{liq} (In_{liq}) and M_f–molecular arsenic, atomic arsenic, gallium (indium) in solid and liquid phases and the film-mask material. In the modified model liquid arsenic is excluded from consideration and the liquid phase is suggested only for gallium (indium). Following [13], the main processes included in the model are: Ga (In) and As_2 adsorption, molecular arsenic dissociation and creation on the surface, gallium and arsenic surface diffusion, As_2, Ga (In), Ga_{liq} (In_{liq}) desorption, transformation of solid gallium (indium) into the liquid state, atomic arsenic dissolution in liquid gallium (indium) and diffusion through a liquid drop, Ga_{liq} (In_{liq}) crystallization at the drop-GaAs (InAs) interface. The processes considered in the drop and at the liquid-crystal interface are illustrated in figure 1.

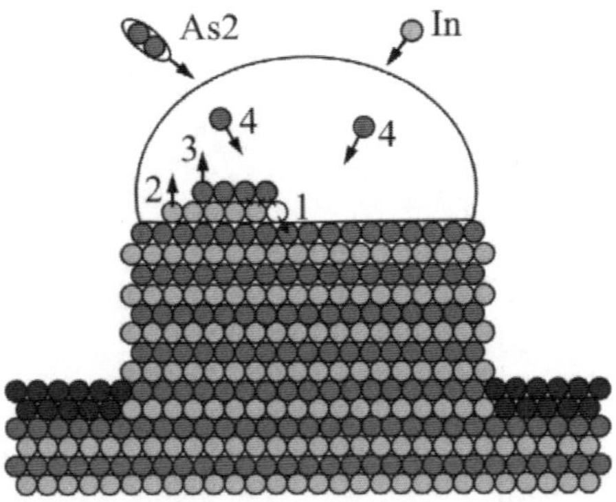

Figure 1. A scheme of the model system and the processes considered in the drop and drop-crystal interface: crystallization (1) and dissolution (2) of Ga (In), As dissolution (3) in the a liquid drop, As diffusion through a metal drop (4). The catalyst material Ga_{liq} (In_{liq}) is marked in white, Ga (In) - in light gray, As - in gray, the material mask - in black colors.

The liquid phase in the lattice model of VLS NW growth is imitated similar to the model of droplet epitaxy growth [15]. The drop surface tension is controlled by covalent binding energies between liquid atoms and wetting – by the binding energies between drop and substrate atoms. The catalytic action of liquid indium or liquid gallium on the growth process was realized by decreasing the energy barrier for molecular arsenic dissociation on the Ga (In) drop surface. The activation energy values of elementary events for the GaAs system were chosen according to [13, 15]. The value of liquid In surface tension of 0.5 J/m^2 is 1.3 times less than the liquid Ga surface tension being of 0.7 J/m^2 [16]. The same relation is chosen for the covalent binding energy of indium and gallium atoms: $E_{Ga\text{-}Ga} = 1.3\ E_{In\text{-}In}$. The energy barrier for In desorption was estimated by fitting the simulation results with the experimental data of temperature dependence of equilibrium In pressure above liquid indium [17]. According to this fitting, we obtained $E_{sub(In)} = 0.7$ eV, where $E_{sub} = E_{des} - E_{dif}$ (desorption energy E_{des} in the model is equal to $E_{des} = E_{dif} + E_{sub}$). The important parameters that determine $A^{III}B^{V}$ nanostructure growth are activation energies of molecular arsenic dissociation and creation on the surface. The first approximation values of these energies, as well as the As_2 desorption barrier for InAs, are the same as for GaAs NW growth. Only on the liquid drop surface, the energy barriers for molecular arsenic dissociation are assumed to be different for In and Ga. In this manner we imitate a different catalytic role of indium and gallium drops.

A cardinal change (in comparison with [13]) has been realized with a description of the processes in the liquid drop and at the energy barriers drop/crystal interface since we excluded liquid arsenic. Now arsenic dissolution and diffusion through a liquid drop are described by the unified exchange reaction: Ga_{liq} (In_{liq}) + As → As + Ga_{liq} (In_{liq}). The reactant neighbourhood specifies the process: dissolution in liquid metal or exchange diffusion through the drop. The activation energy barrier of arsenic dissolution rate higher than the energy barrier of diffusion through the drop can be reached by increasing the activation energy of exchange reaction at the liquid-crystal interface. The presence of solid gallium (indium) at the interface ensures this increase of activation energy barrier. Overestimation of the energy parameters responsible for the phase transformation was required after the model modification. The diffusion activation energy value of 0.3 eV in liquid In and of 0.7 eV in liquid Ga are taken according to the experiment [18] and [19], respectively. The activation energies of Ga (In) dissolution in the liquid drop and Ga (In) crystallization at the drop-crystal interface in the form of GaAs (InAs) are chosen fitting the simulated and experimental solubility of GaAs (InAs) in liquid Ga (In) [20]. The results of such fitting are shown in figure 2. The activation energy of Ga (In) crystallization is found to be equal to 1.2 (1.5) eV and 1.7 (1.85) eV for Ga (In) dissolution. Using the specified energy parameters, the self-catalyzed growth of InAs and GaAs nanowires was carried out.

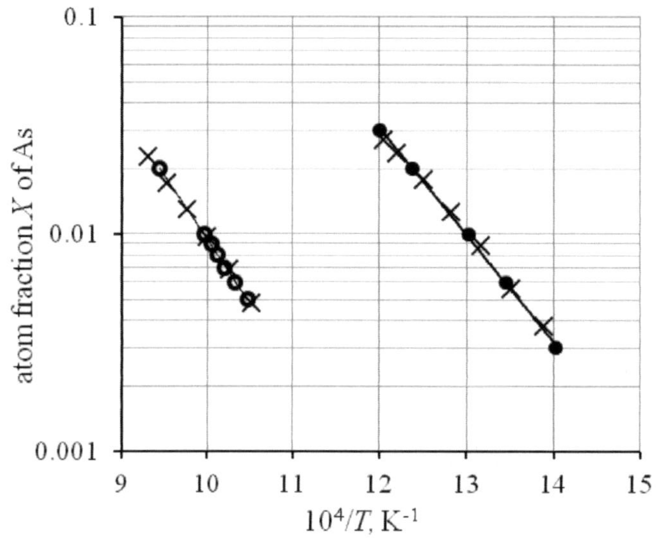

Figure 2. Temperature dependence of equilibrium arsenic concentration in liquid metal (In – solid circles, Ga – open circles from experimental data [20], crosses - simulation results).

3. Simulation results

The initial substrate is a semi-infinite InAs or GaAs crystal with the (111)B surface orientation covered with a film mask having a round through-hole containing an In or Ga drop. The temperature range for growth is chosen according to experimental works [6, 21]: for InAs $T = 740 - 840$ K and for GaAs $T = 800 - 950$ K. The reported data about flux intensities are worthless since our model surface is much smaller than the real one. That means that the Ga diffusion component of NW growth in the experiment is higher than that in simulation. So, the intensities of gallium and indium fluxes in the simulation are chosen higher than those in the experiment. A small model surface size means a significant readsorption effect for the NW growth rate [22]. The input flux intensities ranges used for modelling are: $F_{In} = (0.3–0.7)$ monolayer per second (ML/s), $F_{Ga} = (0.1–0.5)$ ML/s and fluxes ratio $F_{As2}/F_{In} = 0.7–1.7$. InAs and GaAs NW growths are carried out at the same As_2 flux. The NW growth rate dependences on temperature, In (Ga) and As_2 fluxes are obtained.

In a wide range of arsenic fluxes, the InAs NW growth rate was found to be linearly dependent on arsenic flux intensity that is typical for the self-catalyzed growth [23]. The temperature dependences of InAs and GaAs NW growth rates are demonstrated in figure 3. The maximal growth rate of InAs NWs corresponds to 820 K and that of GaAs NW – to 900 K. This result is mainly due to the different arsenic solubility in liquid gallium and indium and different In and Ga volatility. The stable growth of InAs and GaAs NWs was observed around these optimal temperatures. At a lower temperature, the growth rate slows down since the temperature decelerates all processes responsible for NW growth, such as In(Ga) and As_2 desorption, arsenic diffusion through the drop, crystallization at the drop-crystal interface, diffusion of all species along the surface. The NWs at temperatures lower than the optimal temperature are thicker and, at $T < 740$ K, In drops are formed at NW side walls resulting in the following NW branching. At temperatures higher than the optimal one, the growth rate decrease is due to the enhanced In(Ga) and arsenic desorption. At temperatures higher than 850 K (950 K), the NW growth stopped because of indium (gallium) drop evaporation. The morphology of NWs grown at different temperatures is demonstrates in figure 4.

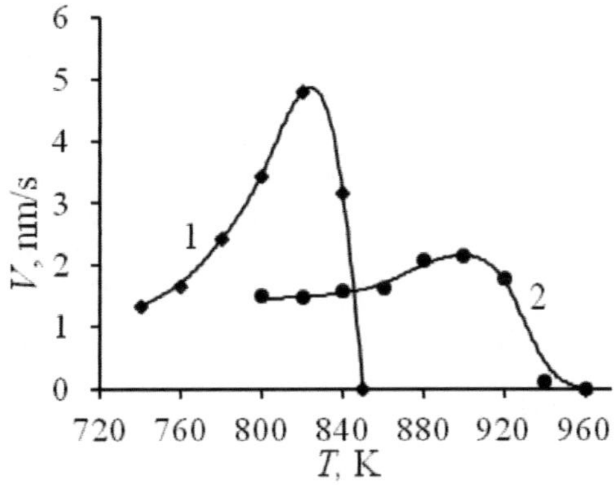

Figure 3. Temperature dependences of NW growth rate: for InAs (1), GaAs (2); F_{As2} = 0.5 ML/s, F_{Ga} = 0.3 ML/s ; F_{In} = 0.5 ML/s.

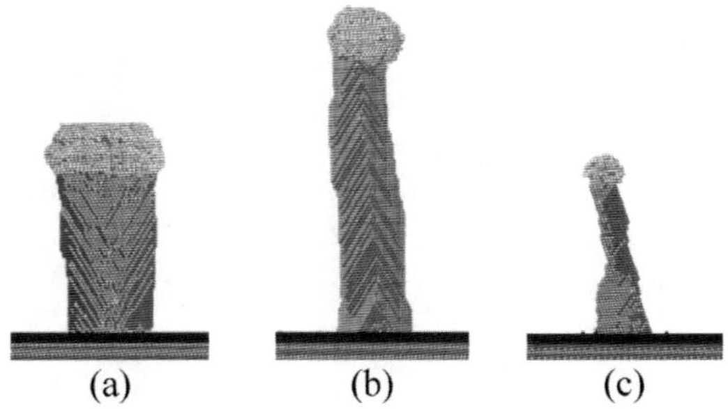

Figure 4. Cross-sections of InAs NWs at the same growth time (t = 10 s) at different temperatures: T = 760 K (a), T = 800 K (b) and T = 840 K (c). As is marked in gray, In – in light gray, In_{liq} – in white, film – in black colors.

It is worth noting that the growth rate presented in figure 3 corresponds to the average growth rate determined for a certain deposition time. The dependence of NW length on deposition time for different growth temperatures is shown in figure 5. For all temperatures one can single out three time intervals. The first interval corresponds to drop gathering (the incubation period), the second interval corresponds to stable growth with a practically constant growth rate and the third interval – with the increasing growth rate. This growth rate increase during the third time interval is associated with the metal drop size decrease. The growth rate decrease in the high temperature region in figure 3 may be due to the untimely metal drop consumption or to the prolonged incubation period. In our model system, with a significant readsorption effect, the In(Ga) drop is rapidly consumed by the enlarged arsenic flux due to readsorption at high temperatures. In figure 6 one can see the difference in the average growth rate values calculated at the same time of growth (curve 1) and at the moment of In drop consumption (curve 2).

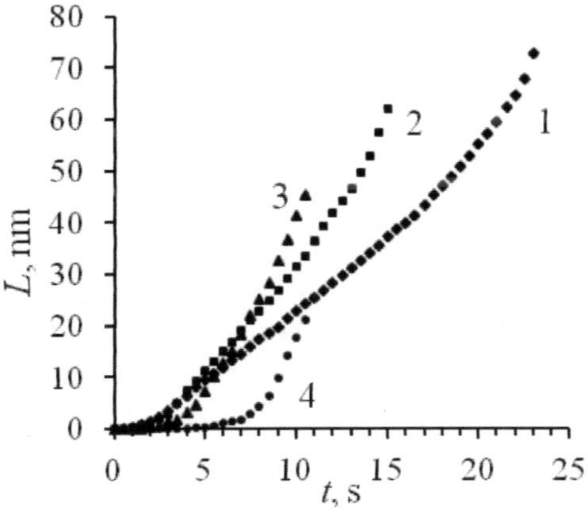

Figure 5. Dependence of NW length on time for different growth temperatures: T = 780 K – (1), T = 800 K – (2), T = 820 K – (3), T = 840 K – (4).

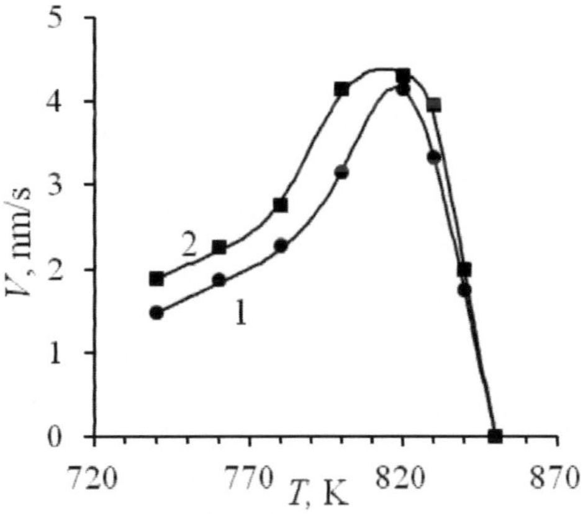

Figure 6. Temperature dependence of InAs NW growth rate determined at t =10 s (1) and at the moment of drop consumption (2) F_{As2} = 0.5 ML/s, F_{In} = 0.5 ML/s.

We analyzed the influence of catalytic indium drop properties on the temperature dependence of NW growth rate. We considered two different energy barriers for molecular arsenic dissociation on the liquid indium drop surface. The first variant – catalytic influence of liquid indium on As_2 dissociation is assumed to be equal to the influence of liquid gallium (curve 1 in figure 7). The second variant – molecular arsenic dissociates on the liquid indium surface that is more difficult than on the liquid gallium surface (curve 2). The reduction of growth rate for curve 2 is predominately due to the enhanced As_2 evaporation. This statement is illustrated in figure 8. Here one can see the cross-sections of InAs NWs at the initial growth stage. The growth time is equal for both NWs, but the NW grown according to curve 1 (figure 8 (a)) is longer and that indicates a higher arsenic inflow at the drop-vapor interface.

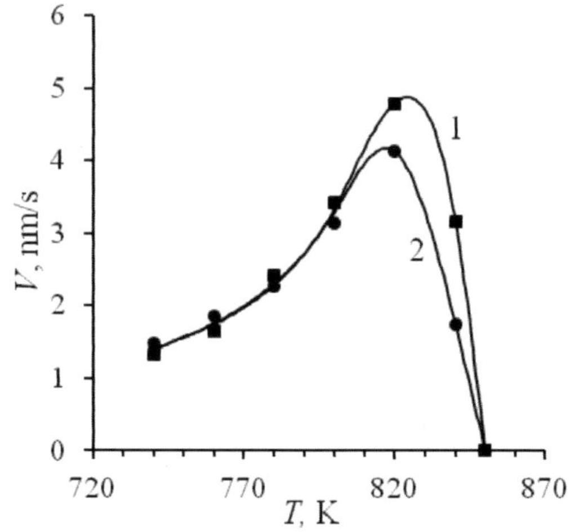

Figure 7. Temperature dependence of InAs NW growth rate: As_2 dissociation on the In_{liq} surface is easier for curve 1 in comparison with curve 2; $F_{As2} = 0.5$ ML/s, $F_{In} = 0.5$ ML/s.

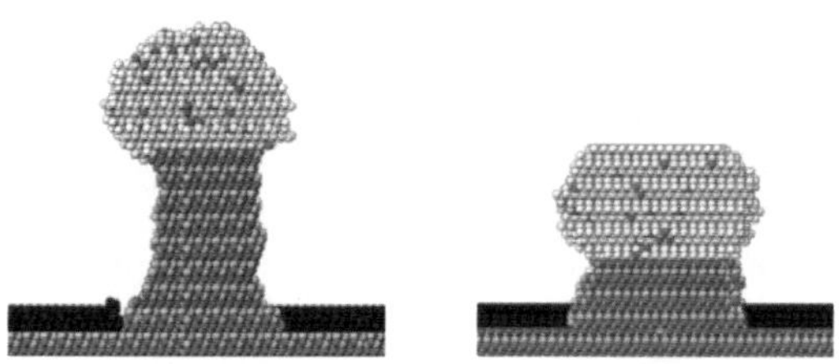

Figure 8. InAs NWs cross-sections at the initial stage of growth (t = 4 s) at T = 820 K; As_2 dissociates on the liquid indium surface is easier in (a) than in (b).

The optimal temperature value depends on arsenic to gallium (indium) flux ratio. The increase of As_2/Ga(In) flux ratio results in the optimal temperature shift to the low temperature region.

4. Conclusions

A modified version of the lattice Monte Carlo model of NW growth according to the vapor-liquid-solid mechanism was presented. The characteristics of self-catalyzed indium arsenide and gallium arsenide nanowire growth on (111)B substrates were analyzed using Monte Carlo simulation. The dependence of NW morphology on growth conditions was demonstrated. The temperature dependence of axial NW growth rate was analyzed. The influence of catalytic properties of the metal drop on the NW growth rate was demonstrated. The difference in optimal conditions for InAs and GaAs NWs self-catalyzed growth is due to different gallium and indium volatility and different arsenic solubility in liquid gallium and indium.

Acknowledgments

This work was supported by the RFBR (grant 14-02-00776) and by the Russian Academy of Sciences Programs.

References

[1] Krogstrup P, Jorgensen H I, Heiss M, Demichel O, Holm J V, Aagesen M, Nygard J and Fontcuberta i Morral A 2013 *Nature Photonics* **7** 306–310
[2] Colombo C, Heiβ M, Gratzel M and Fontcuberta i Morral A 2009 *Appl. Phys. Lett.* **94** 173108
[3] Chuang S, Gao Q, Kapadia R, Ford A C, Guo J and Javey A 2013 *Nano Lett.* **13** 555–558
[4] Tchernycheva M, Travers L, Patriarche G, Glas F, Harmand J-C, Cirlin G E and Dubrovskii V G 2007 *J. Appl. Phys.* **102** 094313.
[5] Colombo C, Spirkoska D, Frimmer M, Abstreiter G and Fontcuberta i Morral A 2008 *Phys. Rev. B* **77** 155326
[6] Grap Th, Rieger T, Blomers Ch, Schapers Th, Grutzmacher D and Lepsa M I 2013 *Nanotechnology* **24** 335601
[7] Yamaguchi M, Horiuchi I, Paek J H and Sawaki N 2011 *AIP Conf. Proc.* **1399** 229–230
[8] Glas F, Ramdani M R, Patriarche G and Harmand J-C 2013 *Phys. Rev. B* **88** 195304
[9] Krogstrup P, Jorgensen H I, Johnson E, Madsen M H, Sorensen C B, Fontcuberta i Morral A, Aagesen M, Nygard J and Glas F 2013 *J. Phys. D: Appl Phys.* **46** 313001
[10] Martelli F, Rubini S, Jabeen F, Felisari L and Grillo V 2011 *J. Cryst. Growth* **323** 297–300
[11] Hertenberger S, Rudolph D, Becker J, Bichler M, Finley J J, Abstreiter G and Koblmüller G 2012 *Nanotechnology* **23** 235602
[12] Nastovjak A G, Neizvestny I G and Shwartz N L 2010 *Pure Appl. Chem.* **82** 2017–2025
[13] Knyazeva M, Nastovjak A, Neizvestny I and Shwartz N 2015 *Semiconductors* **49** 60–68
[14] Zverev AV, Zinchenko KYu, Shwartz N L and Yanovitskaya Z Sh 2009 *Nanotechnologies in Russia* **4** 215–224
[15] Vasilenko M A, Neizvestny I G and Shwartz N L 2015 *Comput. Mat. Sci.* **102** 286–292
[16] Chentsov V P, Shevchenko V G, Mozgovoj A G and Pokrasin M A 2011 *Perspektivnie materiali, in Russian* **3** 46–52
[17] Nesmeyanov A 1961 *Vapor pressure of chemical elements, in Russian* (Moscow: USSR Academy Sciences Publishers)
[18] Dedegkaev T T, Gorokhov V A, Ilyin Yu L, Moshnikov V A, Pustovov V A, Sosov Yu M and Yaskov D A 1986 *Cryst. Res. Technol.* **21** K95–97
[19] Gorokhov V A, Dedegkaev T T, Ilyin Yu L, Moshnikov V A, Petrov A S, Sosov Yu M and Yaskov D A 1984 *Crystal Res. Technol.* **19** 1465–1468
[20] Hall R N 1963 *J. Electrochem. Soc.* **110** 385
[21] Rieger T, Heiderich S, Lenk S, Lepsa M I and Grutzmacher 2012 *J. Cryst. Growth* **353** 39–46
[22] Li A, Sibirev N V, Ercolani D, Dubrovskii V G and Sorba L 2013 *Cryst. Growth Des.* **13** 878
[23] Ramdani M R, Harmand J C, Glas F, Patriarche G and Travers L 2013 *Cryst. Growth Des.* **13** 91–96

Electronic structure, optical and magnetic properties of tetraphenylporphyrins-fullerene molecular complexes

M A Elistratova[1], I B Zakharova[1], N M Romanov[1,2], O E Kvyatkovskii[2,1], I Zakharchuk [3], E Lahderanta[3] and T L Makarova[2]

[1] St. Petersburg State Polytechnical University, 29 Polytechnicheskaya Street, St. Petersburg 195251, Russia
[2] Ioffe Institute, 26 Polytechnicheskaya Street, St. Petersburg, 194021, Russia
[3] Lappeenranta University of Technology, 34 Skinnarilankatu, Lappeenranta, 53850, Finland

E-mail: elistratovamari@yandex.ru

Abstract. This work presents optical, magnetic properties and quasi-equilibrium condensation processes of thin composite films based on C_{60}-MeTPP complexes. The elemental composition, surface motphology and distribution uniformity of the elements were investigated by SEM and energy-dispersive analysis. Different results were obtained for various types of MeTPP. Photoluminescence results showed decrease of intensity with C_{60} adding. The temperature dependence of magnetization M(T) and the field cooling (FC) and zero-field cooling (ZFC) curves for powder samples of CuTPP and CuTPP-C_{60} were measured to determine the changes in the magnetic behavior. Quantum-chemical calculations were performed to elucidate the experimental results.

1. Introduction

One of the most promising materials for organic optoelectronic devices, such as artificial photosynthesis systems and solar cells, are the molecular complexes of fullerenes with organic donors (e.g. porphyrins, phthalocyanines, pentacen) [1-3]. Fullerene is known as an acceptor with strong π-electron accepting ability, thus fullerene-based structures may be important for the study of photoinduced electron transfer. There are two basic approaches to the creation of such systems: synthesis of dyad molecules with C_{60} covalently bonded to porphyrins [4] and developing the self-assembled non-covalent complexes [2, 5, 6].

Intensively developed area of electronic devices based on carbon requires as physical as technological research of new materials. Composite materials and devices based on porphyrins attract a particular interest. Moreover, such devices can replace (or surpass in characteristics) usual silicon-based electronic devices. Tetraphenylporphyrin (TPP, the chemical formula is $C_{44}H_{30}N_4$) and its complexes with metal are p-type semiconductors and have high hole mobility, which determines its great application perspectives in organic optoelectronics [6]. The main type of these materials is fullerene (acceptor) and MeTPP complexes, which can be used in photovoltaic devices [6-8]. Noncovalent and covalent bonded fullerene complexes with various types of porphyrins have a good ability of self-organization because the energy minimum corresponds to the position where fullerene is located under TPP macrocycle. It became possible to create supramolecular structures, but usually they are studied only in solutions [8] or theoretically [9].

This work is devoted to experimental study of tetraphenylporphyrins –fullerene MeTPP-C_{60} complexes (in which Me is Zn(II), Cu(II), Fe(II)).

2. Obtaining and investigation methods

MeTPP and MeTPP-C_{60} thin films were produced by vacuum deposition in quasi-equilibrium conditions on Si, KBr and glass substrates. The film thickness was 200-500 nm. Surface morphology and phase composition of prepared films were investigated by secondary electron microscopy with an energy-dispersive X-ray console. Porphyrins: Zn(II)TPP, Cu(II)TPP, coordinated complex FeClTPP and its composites with fullerene were made by co-deposition in quasi-closed volume and by crystallization from solutions. It was shown that ZnTPP-C_{60} and FeClTPP–C_{60} films have a homogeneous structure with crystallite sizes in the lateral plane of about 1 mkm (see figure 1 (a)). In contrast, CuTPP-C_{60} films have two different phases: CuTPP nanowire (of 20 nm in diameter) and C_{60} cubic crystallites (of about 100 nm) (see figure 1 (b)). In this case, C_{60} crystallites grow only over porphyrin nanowires that leads to the nanostructured films formation.

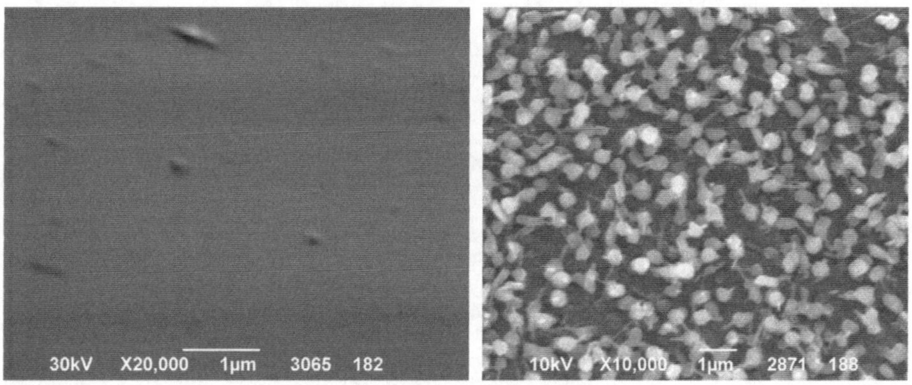

1 (a) 1 (b)

Figure 1. SEM results. (a) – surface morphology of ZnTPP-C_{60} film, (b) – CuTPP-C_{60} film

The method of crystallization from solutions gave the same result. Composition analysis by an energy-dispersive X-ray console showed that at most the composition does not change for ZnTPP, CuTPP –based films and corresponds to the stoichiometry. A depletion of Cl atoms (approximately halved) was found for FeClTPP-based films as compared with the original composition. In the same time, there was an indication of an appreciable amount of oxygen (about 3 wt% when background is 0.7%). Quantum-chemical calculations showed a weak bond of Cl atoms with the core of the porphyrin macrocycles. It can be assumed that Cl atoms partly detached from complexes during the thermal evaporation. Wherein, vacant Fe atoms bond with oxygen. The possible transition from Fe(II) to Fe(III) can lead to changes in magnetic properties. This fact requires further study.

Photoluminescence (PL) spectra for films of both types were measured. PL spectra are shown in figure 2. The reduction in emission intensity in MeTPP-C_{60} as compared with pure MeTPP films is due to the photoinduced charge transfer between C_{60} and MeTPP molecules. Fullerene adding leads to negligible changes in vibrational spectra of MeTPP because the distance between molecules in donor-acceptor couple is big enough (about 3Å). Emission in porphyrins is possible from various types of energy levels (for example, $S_1 \rightarrow S_0$ or $S_2 \rightarrow S_0$, which are prohibited for fullerenes). Due to these facts, the photoinduced charge transfer decreases a possibility of transfer with photon emission.

The temperature dependence of magnetization M(T) and the field cooling (FC) and zero-field cooling (ZFC) curves for powder samples of CuTPP and CuTPP-C_{60} were measured to determine the changes in the paramagnetic behavior of (Cu^{+2}) under the complex formation. The intensity of M(T)

and M(H) dependences was found to decrease for the CuTPP-C_{60} sample. Decrease in magnetization can be explained by interaction between CuTPP and C_{60} after molecular complexes formation.

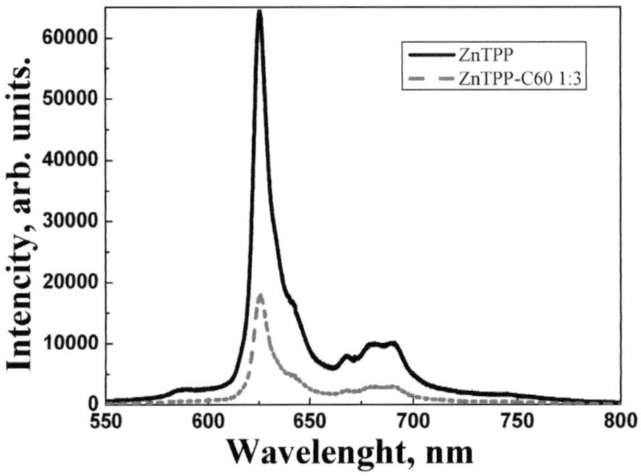

Figure 2. Photoluminescence spectra of ZnTPP (solid line) and ZnTPP-C_{60} (dash line).

Density functional theory DFT/B3LYP calculations for MeTPP, C_{60} and MeTPP-C_{60} complexes were performed to elucidate some experimental results. The electronic structures of electron densities with energy levels at HOMO and LUMO were investigated. The calculations give the optimized geometry, formation energy and HOMO-LUMO gap for molecular complexes. Formation energy E(C_{60})+E(MeTPP)–E(MeTPP-C_{60}) equals to 0.21 eV for ZnTPP and 0.17 eV for CuTPP and HOMO-LUMO gap equals to 1.8–1.6. eV. Charge transfer between the components of the complex in the ground state is very small being of the order of 0.03 qe. Good agreement was found between the experimental measurements, the results of our DFT calculation and the data from [7].

3. Conclusions

MeTPP-based composite structures were obtained by two techniques. Complex experimental and theoretical analyses were performed for these samples. Energy-dispersive analysis found changes in FeClTPP composition after heating (i.e. during the thermal evaporation process). There was a Cl amount depletion. Fe showed active interaction with oxygen after destroying of bonds with Cl. For other MeTPP the composition did not change. Various types of MeTPP have different surface morphology. CuTPP-based films grow as nanowires with phase separating on C_{60} crystallites and CuTPP wires. ZnTPP and FeClTPP samples have a smooth and homogenous surface. Photoluminescence and magnetic measurements showed intensity decrease for samples with fullerene. It can be explained by interaction between MeTPP and C_{60} molecules. Quantum-chemical calculations confirm this conclusion.

Acknowledgements

This work was supported by European FP7 IRSES projects 295180 MagNonMag.

References

[1] Suzuki A, Nishimura K and Oku T 2014 *Electronics* **3** 1 112
[2] Smerdon J A, Rankin R B, Greeley J P, Guisinger N P and Guest J R 2013 *ACS Nano* **7** 3086
[3] Oku T, Noma T, Suzuki A, Kikuchi K and Kikuchi S 2010 *J. Phys. Chem. Solid.* **71** 551
[4] Armaroli N Marconi G, Echegoyen L, Bourgeois J-P, Diederich F 2000 *Chem. Eur. J.* **6** 1629

[5] Konarev D V, Neretin I S, Slovokhotov Y L, Yudanova E I, Drichko N V, Shul'ga Y M, Tarasov B P, Gumanov L L, Batsanov A S, Howard J A K and Lyubovskaya R N 2001 *Chem. Eur. J.* **7** 12 2605
[6] Yoshimoto S and Itaya K 2007 *Journal of Porphyrins and Phthalocyanines* **11** 313
[7] Reid I, Zhang Y, Demasi A, Blueser A, Piper L, Downes J E, Matsuura A, Hughes G, Smith K E 2009 *Applied Surface Science* **256** 720
[8] Mukherjee P and Bhattacharya S 2012 *Spectrochimica Acta Part A: Molecular and Biomolecular Spectroscopy* **90** 186
[9] Buldum A and Reneker D H 2014 *Nanotechnology* **25** 23 235201.

Ge clusters and wetting layers forming from granular films on the Si(001) surface

M S Storozhevykh, L V Arapkina and V A Yuryev
A. M. Prokhorov General Physics Institute of the Russian Academy of Sciences, 38 Vavilov Street, Moscow, 119991, Russia

E-mail: storozhevykh@kapella.gpi.ru

Abstract. The report studies the transformation of a Ge granular film deposited on the Si(001) surface at room temperature into a Ge/Si(001) heterostructure as a result of rapid heating and annealing at 600 °C. As a result of the short-term annealing at 600 °C in conditions of a closed system, the Ge granular film transforms into a usual wetting layer and Ge clusters with multimodal size distribution and Ge oval drops having the highest number density. After the long-term thermal treatment of the Ge film at the same temperature, Ge drops disappear; the large clusters increase their sizes at the expense of the smaller ones. The total density of Ge clusters on the surface drastically decreases. The wetting layer mixed $c(4 \times 2) + p(2 \times 2)$ reconstruction transforms into a single $c(4 \times 2)$ one which is likely to be thermodynamically favoured. Pyramids or domes are not observed on the surface after any annealing.

1. Introduction

This paper presents an experimental study of transformation of a disordered Ge film on Si(001) into a Ge/Si(001) heterostructure. We deposited Ge at room temperature on the Si(001) layer grown on the Si substrate by means of the molecular-beam epitaxy (MBE) and explored crystallization of the obtained Ge granular film as a result of rapid heating and isothermal treatments at 600 °C. This experiment has given the following results. First of all, we have demonstrated that the Ge/Si(001) heterosructure formed as a result of a short-term annealing under the conditions of a closed system consists of a usual patched wetting layer and large clusters of Ge with multimodal size distribution rather than pyramids or domes which appear when a film is deposited in a flux of Ge atoms arriving on its surface [1–5]. Ge oval drops with the lateral dimensions of about a hundred nanometers have the highest number density among the detected clusters. Then, we detected a mixture of $c(4\times 2)$ and $p(2\times 2)$ reconstructions on the surface of the formed wetting layer whereas the simultaneous presence of both these structures in comparable proportions on wetting layer patches is a distinctive feature of the low-temperature mode of the wetting layer growth (at $T_{\mathrm{gr}} < 600$ °C) in the MBE process [6, 7]. And finally, we have shown that the Ge drops disappear from the surface as a result of long-term isothermal annealing of the original Ge film under the conditions of an isolated system; the sizes of the large clusters increase at the expense of the smaller ones and the total density of Ge clusters on the surface decreases by several orders of magnitude. The wetting layer retains all main features of the one forming as a result of the short-term annealing except for the reconstruction of its patches. The latter changes to a pure $c(4\times 2)$ one which is a characteristic feature of the wetting layers grown by MBE at high temperature (at $T_{\mathrm{gr}} > 600$ °C) [8].

Figure 1. STM image of the Si buffer surface before Ge deposition ($h_{Si} = 1000$ Å, $T_{gr} = 650$ °C).

2. Details of experiments

2.1. Samples

Ge films ($h_{Ge} = 7$ Å) were deposited from molecular beams at room temperature ($dh_{Ge}/dt \approx 0.15$ Å/s) on Si buffers (1000 Å thick) grown in the same cycles on commercial (100)-oriented wafers of Si ($\rho = 12\,\Omega$cm) at the temperature of 650 °C. Details of the pre-growth treatments of Si wafers, which included wet chemical etching and oxide removal by short high-temperature annealing ($T_{an} \sim 900$ °C), can be found in our previous articles [9–12]. After Ge deposition, the samples were heated to 600 °C at the maximum rate achievable for the used infrared heaters (0.24 °C/s), annealed at this temperature for 5 or 125 minutes and cooled to room temperature at the quenching mode (0.4 °C/s) [10, 11].

2.2. Techniques and equipment

The experiments were carried out using an ultrahigh-vacuum (UHV) MBE chamber (Riber EVA 32) connected with a UHV scanning tunnelling microscope (STM) chamber (GPI 300) [7, 13, 14]. The rates of Ge and Si deposition and the coverages of Ge and Si (h_{Ge}, h_{Si}) were measured by a graduated in advance film thickness monitors (Inficon Leybold-Heraeus XTC 751-001-G1) with quartz sensors installed in the MBE chamber. During annealing, samples were heated from the rear side by radiators of tantalum. The temperature was monitored with a tungsten-rhenium thermocouple mounted in the vacuum near the rear side of the samples and *in situ* graduated against a specialized pyrometer (IMPAC IS 12-Si). The atmosphere composition in the MBE chamber was monitored using a mass-spectrometer residual gas analyzer (SRS RGA-200) before and during the processes. STM images were obtained at room temperature; they were processed using the WSxM software [15]. Microphotographs of the sample surfaces were obtained with an Axiotech 100 microscope (Carl Zeiss). Additional details concerning the used equipment can be found, e. g., in Refs. [7, 13, 16].

3. Results and discussion

3.1. Si buffer layer

An STM image of the Si(001) surface of the buffer layer before depositing a Ge film is demonstrated in figure 1. Alternate wide terraces ending by S_a and S_b steps [17, 18] with smooth and jagged edges are observed on the surface as well as numerous Si adatom clusters that is characteristic to the Si(001) surfaces cooled at the quenching mode [10, 11, 19].

3.2. Ge film deposited at room temperature

Films of Ge deposited at room temperature on the buffer layers of Si are composed of densely packed grains (figure 2). The sizes of the grains vary from less than a nanometer to a few nanometers. The films are similar in their structure to the films which are formed at room

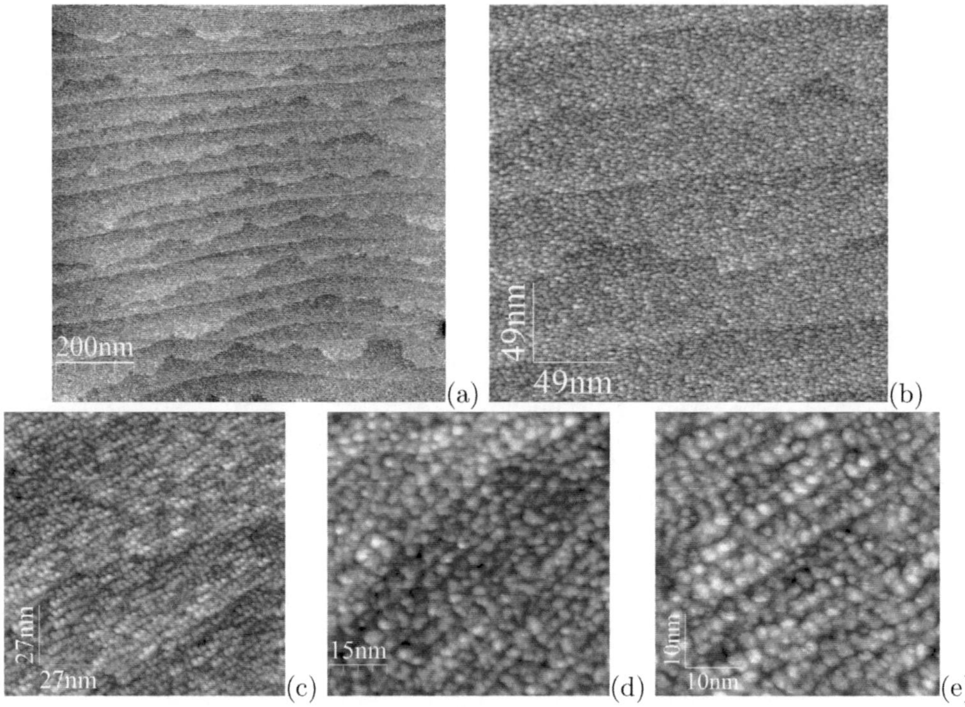

Figure 2. STM images of the sample surface after Ge deposition ($h_{\mathrm{Ge}} = 7$ Å) at room temperature on Si(001): (a, b) a general view and Si terraces under the Ge film, (c, d, e) grains of the Ge film.

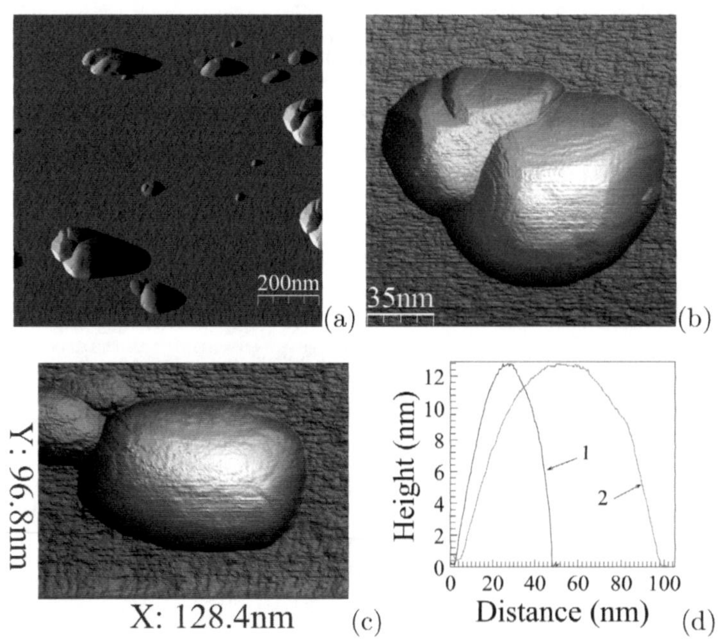

Figure 3. STM images of Ge clusters forming on the Ge/Si(001) wetting layer ($h_{\mathrm{Ge}} = 7$ Å) after Ge deposition at room temperature followed by annealing at 600 °C for 5 min: (a) a derivative image demonstrating numerous clusters of different sizes distributed over the surface; (b, c) typical oval clusters; (d) profiles of the oval cluster shown in panel (c) across (1) and along (2) the cluster (height from the wetting layer vs distance along the wetting layer).

temperature directly on the clean (001) surface of a silicon substrate [20]. The film is seen to be partially ordered; it is composed of relatively long nearly parallel chains consisting of tens of grains. Terraces and steps of the underlaying Si buffer (both $\mathrm{S_a}$ and $\mathrm{S_b}$) are clearly seen in the

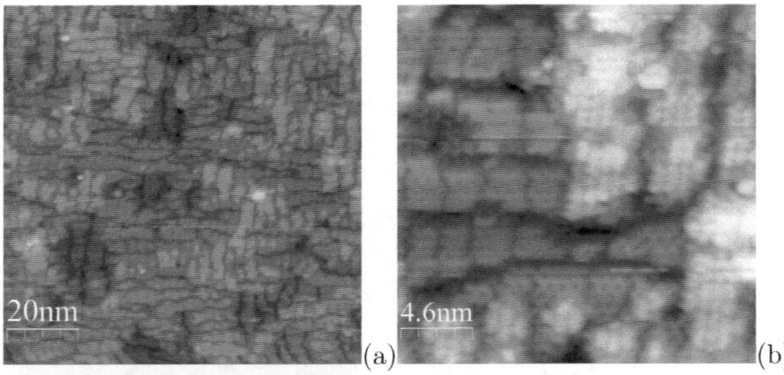

Figure 4. STM images of the Ge/Si(001) wetting layer after Ge deposition at room temperature (h_{Ge} = 7 Å) followed by annealing at 600 °C for 5 min: (a) a general view and (b) a mixed reconstruction of patches.

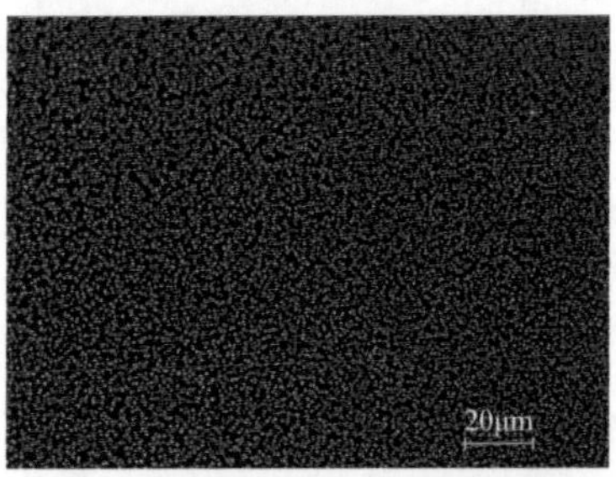

Figure 5. A microphotograph of the sample surface after Ge deposition at room temperature ($h_{\mathrm{Ge}} = 7$ Å) followed by annealing at 600 °C for 5 min; the scale mark is 20 µm.

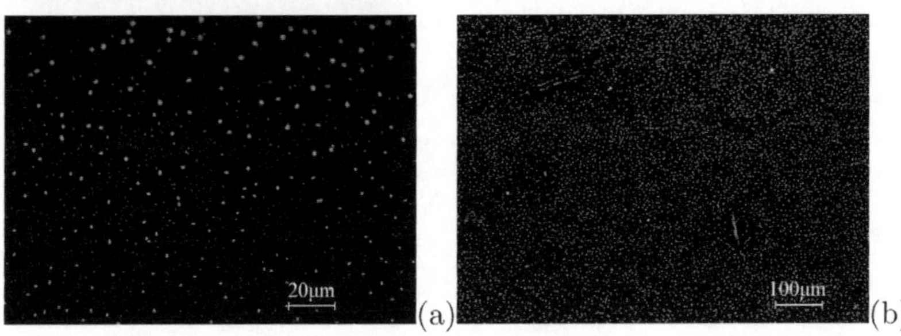

Figure 6. Microphotographs of the sample surface after Ge deposition at room temperature (h_{Ge} = 7 Å) followed by annealing at 600 °C for 125 min; the scale marks are (a) 20 and (b) 100 µm.

STM images presented in figure 2: the granular film strictly follows the relief of the underlying surface and being smooth enough does not conceal the details of this minor roughness. This observation allows us to conclude that the obtained granular film does not reconstruct the Si buffer at least on the level of the structure of its steps.

3.3. Ge layer annealed for 5 minutes at 600 °C

After rapid heating and isothermal annealing of the samples at 600 °C for 5 minutes, i.e. as a result of thermal treatment in the conditions of a closed system and in the conditions of an isolated system for the last 5 minutes, the deposited at room temperature Ge granular film reconstructs into a usual wetting layer with oval (partially faceted) Ge mounds (figures 3 and 4). The final structure coincides with that observed in the reference [20] after the same thermal processing of similar films deposited directly on the clean (001) surfaces of silicon wafers. The number density of the Ge droplets estimated from the STM images is (1.5 to 2) $\times\ 10^9\ \mathrm{cm}^{-2}$ that slightly exceeds the estimates obtained in the reference [20] (figure 3). The drops sizes usually exceed 100 nm in lateral dimensions and reach 15 nm in height. Note that there are many coalescent clusters observed on the surface; in addition, small and relatively large drops are observed simultaneously.

The wetting layer structure resembles the one typical for the wetting layers formed at low

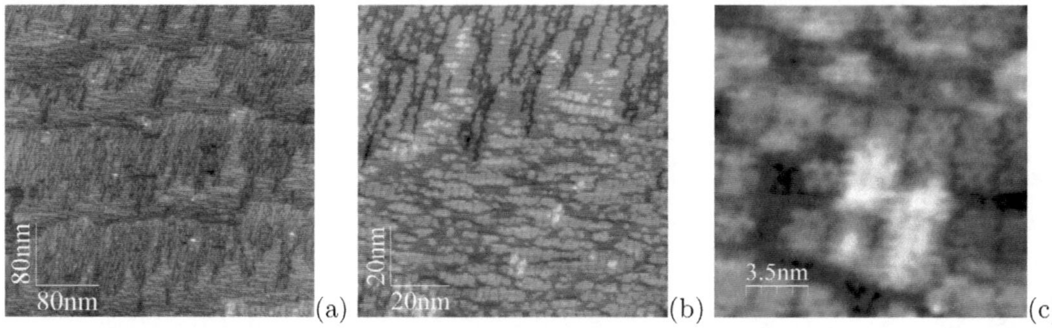

Figure 7. STM images of the Ge/Si(001) layer ($h_{\mathrm{Ge}} = 7$ Å) after Ge deposition at room temperature followed by annealing at 600 °C for 125 min.

temperatures (around 350°C) during MBE [6, 21–23]. Tops of the patches are composed of $c(4 \times 2)$ and $p(2 \times 2)$ reconstructed layers in close proportions (figure 4) [6, 13]. This mixed type of the wetting layer reconstruction is likely a result of a kinetically controlled transformation.

In addition to the oval drops, large clusters of Ge were also detected on the surface using a light microscope (figure 5). Since their number density estimated from the microphotographs was about $5 \times 10^7\,\mathrm{cm}^{-2}$, a probability of their detection with STM was very low. Unfortunately, now we cannot judge about their shapes; their sizes can be estimated as at least several hundreds of nanometers. Their size distribution is at least bimodal; they often form chains.

3.4. Ge layer annealed for 125 minutes at 600 °C

After rapid heating and isothermal annealing of the samples at 600 °C for 125 minutes, the large clusters of Ge increase their sizes but their distribution remains multimodal (figure 6); the number density of these clusters reduces by two orders of magnitude to $\sim 5 \times 10^5\,\mathrm{cm}^{-2}$ (figure 6a) however they retain some tendency to form chains (figure 6b). The largest clusters are seen to consume the substance from the surrounding smaller ones and form an empty space around themselves (figure 6a).

STM images do not show any presence of oval drops on the wetting layer (figure 7). At the same time, no pyramids or domes are also observed on the surface. And evidently, wedges are not observed, either [8]. Patches of the Ge wetting layer are mainly $c(4 \times 2)$-reconstructed. This type of the wetting layer reconstruction is characteristic of the high-temperature mode of the MBE growth of Ge on Si(001) ($T_{\mathrm{gr}} > 600$ °C) [8]. We conclude that the $c(4 \times 2)$ reconstruction forms under the thermodynamic control and is thermodynamically favoured.

We believe that the observed disappearance of the Ge drops from the wetting layer after sample annealing at 600 °C for 125 minutes is explained by Ge redistribution in favour of the growing large clusters. Ge and Si intermixing may also affect the drop disappearance.

4. Conclusion

In summary, as a result of the short-term annealing at 600 °C at the described above conditions of an isolated system, the Ge/Si(001) granular film deposited at room temperature transforms into a usual wetting layer and Ge clusters with multimodal size distribution with the Ge oval drops having the highest number density. After the long-term thermal treatment of the described initial samples at the same temperature, Ge drops disappear; the large clusters increase their sizes at the expense of the smaller ones. The total density of Ge clusters on the surface decreases by about four orders of magnitude. The wetting layer mixed $c(4 \times 2) + p(2 \times 2)$ reconstruction forming after the sort-term annealing transforms into a single $c(4 \times 2)$ one, which is likely to be thermodynamically favoured, as a result of long-term annealing. Pyramids or domes are not observed on the surface after any annealing.

Acknowledgments
We cordially thank Ms N V Kiryanova for her invaluable contribution to arrangement of this research. We express our appreciation to Mr O V Uvarov for obtaining microphotographs. We also thank Ms L M Krylova for chemical treatments of the samples. The Center of Collective Use of Scientific Equipment of GPI RAS supported this research by presenting admittance to its instrumentation. We acknowledge the support of this research.

References
[1] Medeiros-Ribeiro G, Kamins T I, Ohlberg D A A and Williams R S 1998 *Phys. Rev. B* **58** 3533
[2] Brehm M, Montalenti F, Grydlik M, Vastola G, Lichtenberger H, Hrauda N, Beck M J, Fromherz T, Schäffler F, Miglio L and Bauer G 2009 *Phys. Rev. B* **80** 205321
[3] Montalenti F, Raiteri P, Migas D B, von Känel H, Rastelli A, Manzano C, Costantini G, Denker U, Schmidt O G, Kern K and Miglio L 2004 *Phys. Rev. Lett.* **93** 216102
[4] Ross F M, Tromp R M and Reuter M C 1999 *Science* **286** 1931
[5] Medeiros-Ribeiro G, Bratkovski A M, Kamins T I, Ohlberg D A A and Williams R S 1998 *Science* **275** 353
[6] Arapkina L V and Yuryev V A 2011 *J. Appl. Phys.* **109** 104319
[7] Arapkina L V and Yuryev V A 2011 *Nanoscale Res. Lett.* **6** 345
[8] Yuryev V A and Arapkina L V 2012 *J. Appl. Phys.* **111** 094307
[9] Arapkina L V, Shevlyuga V M and Yuryev V A 2008 *JETP Lett.* **87** 215–219
[10] Arapkina L V, Yuryev V A, Chizh K V, Shevlyuga V M, Storojevyh M S and Krylova L A 2011 *Nanoscale Res. Lett.* **6** 218
[11] Arapkina L V, Yuryev V A, Shevlyuga V M and Chizh K V 2010 *JETP Lett.* **92** 310–314
[12] Arapkina L V, Krylova L A, Chizh K V, Chapnin V A, Uvarov O V and Yuryev V A 2012 *J. Appl. Phys.* **112** 014311
[13] Yuryev V A and Arapkina L V 2011 *Nanoscale Res. Lett.* **6** 522
[14] Eltsov K N, Klimov A N, Kosyakov A N, Obyedkov O V, Shevlyuga V M and Yurov V Y 2003 *Chemical state and atomic structure of fcc metal surfaces in chemical reaction with halogens* (*Proc. of General Physics Institute* vol 59) ed Konov V I and Eltsov K N (Moscow, Russia: Nauka) p 45
[15] Horcas I, Fernandez R, Gomez-Rodriguez J M, Colchero J, Gomez-Herrero J and Baro A M 2007 *Rev. Sci. Instrum.* **78** 013705
[16] Arapkina L V and Yuryev V A 2010 *Phys. Usp.* **53** 279–290
[17] Chadi D J 1979 *J. Vac. Sci. Technol.* **16** 1290–1297
[18] Misbah C, Pierre-Louis O and Saito Y 2010 *Rev. Mod. Phys.* **82** 981–1040
[19] Yuryev V A and Arapkina L V 2013 *Nanotechnology VI* (*Proc. SPIE* vol 8766) ed Adelung R (SPIE, Bellingham, WA) p 87660M
[20] Storozhevykh M S, Arapkina L V and Yuryev V A 2015 *Nanoscale Res. Lett.* **10** 295
[21] Yuryev V A, Arapkina L V, Storozhevykh M S, Uvarov O V and Kalinushkin V P 2014 *J. Nanoelectron. Optoelectron.* **9** 196–218
[22] Yuryev V A, Arapkina L V, Storozhevykh M S, Uvarov O V and Kalinushkin V P 2013 *Nanotechnology VI* (*Proc. SPIE* vol 8766) ed Adelung R (SPIE, Bellingham, WA) p 876606
[23] Arapkina L V and Yuryev V A 2013 *J. Appl. Phys.* **114** 104304

High-quality Ge epilayers grown on a Si substrate in one step process via hot wire chemical vapor deposition

S A Denisov[1,2], S A Matveev[2], V Yu Chalkov[1] and V G Shengurov[1]

[1] Physico-Technical Research Institute, Nizhniy Novgorod State University, Nizhniy Novgorod, 23/3 Gagarin Ave., Nizhny Novgorod 603950 Russia
[2] Physics Department, Nizhniy Novgorod State University, Nizhniy Novgorod, 23/3 Gagarin Ave., Nizhny Novgorod 603950 Russia

E-mail: matveevsa.sou@gmail.com

Abstract. High-quality Ge epilayers on Si with a low threading-dislocation density (TDD) were grown by a one step hot wire chemical vapor deposition process at 350°C without cyclic thermal annealing. The Ge layers with threading dislocation density (TDD) of $1\cdot10^{5}$ cm^{-2} for a 1.4 μm thick Ge layer were obtained on Si wafers of a diameter Ø = 5 cm. Root mean square of roughness (RMS) of ~ 0.37 nm is achieved. The Ge layers produced are of high optical quality.

1. Introduction

Epitaxial growth of fully relaxed Ge epilayers on Si substrates is of current interest for a number of new technologies, such as high performance Ge-on-Si photodetectors for infra-red sensing, solar cells, optoelectronics or Ge lasers [1]. These technologies all benefit from being compatible with the Si(001) substrate. Furthermore, successful growth on Si provides a virtual substrate for the integration and fabrication of GaAs-based optical devices on Si. The epitaxial growth is usually performed by ultra-high vacuum chemical vapor deposition (UHV-CVD) or molecular-beam epitaxy (MBE). However, these methods operate under an ultra-high vacuum atmosphere. On the contrary, chemical vapor deposition (CVD) offers such advantage as high throughput.

Growth of epitaxial Ge on Si at low substrate temperature by hot wire (HW) CVD has been reported in [2]. However, at a lower substrate temperature T_S = 350°C Mukherjee could grow only the 300 nm thick epitaxial Ge film. The main advantage is that this technique is relatively simpler than MBE and UHV CVD.

Here we report the growth of relatively thick (with a thickness of about 0.2 – 5 μm and above) epitaxial Ge films on (001) oriented c-Si wafers by HW CVD.

2. Experimental details

The samples were prepared on 5 cm diameter n-type (001) Si wafers with a resistivity of 5 – 10 Ω·cm by HW CVD with a base pressure of $5\cdot10^{-8}$ torr [3]. After standard RCA etching and thermal cleaning of Si substrate at 900°C for 20 min, a 200 nm thick Si buffer layer was grown at 800°C. Then a Ge-film was grown at 350°C. The deposition of Ge was carried out by letting purified germane (GeH_4) inside the growth chamber through a leak valve. The partial pressure of GeH_4 was $4\cdot10^{-4}$ Torr. Germane was cracked pyrolitically on a tantalum wire heated directly by passing the direct electric current through the wire up to 1200 - 1300°C.

Journal of Physics: Conference Series **690** (2016) 012014 doi:10.1088/1742-6596/690/1/012014

The Ge films were grown at the same growth conditions as the Ge/n+-Si(001) structures for the fabrication of the photodiodes. The surface morphology of the Ge/Si(001) layers was examined by Atomic Force Microscopy (AFM) in ambient air using a NT-MDT Solver ProTM instrument. The crystal structure of the Ge/Si(001) films was examined by Double crystal X-Ray Diffraction (DXRD) and by Transmission Electron Microscopy (TEM). The X-ray diffraction experiments have been carried out using a Bruker D8 Discoverer TM X-ray diffractometer with CuK radiation. TEM investigations have been carried out using a JEOLJEM 2100F transmission electron microscope.

3. Results and discussion

Figure 1 shows typical optical image of TDD in a Ge epilayer on a Si substrate. The typical TDD in ~ 1.4 μm thick layers is in a range of $1 \cdot 10^5$ cm^{-2}. Notice that TDD of Ge epilayers of the same thickness obtained by the RP CVD method was 10^7 - 10^8 cm^{-2} even after thermal annealing [4].

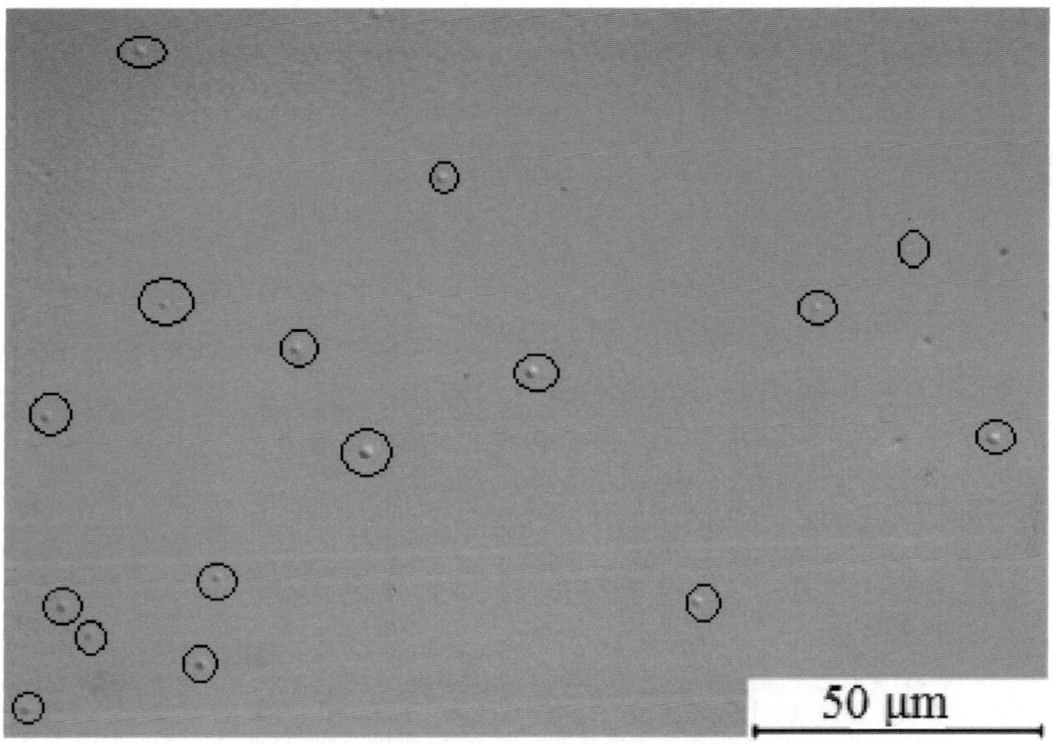

Figure 1. Optical image of a Ge epilayer on a Si substrate after etching by I_2 solution

XRD ω-2θ scans around Si(400) and Ge (400) diffraction of a 1μm thick Ge layer are shown in figure 2. The crystallity of the Ge layer was approved by the presence of an intensive Ge(004) peak. The sample shows a nearly symmetric Ge (400) peak including no detectable diffusion between Ge and Si. The full width at half maximum (FWHM) of the Ge peak is 0.052°.

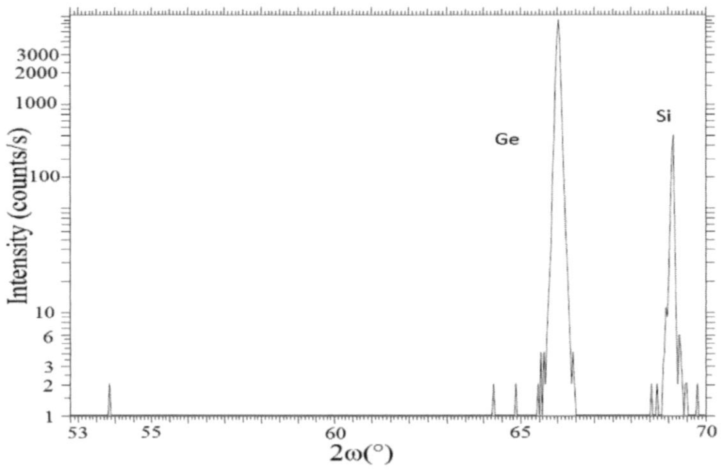

Figure 2. HR XRD 2θ scan profile of a Ge film on a Si substrate, grown at 350°C substrate temperature.

Cross-sectional TEM of the heteroepitaxial Ge films of ~ 1 μm shows a high density of threading dislocations confined to a region ~ 0.2 μm wide adjacent to the Ge/Si interface (figure 3(a)).

High resolution images were also observed in the interface region between the Si substrate and the Ge film (figure 3(b)). The continuous crystalline structures are clearly seen at the Ge/Si interface and thus the epitaxial growth following the substrate is confirmed.

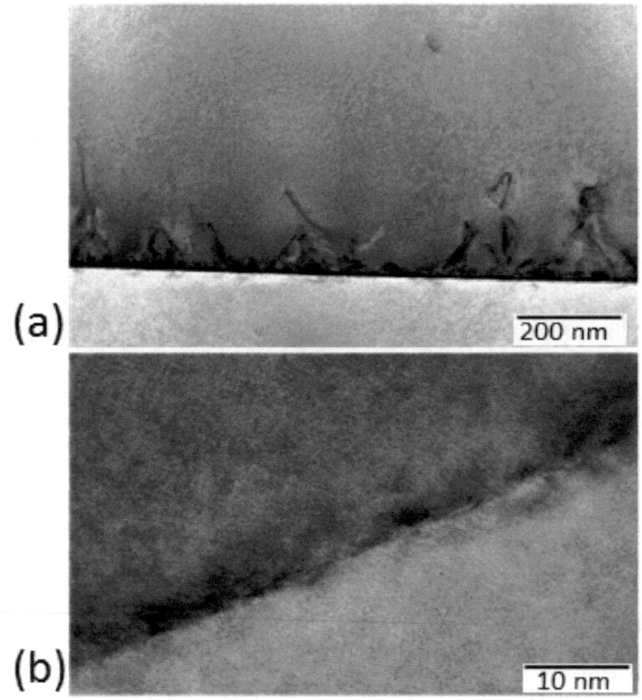

Figure 3. The cross-sectional TEM image of a Ge film on a single crystal Si(100) substrate: (a) TEM image observed in the whole thickness (b) the interface region between the Si(100) substrate and the Ge film.

Figure 4 shows the atomic force microscope (AFM) image of a Ge epilayer on a Si substrate. The surface is found to consist of step and terrace structures, indicating the two-dimensional layer-to-layer growth. A small RMS surface roughness of 0.37 nm is obtained.

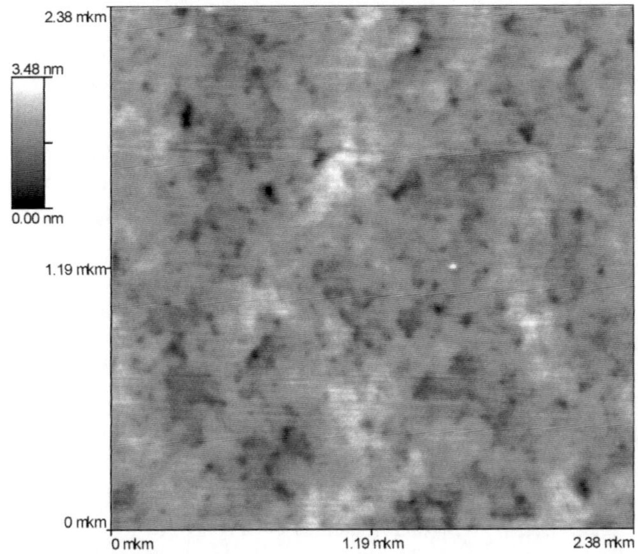

Figure 4. A morphology of a Ge epilayer on a ~ 0.25 μm thick Si substrate.

A spectral dependence of the external photoelectric quantum efficiency η of a photodiode based on a Ge/Si heterostructure is presented in figure 5. The photodiode demonstrated a good photoresponse in the 1.1 - 1.5 μm wavelength band.

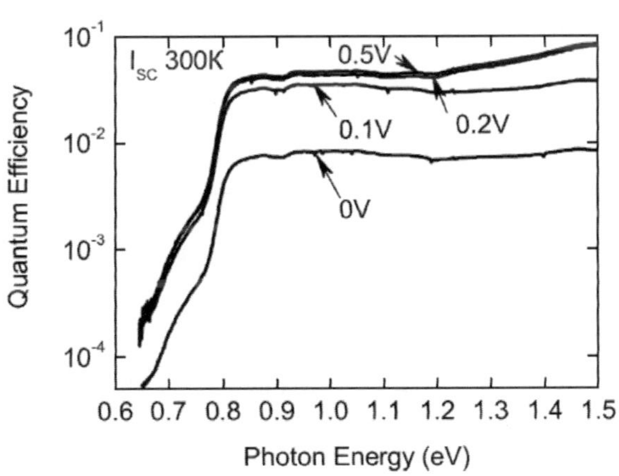

Figure 5. The spectra of the external photoelectric quantum efficiency (300 K) of a photodiode based on a Ge/Si(001) heterostructure measured at different values of reverse bias U_b.

Although this value is ≈ 4 times lower than the highest ones reported in the literature, there is still a wide room for further improvement of the quantum efficiency. High crystal quality of Ge layers on Si(100) substrates can be explained.

When using the low temperature HW CVD method, the hydrogen injection into a Ge layer with formation of clusters occurs. Low diffusion barrier promotes the diffusion and surface mobility of adatoms, which facilitates Ge relaxation and reduces the density of defects caused by the lattice mismatch.

4. Conclusion

In summary, a virtual Ge template composed of a Ge epilayer on a Si substrate can be obtained in one step process via the HW CVD method without UHV condition. The high-quality 0.2 – 5 μm thick Ge epilayers on Si with a surface RMS roughness of about 0.37 nm and TDD of $1\cdot10^5$ cm^{-2} were obtained. HW UHV CVD has been demonstrated to be a promising method for fabrication of the Ge/Si(001) photodetectors used in the Si-based electronic-photonic integrated circuits. Those results suggest that Ge will play a significant role as an enabler for integrated active photonic devices on Si.

Acknowledgments
This work was supported by RFBR (project no. 14-02-31370 mol_a) and grant of the President of the Russian Federation (no. MK-7021.2015.2)

References

[1] Shah V A, Dobbie A, Myronov M and Leadley D R 2011 *Thin Solid Films* **519** 7911
[2] Mukherjee C, Seitz H and Schroder B 2001 *Appl. Phys. Lett.* **78** 3457
[3] Matveev S A, Denisov S A, Guseinov D V, Trushin V N, Nezhdanov A V, Filatov D O and Shengurov V G 2014 *Journal of Physics: Conf. Series* **541** 012026
[4] Hartmann J M, Abbadie A, Papon A M, Holliger P, Rolland G, Billon T, Fedeli J M, Rouviere M, Vivien L and Laval S 2004 *Journal of Applied Physics* **95** 5905

Investigation of delta-doped pHEMT InGaAs/GaAs/AlGaAs structures by the electrochemical capacitance-voltage technique

G Yakovlev, D Frolov and V Zubkov
Department of micro- and nanoelectronics, St. Petersburg Electrotechnical University "LETI", Prof. Popov str. 5, St. Petersburg 197376, Russia

E-mail: gy@etu.ru

Abstract. In pHEMT devices, a two dimensional (2D) conducting channel representing a quantum well (QW) and adjacent layers of wide-gap semiconductors are fabricated on an undoped material which provides a high mobility of free carriers in the channel. During the development of pHEMT structures it is always very important to estimate and control the free charge carriers concentration, that is not a trivial task because of its redistribution which causes changing of the electric field distribution and, respectively, the energy spectrum of states in the 2DEG channel. By using the modern ECV technique, the pHEMT InGaAs/GaAs/AlGaAs structures were investigated. Free charge carriers concentration profiles were obtained, and the location of δ-layers and the carrier concentration in the QW were estimated.

1. Introduction

Capacitance-voltage (CV) profiling is widely used to characterize semiconductor materials and structures. It is a well-known technique with highly reproducible results. Conventional CV measurements are performed by forming a metal Schottky contact on a semiconductor surface and sweeping the applied biases. However, in some semiconductor structures like pHEMT, which have top highly doped layers, the use of this technique will result in a concentration profile at a very limited depth (of around 20 nm). Because of the breakdown of the structure at high applied voltages, the necessary information about a deeply located QW stays unknown. In this case, we prefer using a modification of conventional CV, i.e. the electrochemical capacitance-voltage (ECV) profiling. Here an electrolyte is used as a rectifying contact, which could be used in a conventional CV regime, and also for controlled dissolution of the semiconductor [1] that helps to overcome the limitations mentioned above and obtain the full information about the concentration distribution [2].

During the development of pHEMT structures for the goal of enhancing the channel conductivity, the doping level and thicknesses of functional layers in its vicinity are typically been increased, however, it could lead to the shunt of the channel by these highly doped layers. Therefore, it is very important to control the redistribution of free charge carriers between the 2DEG channel and non-uniformly doped wide-band gap barriers [3]. The problem is complicated by the fact that redistribution of free carriers in a complex multilayer pHEMT structure does change, by itself, the electric field over the structure, which in turn causes the change in the energy spectrum of the 2DEG channel and, hence, alters the carrier concentration in the channel. Commonly, the charge carrier concentration in pHEMT channels is determined experimentally by the Hall technique [4]. However, such measurements include significant errors due to difficulties in separating the contributions to the conductance of the parallel connected 2DEG channel and heavily doped regions of the wide-band gap barrier [5]. That is why we use the ECV technique for accurate measurements of free charge carrier distribution, and for estimating the true location of δ-layers and a quantum well as well as the free charge carrier concentration in their vicinity [6].

2. Samples and experimental technique

ECV concentration profiles were measured on typical InGaAs/GaAs/AlGaAs pHEMT structures, some of them were fabricated without top highly doped layers. The structures were grown on (100) semi-insulating GaAs substrates by the MBE technique. Sample #1 has two-side δ-doping, while samples #2 and #3 have only one δ-layer with a lower doping level, other specification is the same. The $In_{0.22}Ga_{0.78}As$/GaAs-QW is located at a depth of 54.5 nm and is of 12 nm width.

The measurements were performed at room temperature using a ECVPro profiler (Nanometrics). The solution of 0.2M NaOH with EDTA was used to form the electrolyte barrier with a nominal contact area of 0.1 cm^2. During the ECV profiling the structures were etched gradually with a 1 nm step, and the total thickness of the removed material was ~10 nm. Atomic force microscopy (AFM Solver NEXT, NT-MDT) was used to verify the etching depth and surface quality. It should be noted that for modern multilayer pHEMT structures it is ultimately necessary to make careful selection of etching parameters, as well as to consider the impact of series resistance of lightly doped layers on the capacitance measurements [7]. The ac voltage frequency and amplitude of the test signal were 300 Hz and 10 mV, respectively. To simulate the potential lineup we used our self-consistent Poisson-Schroedinger algorithm [8].

3. Experimental results and discussion

The experimental capacitance-voltage curves and derived free charge carrier distributions for sample #1 are shown in figures 1 and 2, respectively. The CV characteristics were measured after four etching steps at the depth of 0, 3, 7 and 10 nm, respectively. Note that the biases in figure 1 were measured relative to a Pt electrode. The simulated potential lineup at the equilibrium for sample #1 is also shown in figure 2.

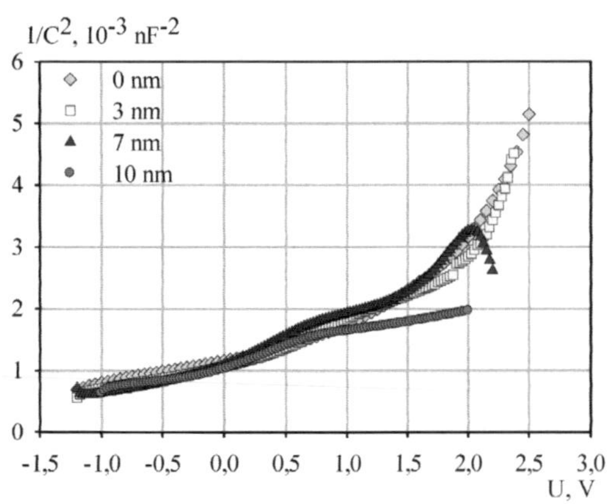

Figure 1. CV profiles of pHEMT sample #1 measured by the ECV depletion technique.

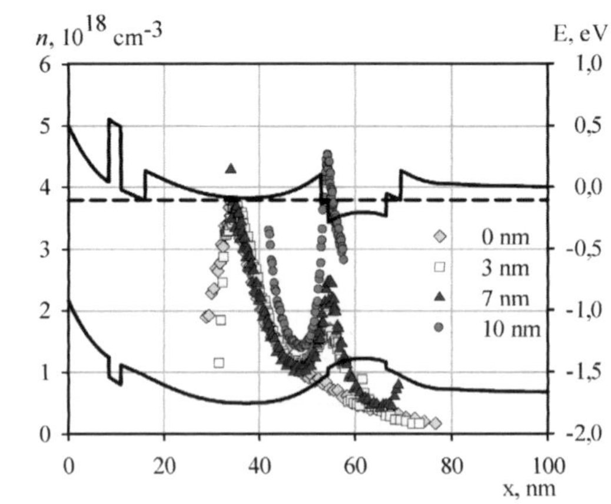

Figure 2. Apparent charge carrier profiles and the potential lineup in pHEMT sample #1.

We could clearly distinguish the δ-layer and the 2DEG channel. In the carrier concentration profile measured without etching (the green curve in figures 1 and 2), no response from the QW was noticed. During each etching step the thickness of the cap layer decreased, making the concentration peak from the QW more pronounced. So, we may conclude that this technique can be used for optimization of cap layer thickness under the gate contact of pHEMT devices to increase the QW filling by carriers. From the obtained data it follows that the etching has no influence on the peak concentration associated with a δ-layer, while the QW response changes significantly.

The δ-layer in sample #1 locates at 35 nm under the surface and has the charge carrier concentration in the peak of $3.7\cdot10^{18}$ cm^{-3}. The parameters characterizing the carrier distribution in the QW are summarized in Table 1. All profiles are tailing on the Debye length. From the detailed investigation of the obtained data one could conclude that there exists a small (no more than 6 nm) mismatch between the measured and expected QW locations. This could be explained, for instance, by errors in etch depth determination or by QW skewness [9].

Table 1. Parameters of free carrier concentration profiles near InGaAs/GaAs/AlGaAs-QW in sample #1, obtained from ECV

Etching depth, nm	QW carrier concentration in maximum, cm^{-3}	2DEG density, cm^{-2}	QW peak FWHM, nm
0	-	-	-
3	$1.7\cdot10^{18}$	$2.3\cdot10^{12}$	8
7	$2.5\cdot10^{18}$	$2.6\cdot10^{12}$	6
10	$4.5\cdot10^{18}$	$4.1\cdot10^{12}$	4

It should be noted that in such complicated multilayer structures the ECV technique (the standard etching mode) does not give enough resolution, particularly because all measurements are usually performed at a fixed bias. We believe that for pHEMT structures it is necessary to measure the full CV characteristic after each etch step. The resulting carrier concentration profile can be calculated in this case by superposition of CV characteristics at different etching steps, one is represented in figure 3 (the red curve).

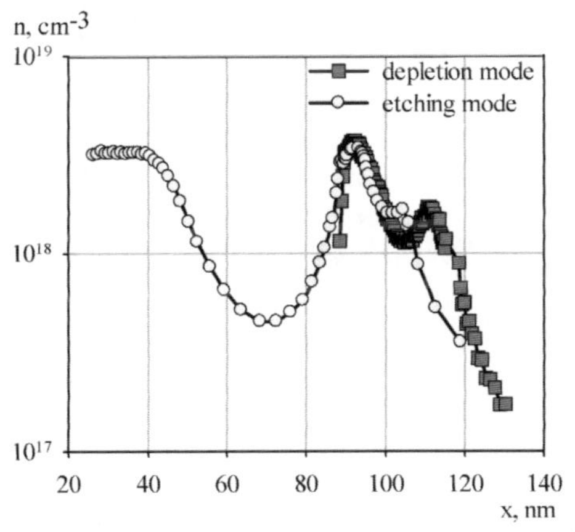

Figure 3. Apparent free charge carrier profiles obtained by ECV in etching end depletion modes. Profile of sample 1 (the red curve) is shifted by the removed material thickness (55 nm).

Another structure investigated by ECV had an additional highly doped layer, in contrast to sample #1. Its carrier concentration profile obtained by the standard ECV etching mode is also shown in figure 3 (the yellow circles). As it could be seen, the standard ECV etching mode provides the possibility of getting the concentration distribution over a wide range but with some loss in selectivity in important points (one couldn't distinctly distinguish the δ-layer and 2DEG channel responses). In this case the integration of CV curves at each etch step is preferable for improved accuracy.

Journal of Physics: Conference Series **690** (2016) 012015 doi:10.1088/1742-6596/690/1/012015

The experimental CV curves and free charge carrier concentration profiles for single δ-doped samples (#2 and #3 grown in equivalent conditions) are shown in figures 4 and 5, respectively.

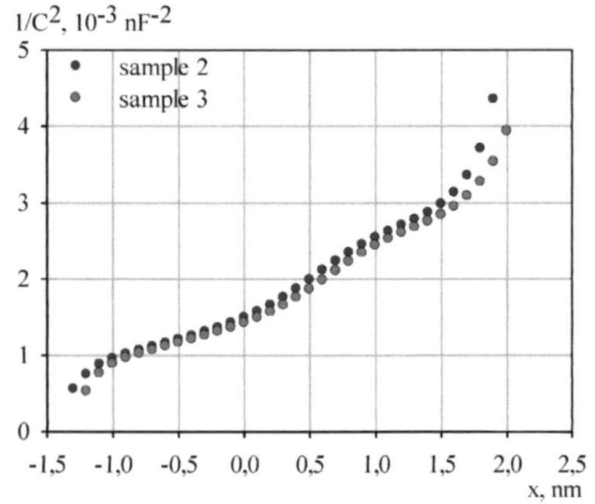

Figure 4. Experimental CV profiles of pHEMT samples #2 and #3, measured by the ECV technique.

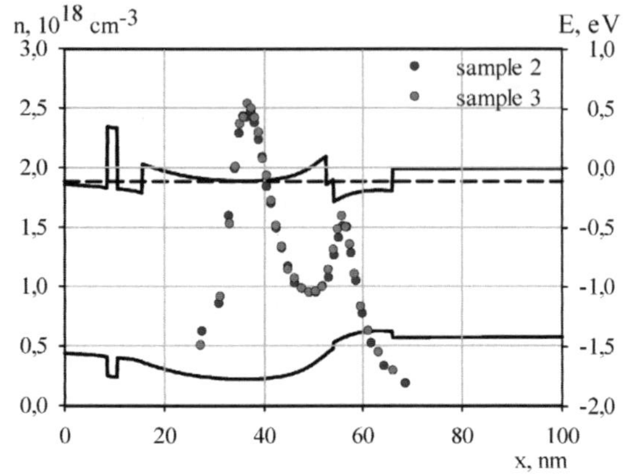

Figure 5. Apparent profiles of free charge carriers of pHEMT samples #2 and #3, measured by the ECV technique.

The concentration profiles presented in figure 5 show a very good coincidence that confirms the perfect reproducibility of structures during the growing process. The Delta-layer peak carrier concentration is $2.5\cdot10^{18}$ cm^{-3}, and for the QW it is $1.5\cdot10^{18}$ cm^{-3} (2DEG density equals to $2.2\cdot10^{12}$ cm $^{-2}$). As one can see, in the case of two-side δ-doping (sample #1), if the pre-etching process takes place, the 2DEG concentration in the QW is almost twice higher than that for single-doped samples #2 or #3.

4. Conclusion

In complicated multilayer pHEMT structures with highly-doped δ-layers it is necessary to measure CV characteristics after each etch process and make the superposition of CV characteristics for getting the total charge carrier concentration profile with high precision. In this study several pHEMT InGaAs/GaAs/AlGaAs structures were investigated using the ECV technique. Free charge carrier concentration profiles were obtained, the location of δ-layers and the carrier concentration in the QW were estimated. It was shown that by using the ECV technique one could optimize the etched thickness of a cap layer to obtain the best concentration of free carriers in the QW of a pHEMT device. Particularly, in the case of two-side δ-doping, the ECV technique can help to determine the optimal cap layer thickness, and the 2DEG concentration in the QW can be increased significantly in contrast to the single-doped case.

References

[1] Blood P 1986 *Semicond. Sci. Technol.* **1** 7–27
[2] Zubkov V, Kucherova O, Frolov D and Zubkova A 2013 *Phys. Status Solidi* C **10** 342–45
[3] Look D et al 1996 *J. Appl. Phys.* **79** 540-44
[4] Egorov A, Gladyshev A, Nikitina E, Denisov D, Polyakov N, Pirogov E and Gorbazevich A 2010 *Semiconductors* **44** 950-53
[5] Brunkov P et al 2011 *Semiconductors* **45** 829-35
[6] Ambridge T and Faktor M 1975 *J. Appl. Electrochem.* **5** 319-28
[7] Jogai B and Stutz C 1997 *J. Electron. Mater.* **7** 863-867
[8] Zubkov V 2006 *Semiconductors* **40** 1204-8
[9] Vasilevskii I, Galiev G, Klimov E, Mokerov V, Shirokov S, Imamov R and Subbotin I 2008 *Semiconductors* **42** 1084-91

Journal of Physics: Conference Series **690** (2016) 012016 doi:10.1088/1742-6596/690/1/012016

Eigenstate modelling in arbitrary shaped nanostructres with gradual heterointerfaces

Yu D Sibirmovsky, I S Vasil'evskii and N I Kargin
National Research Nuclear University "MEPhI", 115409, Moscow, Russia
E-mail: sibirmovsky@gmail.com

Abstract. We consider the problem of evaluating the energy levels and envelope wavefunctions of electrons in semiconductor nanostructures with a smooth confining potential, such as inter-diffused quantum dots or quantum wells. We show how Gaussian functions can be used to model various confinement potentials in 1D, 2D and 3D. We develop a fast and accurate numerical algorithm for solving the one-particle Schrodinger equation with a position-dependent effective mass for this type of potential. We examine the convergence and accuracy of our algorithm. This method can be used to analyze experimental data, such as the PL spectra of quantum wells, wires and dots and the charge carrier transport through various nanostructures. $GaAs/Al_{0.3}Ga_{0.7}As$ system is used in calculations.

1. Introduction

The problem of accurate evaluation of bound states of electrons and holes in semiconductor nanostructures is not trivial. For practical purposes such as modelling the photoluminescence spectra, tunnelling probability or charging and ionization of quantum dots the computational algorithm should be accurate and easy to use without requiring excessive computing power.

The Schrodinger equation can be solved analytically only for a handful of potentials. In 2D and 3D they are either symmetric or separable in certain coordinate systems [1]. For semiconductor heterostructures with varying composition the effective mass is position-dependent further complicating the problem. Numerical algorithms for one-dimensional problems are fairly straightforward, such as the transfer matrix method [2]. In two and three dimensions however, it is not as easy to correctly define the boundary conditions and solve the eigenvalue problem for an arbitrary non-separable confining potential. Various methods continue to be developed [3,4] but most of them are limited either by required computing power or by the amount of information which can be obtained.

In this work we show that Gaussian functions can be used to model arbitrary shaped nanostructures in 1D, 2D or 3D and develop a fast and accurate numerical algorithm for solving the Schrodinger equation for this type of potential. We use the effective mass, one particle, and envelope function approximation. Nonparabolicity and particle interaction are not considered here, but the method can be modified to include these effects. We use $GaAs/Al_{0.3}Ga_{0.7}As$ system in all calculations in this paper. Both the conduction band offset and effective mass depend linearly on the AlAs mole fraction t for $t < 0.45$.

Journal of Physics: Conference Series **690** (2016) 012016 doi:10.1088/1742-6596/690/1/012016

2. Gaussian potential

We propose the following general potential for modelling the confinement in real nanostructures:

$$U(\vec{r}) = -\sum_n U_{0n} \exp\left(-\frac{(x-x_{0n})^2}{2a_n^2} - \frac{(y-y_{0n})^2}{2b_n^2} - \frac{(z-z_{0n})^2}{2c_n^2}\right) \tag{1}$$

This potential is very convenient. It is smooth, local (vanishes at infinity) and each term can be represented as a product $U(x)U(y)U(z)$, which makes the resulting Schrodinger equation easy to solve. This potential is also very flexible, since it does not rely on any type of symmetry and thus can be used to model real, asymmetric nanostructures. Due to its locality, it can also be used to model nanostructure complexes (superlattices, nanowire and quantum dot arrays).

Figure 1 shows how the rectangular quantum well can be modelled very accurately by a sum of 9 Gaussian functions. The energy levels and wavefunctions in the latter case are calculated by our method described below. Their differences in this case are $\Delta E_1 \approx 0.3\ meV$ and $\Delta E_2 \approx 1.5\ meV$.

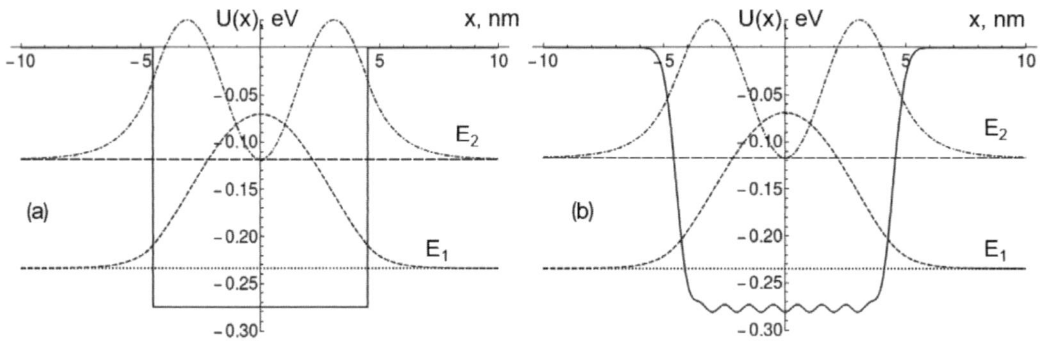

Figure 1. Modelling a rectangular potential well (a) using 9 Gaussian functions (b).

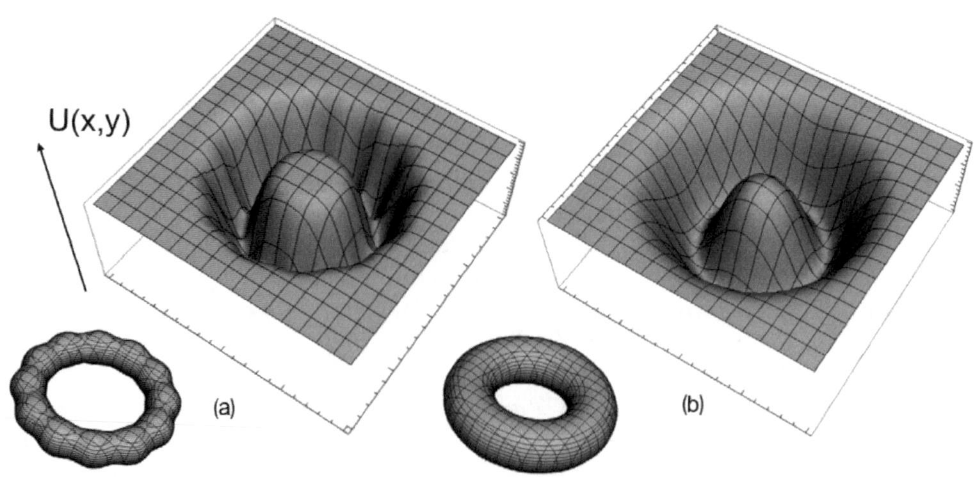

Figure 2. Modelling a quantum ring using 12 Gaussian functions with half-width to radius ratio 0.2 for (a) and 0.3 for (b). 2D potentials and equipotential surfaces in 3D space are shown.

In figure 2 a sum of 12 Gaussian functions is used to model a quantum ring in 3D. The half-width of Gaussian functions has to be chosen correctly if we want to obtain the smooth potential surface. On the other hand, we can create smooth potential with abrupt boundary by increasing the number of functions used.

For semiconductor heterostructures we also need to consider the position dependence of the effective mass. If its dependence on AlAs mole fraction is linear, the effective mass for a single Gaussian well is defined as follows (m_b^* - the effective mass of barrier material at infinity, m_w^* - the effective mass in the center of the well):

$$m^*(\vec{r}) = m_b^* - (m_b^* - m_w^*)\exp\left(-\frac{(x-x_0)^2}{2a^2} - \frac{(y-y_0)^2}{2b^2} - \frac{(z-z_0)^2}{2c^2}\right) \tag{2}$$

3. The method in general

We consider the Schrodinger equation with the position-dependent effective mass [5]:

$$-\frac{\hbar}{2}\left(\nabla\frac{1}{m^*(\vec{r})}\nabla\right)\Psi(\vec{r}) + U(\vec{r})\Psi(\vec{r}) = E\Psi(\vec{r}) \tag{3}$$

We then use the spectral method first proposed in [6] for nanowires and further developed in [7-10] for quantum dots with abrupt interfaces. We expand the wavefunction into a complete orthonormal basis of eigenfunctions of a simple potential, such as a quantum box or a harmonic oscillator, both of which are separable in Cartesian coordinates in 3D.

$$\Psi(\vec{r}) = \sum_{j,k,l}^{\infty}\Psi_{jkl} = \sum_{j,k,l}^{\infty} C_{jkl}\psi_j(x)\psi_k(y)\psi_l(z) \tag{4}$$

Then we can substitute (4) into (3), multiply (3) by Ψ_{pqs}^* on the left and integrate over the whole space, obtaining a matrix eigenvalue problem:

$$< \Psi_{pqs}^* \mid -\frac{\hbar}{2}\left(\nabla\frac{1}{m^*(\vec{r})}\nabla\right) + U(\vec{r}) \mid \Psi_{jkl} > C_{jkl} = E\delta_{pj}\delta_{qk}\delta_{sl}C_{jkl} \tag{5}$$

If we evaluate matrix elements on the left side we can solve this problem numerically, obtaining the complete set of eigenvectors $\{C_{jkl}\}_n$ and eigenvalues E_n. As we can only diagonalize a finite matrix, we have to truncate the expansion (4) and the result now depends on the number of basis functions N_x, N_y, N_z and their exact form (for example quantum box dimensions L_x, L_y, L_z or harmonic oscillator frequencies $\omega_x, \omega_y, \omega_z$). To obtain the best possible agreement with the exact result for certain N_x, N_y, N_z, we choose the parameters variationally (so as to minimize the ground state energy).

Unlike other numerical methods, such as a widely used finite element method, this algorithm does not require discretization in space or energy. It outputs normed continuous wavefunctions and all energy levels, including degenerate ones. This method is general and can be used with any potential with addition of electric and magnetic fields. The goal of this work is to create a fast, simple and accurate algorithm for the chosen family of potentials. By evaluating all matrix elements analytically, we will ensure that there is no numerical integration, which can be costly and introduce additional error.

4. Matrix elements calculation

We will first obtain the explicit expressions for matrix elements in the case of a single Gaussian well in one dimension. It will be easy to extend them to 2 and 3 dimensions. In this section we also consider the effective mass to be constant in space.

$$U(x) = -U_0\exp\left(-\frac{x^2}{2a^2}\right) \tag{6}$$

We will use two basis sets. First is the set of solutions of harmonic oscillator (HO) potential. They have the following form (H_j - Hermite polynomials).

$$\psi_j(x) = \frac{1}{\sqrt{\sqrt{\pi}\ 2^j j!\ L}} \exp\left(-\frac{x^2}{2L^2}\right) H_j\left(\frac{x}{L}\right) \tag{7}$$

Here $j = 0, 1, 2, \ldots$ Then matrix elements will have the form:

$$M_{jp} = -\frac{\hbar}{2m^* L^2}\left(\frac{1}{2}\sqrt{(j+1)(j+2)}\delta_{j+2,p} - \left(j+\frac{1}{2}\right)\delta_{j,p} + \frac{1}{2}\sqrt{j(j-1)}\delta_{j-2,p}\right) - \\ -U_0\sqrt{\frac{j!\ p!}{2^{j+p}}} g^{j+p+1} \sum_{n=0}^{[j/2]} \sum_{l=0}^{[p/2]} \left(-\frac{L^2}{2a^2}\right)^{n+l} \frac{2^{j-2n}}{n!\ l!\ (p-2l)!}\delta_{j-2n,p-2l} \tag{8}$$

$$g = \frac{a}{\sqrt{a^2 + L^2/2}} \tag{9}$$

Here we used the relation for $H_j(gx)$ from [11]. The second basis is the set of solutions of the quantum box (QB) problem:

$$\psi_j(x) = \frac{1}{\sqrt{L}} \sin\left[\frac{\pi j}{2}\left(1 + \frac{x}{L}\right)\right] \tag{10}$$

Here $j = 1, 2, 3, \ldots$ To obtain matrix elements for the potential analytically we need to integrate:

$$\int_{-L}^{L} \sin\left[\frac{\pi j}{2}\left(1+\frac{x}{L}\right)\right] \exp\left(-\frac{x^2}{2a^2}\right) \sin\left[\frac{\pi p}{2}\left(1+\frac{x}{L}\right)\right] dx \tag{11}$$

Now let us assume that $L \gg a$. Then we can move the integration boundaries to infinity and (11) is reduced to the Euler-Poisson integral. While this assumption may lead to additional error, when we compare the result to the HO basis calculation (figure 3) they show very good agreement.

$$M_{jp} = \frac{\pi^2 \hbar}{8m^* L^2} j^2 \delta_{j,p} + \sqrt{\frac{\pi}{2}} U_0 \frac{a}{L}\left(\exp\left(-\frac{\pi^2 (j+p)^2 a^2}{8L^2}\right) \cos\left[\frac{\pi}{2}(j+p)\right] - \\ - \exp\left(-\frac{\pi^2 (j-p)^2 a^2}{8L^2}\right) \cos\left[\frac{\pi}{2}(j-p)\right]\right) \tag{12}$$

The algorithm was implemented in Wolfram Mathematica 10. In our experience, the most time-consuming operation is calculation of the matrix elements. Numerical matrix diagonalization is performed much faster. Both operations scale with number of basis functions as N^{2d} with d being the number of dimensions. In one dimension the calculations are fast and take several seconds with a personal computer even for $N = 100$. Thus, we were able to analyze the continuous dependence of the results on the parameter L for HO and QB bases (figure 3). We can see definite intervals of convergence where the energy does not depend on L. These intervals become longer as N increases. For two dimensions the region of convergence was found. For simple potentials it is almost square, meaning we can choose L_x, L_y and L_z independently in their own intervals of convergence.

The accuracy of the algorithm can also be checked by comparing the results with analytically solvable potentials (figure 1). A good example is modified Poschl-Teller potential which has the form $-U_0/\cosh^2(x/\sqrt{2}a)$ and is very close in shape to the Gaussian potential. Indeed, for $a = 5$ nm their ground state energies are almost equal ($\Delta E_1 \approx 1\ meV$). For wider wells ΔE_1 decreases. We also used exactly solvable symmetrical potentials in 2D and 3D to verify the results of our method in several dimensions.

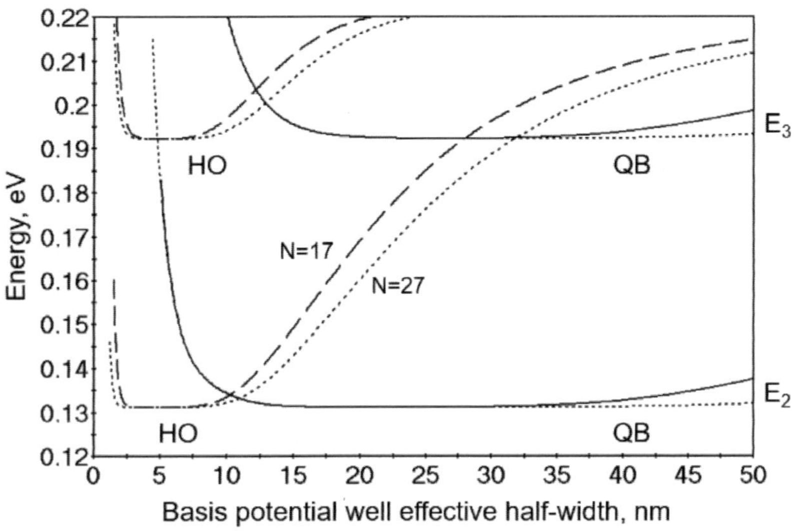

Figure 3. Dependence of electron energy levels in a Gaussian well on the basis function parameters. HO - harmonic oscillator, QB - quantum box. For HO the effective half-width of ground state wavefunction is shown. The Gaussian well half-width is 5 nm. N is the number of basis functions used in the expansion.

5. Position dependent effective mass

Now we allow the effective mass to be position dependent. For the QB basis the matrix elements can still be calculated semi-analytically (using the Taylor expansion).

$$\frac{1}{m^*(x)} = \frac{1}{m_b^* - (m_b^* - m_w^*)\exp\left(-\frac{x^2}{2a^2}\right)} \tag{13}$$

$$\frac{1}{m^*(x)} = \frac{1}{m_b^*}\left[1 + \sum_{n=1}^{\infty} \mu^n \exp\left(-\frac{nx^2}{2a^2}\right)\right] \tag{14}$$

$$\mu = 1 - \frac{m_w^*}{m_b^*} < 1 \tag{15}$$

$$-\frac{{}^{-}\hbar}{2} < \psi_p(x) \mid \frac{\partial}{\partial x}\frac{1}{m^*(x)}\frac{\partial}{\partial x} \mid \psi_j(x) >= \frac{\pi^{2-}\hbar}{8m_b^* L^2} j^2 \delta_{j,p} +$$
$$+\frac{\pi^{2-}\hbar}{8m_b^* L^2}\sqrt{\frac{\pi}{2}\frac{a}{L}}jp\Bigg[\cos\left[\frac{\pi}{2}(j+p)\right]\sum_{n=1}^{\infty}\frac{\mu^n}{\sqrt{n}}\exp\left(-\frac{\pi^2(j+p)^2a^2}{8L^2n}\right) +$$
$$+\cos\left[\frac{\pi}{2}(j-p)\right]\sum_{n=1}^{\infty}\frac{\mu^n}{\sqrt{n}}\exp\left(-\frac{\pi^2(j-p)^2a^2}{8L^2n}\right)\Bigg] \tag{16}$$

The sum in (16) converges very fast, giving at least 5 correct digits for 20 terms. We compare calculation results for the constant and position dependent effective mass (figure 4). We can see that in the latter case the energy levels decrease ($\Delta E_1 \approx -2\ meV$, $\Delta E_2 \approx -3\ meV$, $\Delta E_3 \approx -5\ meV$) and electron becomes more localized. It can be also understood qualitatively, since $m_b^* > m_w^*$, thus the mean effective mass increases when we account for m_b^*.

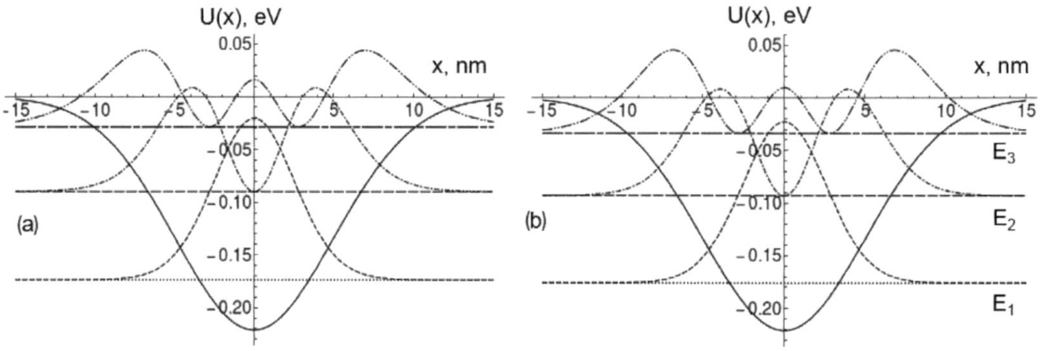

Figure 4. Energy levels and wavefunctions for the electron in a Gaussian potential well with the constant (a) and position-dependent (b) effective mass.

In conclusion, we have shown that Gaussian functions can be used to model arbitrary shaped confinement potentials in any number of dimensions. We have developed a fast and accurate algorithm for evaluating energy levels and wave functions for this family of potentials and examined its convergence. The electric and magnetic fields can be included in straightforward way. Note that this method can be modified to include the effects of nonparabolicity and particle interaction (for example, by an iterative algorithm).

Acknowledgments
The study was supported by the Ministry of Education and Science of the Russian Federation on the agreement No. 14.578.21.0062 dated 20.10.2014 (unique identification number of the work is RFMEFI57814X0062).

References

[1] Khare A and Bhaduri R K 1994 *Am. J. Phys.* **62** 1008
[2] Jonsson B and Eng S T 1990 *IEEE. J. Quant. Electron.* **26** 2025
[3] Strickland M and Yager-Elorriaga D 2010 *J. Computat. Phys.* **229** 6015
[4] Tichy V, Skala L and Hudec R 2014 *Cent. Eur. J. Phys.* **12** 730
[5] Levy-Leblond J-M 1995 *Phys. Rev.* A **52** 3
[6] Gershoni D, Temkin H, Dolan G J, Dunsmuir J, Chu S N G and Panish M B 1988 *Appl. Phys. Lett.* **53** 995
[7] Marzin J-Y and Bastard G 1994 *Solid State Comm.* **92** 437
[8] Califano M and Harrison P 2000 *Phys. Rev.* B **61** 10959
[9] Ngo C Y, Yoon S F and Fan W J 2006 *Phys. Rev.* B **74** 245331
[10] Tablero C 2009 *J. Appl. Phys.* **106** 074306
[11] Feldheim E 1942 *Acta Mathematica* **75** 117

Intersubband light absorption in double GaAs/AlGaAs quantum wells under lateral electric field

A A Shumilov, M Ya Vinnichenko, R M Balagula, I S Makhov, D A Firsov and L E Vorobjev

Peter the Great St. Petersburg Polytechnic University, 29 Polytechnicheskaya str., St. Petersburg 195251, Russia

e-mail: proshikote@gmail.com

Abstract. Spectra of variation of intersubband light absorption induced by lateral electric field and lattice temperature change were studied in the structure with double GaAs/AlGaAs quantum wells. The calculated positions of energy levels have been verified with interband photoluminescence measurements. The change in intersubband light absorption under lateral electric field was explained by the redistribution of hot charge carriers between the two lowest electronic states and by the subsequent emergence of space charge between quantum wells.

1. Introduction

Nowadays, investigation of intersubband light absorption in quantum wells attracts the scientific community's attention. A lot of publications describe the influence of the electric field on the optical properties of the quantum wells, for example see Refs. [1–3]. The majority of papers are devoted to the analysis of transverse electric field effects. For example, Refs. [2–4] report the investigation of the effect of transverse electric field and temperature on intersubband light absorption in tunnel-coupled quantum wells. The electron redistribution between the size-quantization levels and a variation of the electron energy spectrum of quantum wells is stated as a cause of the observed absorption modulation. However, the effect of lateral electric field on the optical properties of tunnel-coupled quantum wells has not been adequately studied while it can heat electrons and, in turn, change intersubband absorption. The observed change in the intersubband absorption magnitude and spectra under strong heating lateral electric field in the single rectangular quantum wells was explained in Ref. [5, 6] by the real space transfer to the barrier and change of the electron exchange interaction due to their heating. The influence of nonparabolicity can also cause these changes. The electron heating effect on the photoluminescence spectrum was investigated in [7] under lateral electric field and intense optical pumping. However, the effect of lateral electric field on the optical properties of the double tunnel-coupled quantum well has not been extensively studied. This paper presents the original results on the stated phenomenon in double GaAs/AlGaAs tunnel-coupled quantum wells.

2. Sample and experiment

In order to estimate the electric field heating effect, the complex semiconductor structure was grown by molecular-beam epitaxy on a semi-insulating GaAs substrate. It contained 100 pairs of $GaAs/Al_{0.38}Ga_{0.62}As$ tunnel-coupled quantum wells separated by 12 nm tunnel-nontrasparent barriers. Each pair of tunnel-coupled quantum wells was formed by 5.6 nm and 4.0 nm GaAs layers separated by a 1.2 nm tunnel-transparent $Al_{0.38}Ga_{0.62}As$ barrier (see figure 1). Quantum well layers were uniformly doped with silicon up to the donor impurity concentration $N = 5{\cdot}10^{17}$ cm^{-3} (the surface impurity concentration in the quantum wells was $N_d = 6{\cdot}10^{11}$ cm^{-2}). Figure 1 schematically shows the potential profile of the conduction band of one period of tunnel-coupled GaAs/AlGaAs quantum wells at $T = 77$ K. During postgrowth processing, ohmic indium contacts were deposited on the sample. The contacts were made by annealing strips of indium deposited on the surface of the structure in nitrogen atmosphere. The sample was heated to the temperature $T = 450^{\circ}$ C over a span of 2 minutes, and then cooled to room temperature. The facets of the sample were ground at 45° angle to create a multi-pass sample geometry. This sample geometry allows independent investigation of intersubband absorption of the light of two polarizations. The dimensions of the sample were chosen so that radiation passed 12 times through the quantum well layers. The total optical path length in the quantum-well layers was $L = 11.5$ μm.

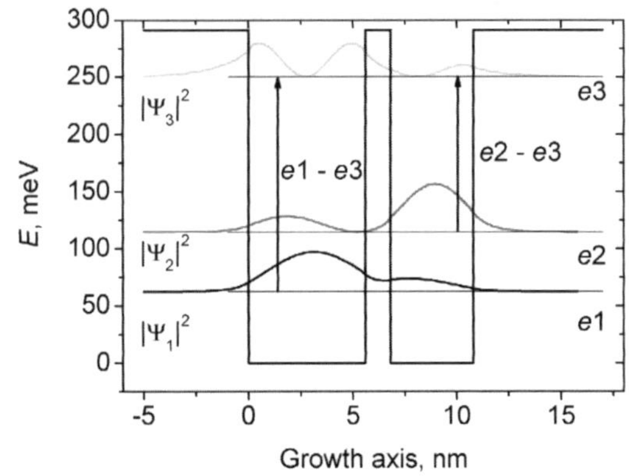

Figure 1. Profile of the conduction band bottom for tunnel-coupled GaAs/AlGaAs quantum wells at $T = 77$ K. The calculated positions of the energy levels and the squared electron wave function Ψ moduli are shown. Arrows show intersubband electron transitions.

In order to perform the structure characterization and verification of the calculated energy level positions in quantum wells, the interband photoluminescence was studied. An automated monochromator Horiba Jobin Yvon FHR640 was used for photoluminescence studies. A liquid nitrogen cooled silicon CCD-camera served as a photodetector. A solid-state laser with a wavelength of 532 nm was used as a source of interband optical pumping. The sample was mounted into the liquid-nitrogen-cooled cryostat which allowed us to set the precise temperature in the range from 77 to 320 K. The input and output cryostat windows were made of fused silica due to its transparency in the near-infrared range.

The intersubband absorption spectra were studied using a Bruker Vertex 80v vacuum Fourier spectrometer. A built-in globar was used as a wideband infrared radiation source. A photovoltaic HgCdTe photodetector was chosen as a detector of radiation. The sample was mounted into the previously described liquid-nitrogen cryostat or into a Janis closed-cycle cryostat which allows the temperature to be maintained in the range of 4–320 K with an accuracy of 0.1°. The input and output cryostat windows made of ZnSe were selected because of ZnSe transparency in the mid-infrared range.

The absorption change under lateral electric field was measured in pulsed mode (electric field pulse duration was 250 ns with a duty cycle from 50 000 to 250 000). Since the signal of absorption change was rather small we used the averaged value derived from a large number of absorption spectra (30 - 100 counts). The results were obtained in two different ways. The necessity of secondary measurements will be discussed further in text. The first method of measurements consisted of acquiring the AC-coupled photodetector signal proportional to the transmission change for *p*-polarized light by an SR250 boxcar integrator synchronized with a pulse generator. The second used method involved measurements of time slices of DC-coupled photodetector signal proportional to the sample transmission. Change in transmission related to the electric field was obtained by subtraction of spectrum measured after the electric field pulse from the spectrum at the moment of the pulse. The transmission spectra were measured with a 25 ns time interval. The spectra measurements began 1.2

μs before and ended 25 μs after the 250 ns electric field pulse. This allowed us to estimate the contribution of lattice and electron heating. The main difference between the methods used is the signal sign sensitivity. For the first method, the transmission change signal can be of both signs. Application of the Fast Fourier Transform technique assumes the positivity of the resulting spectrum, therefore in our case the resulting signal change spectra do not contain absorption variation sign. Thus, the results require phase correction to discover the real modulation spectra. The procedure of phase correction is not a simple task and usually is uncertain. The second method is accurate in the sense of signal sign but is not so reliable in case of small signal values. Our implementation of both methods gave us the identical results and verified the phase correction procedure. The spectral resolution of the transmission spectra was about 3.5 meV in both measurement schemes.

Theoretical calculation of energy levels of electrons and holes and optical matrix elements of intersubband transitions in tunnel-coupled quantum wells was performed (see figure 1). The measured spectra of interband photoluminescence at $T = 300$ K and $T = 77$ K are presented in figure 2. At room temperature one can see the peaks associated with transitions between localized electron states and heavy hole and light hole states *e*1-*hh*1, e2-*hh*2, *e*1-*lh*1. At liquid nitrogen temperature only one peak associated with transitions *e*1-*hh*1 can be observed due to the lack of electrons on the *e*2 level and holes on the *lh*1 level. The temperature shift of photoluminescence peaks is related to the bandgap change. The peak positions are in a good agreement with the results of energy spectra calculations for the structure with parameters stated above.

3. Results and discussions

According to the performed solution of the Schrödinger equation for electron energy levels at $T = 77$ K, the intersubband optical transitions *e*1-*e*3 and *e*2-*e*3 have energies of 187 meV and 135 meV, respectively. The transmission spectra of the structure were measured using multipass geometry for two different light polarizations (*p* and *s*) at different lattice temperatures. According to the Bouguer–Lambert–Beer law, the intersubband absorption spectra were obtained as a ratio of transmission spectra of *p*- and *s*-polarized light. The absorption spectra for the different lattice temperatures are presented in figure 3. The calculated energies of intersubband transitions *e*1-*e*3 and *e*2-*e*3 at $T = 77$ K are shown with arrows. The temperature increase results in a slight red shift of these transition energies. A satisfactory agreement between calculations and experimental results is observed. At low lattice temperatures the charge carriers are mainly located at the ground subband thus only *e*1-*e*3 optical transitions can be observed in the measured spectra (the spectral position of optical transitions *e*1-*e*2 is located outside the photodetector sensitivity range). The electron concentration in the second electron subband *e*2 increases with increase in temperature which, in turn, causes optical transitions *e*2-*e*3. The observed shift of the intersubband absorption peaks to the longwavelength spectral region with temperature increase is connected with the change of the bandgap of the barrier and quantum well materials with the temperature and complies with calculations of the electron energy states. It should also be noted that the ground electron subband *e*1 is "genetically" connected with the wide quantum well while the second electron subband *e*2 is connected with the narrow quantum well. This fact can result in emergence of space charge and electric field between narrow and wide quantum wells which in turn can shift the peak positions.

The light transmission change under lateral electric field was obtained using two different experimental methods: as a spectrum of actual change of the signal, and as a difference of transmission spectra during the electric field pulse and without it. In the first case, transformation of the interferogram into the spectrum included the mandatory procedure of phase correction, because the standard Fourier spectroscopy method does not save information about a signal sign. The phase correction of the transmission change spectra is described more specifically in Refs. [3, 4, 8]. The measurements of the transmission spectra at the moments of the electric field pulse and without electric field were carried out mostly for confirmation of the phase correction results. The three-dimensional time and photon energy dependences of the *p*-polarized light intensity I_p passed through the sample were obtained as a result of time-resolved step scan measurements. Time dependencies of the intensity I_p are presented in figure 4 for two certain photon energies $h\nu$ corresponding to the optical transitions *e*1-*e*3 ($h\nu$ = 187 meV, curve *1*) and *e*2-*e*3 ($h\nu$ = 135 meV, curve *2*), at the lattice temperature T = 77 K and electric field E = 695 V/cm. The first plot illustrates the fact that absorption associated with optical transitions *e*2-*e*3 increases, and absorption related to *e*1-*e*3 transitions decreases under electric field. A few microseconds after electric field pulse, the intensity I_p returns to the equilibrium level which proves that lattice heating is negligible. Each transmission change spectrum was obtained as a difference between the spectrum under electric field in a moment of the maximum change in the transmitted light intensity I_p (t = 0.8 μs in figure 4) and the spectrum without electric field. The 100 ns delay between the electric field pulse maximum and I_p signal maximum can be associated with the response time of the photodetector used. The transmission change spectra of the structure under electric field obtained by both methods are identical which proves the feasibility of the chosen phase correction procedure.

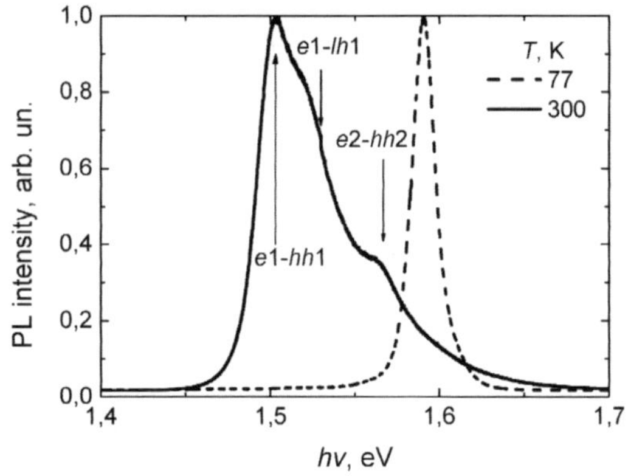

Figure 2. Normalized interband photoluminescence spectra at temperatures 77 K (dashed curve) and 300 K (solid curve). Arrows mark the calculated energies of intersubband transitions *e*1-*e*3 and *e*2-*e*3 at T = 300 K.

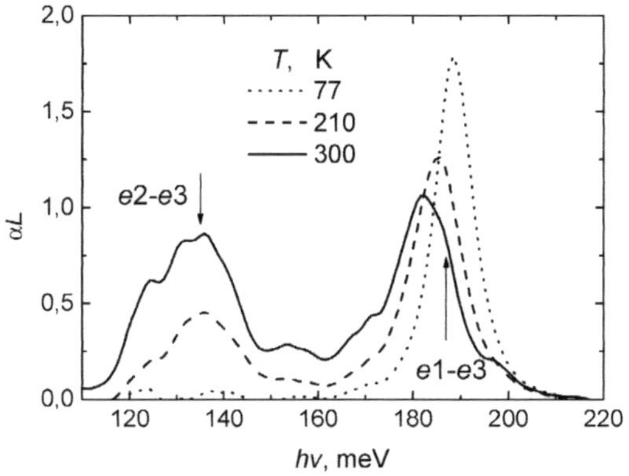

Figure 3. Intersubband absorption spectra at different temperatures T. L is optical path length. Arrows mark theoretically calculated transition energies at T = 77 K.

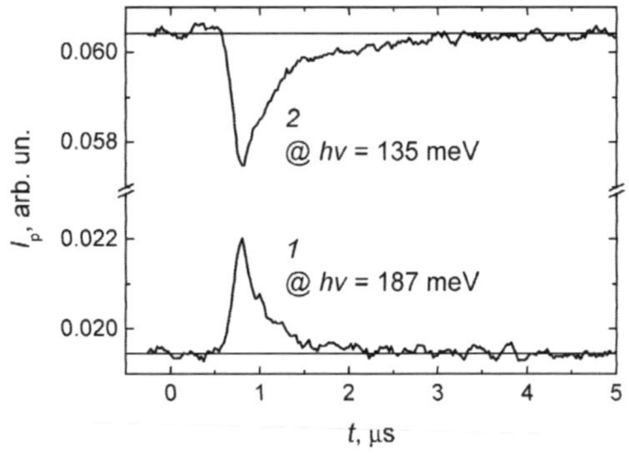

Figure 4. Time dependencies of light intensity passed through the sample at T = 77 K, E = 695 V/cm. Curve *1* and *2* were measured at quantum energies $h\nu$ = 187 meV and $h\nu$ = 135 meV which correspond to electron transitions *e*1-*e*3 and *e*2-*e*3, respectively.

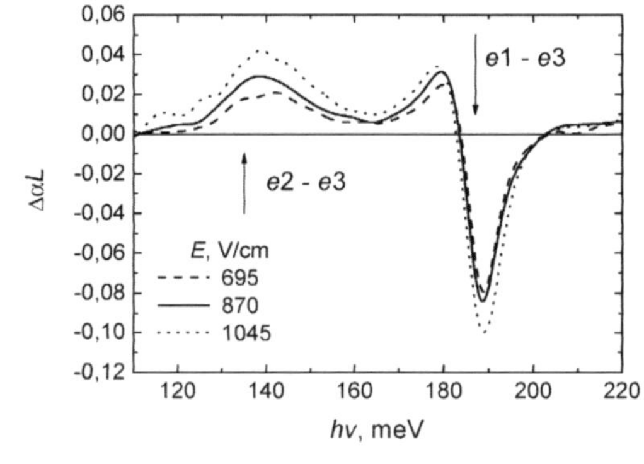

Figure 5. Spectra of intersubband absorption change induced by different lateral electric fields at T = 77 K.

Figure 5 demonstrates the change in intersubband absorption spectra obtained at liquid nitrogen temperature at different lateral electric fields. The observed increase of absorption at the photon energy $h\nu = 135$ meV is attributable to the heating of the first subband $e1$ electrons by lateral electric field. As a result, nonequilibrium electrons occupy the upper electronic states and can populate the level $e2$. An increase of the electron concentration in the second subband $e2$ leads to an increase of intersubband absorption e2-e3 ($h\nu = 135$ meV). The negative sign of modulation at the photon energy $h\nu = 187$ meV can be attributed to the combined action of the absorption decrease and a shift of the $e1$-$e3$ absorption peak to longer wavelengths under lateral electric field. The observed modulation spectra correspond to the heating of the electron gas. Nonequillibrium hot electrons are being redistributed in the real space between two quantum wells. The consequent space charge variation leads to the change of energy level positions. Comparing the values of absorption variation $\Delta\alpha L$ induced by temperature change and $\Delta\alpha L$ induced by applied lateral electric field we were able to estimate the degree of electron heating: the electron temperature was $T_e = 97$ K at $E = 695$ V/cm and $T_e = 103$ K at $E = 1045$ V/cm.

4. Conclusion

The intersubband light absorption in double tunnel-coupled GaAs/AlGaAs quantum wells in mid-infrared spectral range have been investigated under the conditions of applied lateral electric field. The temperature modification of the intersubband absorption spectra was studied as well. The redistribution of hot charge carriers between the two lowest electronic states under lateral electric field and subsequent emergence of space charge in the structure can cause the change in intersubband light absorption spectra. The electron temperature T_e was estimated. The value of the observed absorption variation suggests the application prospects of tunnel-coupled quantum wells as the basis for fast infrared light modulators.

Acknowledgments

The authors are grateful to A.N. Sofronov and V.Yu. Panevin for their help with the experiment preparation. This work was supported by the Russian Foundation for Basic Research (grants 14-02-00336, 14-02-31489) and the Ministry of Education and Science of the Russian Federation (state assignment).

References

[1] Faist J, Capasso F, Sirtori C, Sivco D and Cho A 2000 *Intersubband transitions in quantum wells: physics and device applications.* **62** 1
[2] Dupont E, Delacourt D, Berger V, Vodjdani N and Papuchon M 1993 *Appl. Phys. Lett.* **62** 1907
[3] Firsov D A, Vorobjev L E, Vinnichenko M Ya, Balagula R M, Kulagina M M and Vasil'iev A P 2015 *Semiconductors* **49** 1425
[4] Rybalko D A, Vinnichenko M Ya, Vorobjev L E, Firsov D A, Balagula R M, Panevin V Yu, Kulagina M M and Vasil'iev A P 2014 *J. Phys.: Conf. Ser.* **541** 012081
[5] Towe E, Sun D, Vorobjev L E, Danilov S N, Firsov D A and Zibik E A 1995 *Superlattices and microstructures* **17** 129
[6] Vorob'ev L E, Firsov D A, Shalygin V A and Saidashev I I 1997 *JETP Letters* **65** 549
[7] Vorobjev L E et. al. 2010 *Semiconductors* **44** 1402
[8] Firsov D D and Komkov O S 2013 *Tech. Phys. Lett.* **39** 1071

Theoretical modeling of exciton-light coupling in quantum wells

E S Khramtsov[1], P A Belov[2], P S Grigoryev[1], I V Ignatiev[1], S Yu Verbin[1] and S L Yakovlev[2]

[1]Spin Optics Laboratory, St. Petersburg State University, 7/9 Universitetskaya nab., St. Petersburg, 199034 Russia

[2]Department of Physics, St. Petersburg State University, 7/9 Universitetskaya nab., St. Petersburg, 199034 Russia

E-mail: e.khramtsov@spbu.ru

Abstract. The wave function of excitons in GaAs-based finite square quantum wells (QWs) is calculated by the direct numerical solution of the three-dimensional Schrödinger equation. The precise results for the lowest exciton state are obtained by the Hamiltonian discretization using the fourth-order finite-difference scheme. The radiative decay rate is calculated for QWs of various widths using the obtained exciton wave function.

1. Introduction

The theoretical modeling of excitons in a bulk semiconductor is usually carried out in the framework of the hydrogen model [1]. This model, however, becomes unsuitable for semiconductors with a degenerate valence band. In such a case, the exciton Hamiltonian should be generalized by introduction of the Luttinger terms describing the complex valence band [2]. The terms couple the exciton center-of-mass and intrinsic electron-hole motions. As a result, the variables of the exciton Schrödinger equation (SE) cannot be separated and one has to consider the multi-dimensional SE [3]. A study of the exciton in a quantum well (QW) meets further complications of the problem. Even in the simplest case, when only the diagonal part of the Luttinger Hamiltonian is included into the problem, the presence of the QW potential requires one to consider at least the three-dimensional SE, which cannot be solved analytically.

The calculations of the exciton radiative decay rate (or the oscillator strength) have been carried out in several works using different simplifications of the problem [4-8] The experimental studies of the radiative characteristics are widespread, see, e.g., Refs. [9-14] Recent measurements of the radiative decay rate have been carried out by Poltavtsev *et al.* [15, 16]. These experimental results and lack of the available theoretical calculations motivated us for this study.

In the present paper, we report on the results of calculations of the radiative decay rate for QWs of various widths. The radiative decay rate is obtained using the wave function found from the SE for excitons in QWs with a degenerate valence band. Partial separation (over two variables) of the center-of-mass motion and cylindrical symmetry of the problem allowed us to reduce the initial SE to the three-dimensional one. The numerical solution of the problem has been done for the GaAs/AlGaAs and InGaAs/GaAs QWs, which are widely experimentally and theoretically studied now as model heterostructures [13, 17].

2. Microscopic model

In our study, we assume that the exciton in a QW is described by the SE with the diagonal part of the Luttinger Hamiltonian [2] only:

$$H = \frac{\mathbf{k}_e^2}{2m_e} + \frac{\left(k_{hx}^2 + k_{hy}^2\right)}{2m_{hxy}} + \frac{k_{hz}^2}{2m_{hz}} - \frac{e^2}{\epsilon|\mathbf{r_e} - \mathbf{r_h}|} + V_e(\mathbf{r_e}) + V_h(\mathbf{r_h}). \quad (1)$$

Here, indices e and h denote the electron and the hole, respectively. The relative electron-hole distance is $|\mathbf{r_e} - \mathbf{r_h}|$. The electron charge is denoted by e and the dielectric constant is ϵ. The square QW potentials $V_e(\mathbf{r_e})$ and $V_h(\mathbf{r_h})$ are zero inside the QW and equal to some constant values in the barriers (conduction- and valence-band offsets, respectively). Term $\mathbf{k}_e^2/2m_e$ is the kinetic operator of the electron in the conduction band in the effective-mass approximation [3]. The kinetic terms of the hole in the valence band, the second and third terms in Eq. (1), come from the diagonal part of the Luttinger Hamiltonian [2].

Hamiltonian (1) specifies the six-dimensional SE, $H\Psi = E\Psi$, for the electron and the hole in QW coupled by the Coulomb interaction. The translational symmetry along the QW layer allows us to reduce this equation to the four-dimensional one by separation of the center-of-mass motion in the (x, y)-plane. This motion is described by an analytical part of the complete wave function, Ψ. The relative motion of the electron and the hole in the exciton is described by $\psi(x, y, z_e, z_h)$, where $x = x_h - x_e$, $y = y_h - y_e$. One more dimension is eliminated by taking advantage of the cylindrical symmetry of the problem and introducing the polar coordinates (ρ, ϕ) for description of the relative motion, since the Coulomb potential does not depend on ϕ. Representing the wave function in the form

$$\psi(x, y, z_e, z_h) = \psi(z_e, z_h, \rho)e^{ik_\phi \phi} = \frac{\chi(z_e, z_h, \rho)}{\rho}e^{ik_\phi \phi}, \quad (2)$$

where $k_\phi = 0, 1, 2, \ldots$, we proceed to the three-dimensional equation, which is numerically studied in the present paper. In Eq. (2), we introduce a factor $1/\rho$ in order to fulfill the cusp condition at $\rho = 0$.

Since the light interacts mainly with the ground $1s$ state of the exciton, we study the case when $k_\phi = 0$. In this case, the equation under consideration is written as [18]:

$$\left(K - \frac{e^2}{\epsilon\sqrt{\rho^2 + (z_e - z_h)^2}} + V_e(z_e) + V_h(z_h)\right)\chi(z_e, z_h, \rho) = E_x\chi(z_e, z_h, \rho) \quad (3)$$

where the kinetic term reads:

$$K = -\frac{\hbar}{2m_e}\frac{\partial^2}{\partial z_e^2} - \frac{\hbar}{2m_{hz}}\frac{\partial^2}{\partial z_h^2} - \frac{\hbar}{2\mu}\left(\frac{\partial^2}{\partial \rho^2} - \frac{1}{\rho}\frac{\partial}{\partial \rho} + \frac{1}{\rho^2}\right), \quad (4)$$

and $\mu = m_e m_{hxy}/(m_e + m_{hxy})$ is the reduced exciton mass in the (x, y)-plane. The obtained three-dimensional Eq. (3) cannot be solved analytically because the potential terms do not allow further separation of the variables.

3. Numerical method

We performed the direct numerical solution of Eq. (3) for precise calculations of the exciton ground state energy, E_x, and function $\chi(z_e, z_h, \rho)$. The exponential decrease of the exciton wave function at large values of variables allows us to impose the zero boundary conditions at the boundary of some rectangular domain. The size of this domain varies from dozens of

QW widths (for small widths) down to several QW widths (for large widths). Therefore, the studied boundary value problem (BVP) is formed by Eq. (3) and the zero boundary conditions at $\rho = 0$, at some large value of ρ, and at large positive and negative values of variables $z_{e,h}$. The direct numerical solution of the BVP is feasible using available computational facilities. For this purpose, we employed the fourth-order finite-difference (FD) approximation [19] of the derivatives in Eq. (3) on the equidistant grids over three variables. We use the central fourth-order FD formula for approximation of the second partial derivative of $\chi(z_e, z_h, \rho)$ with respect to z_e:

$$\frac{-\chi_{k-2,l,m} + 16\chi_{k-1,l,m} - 30\chi_{k,l,m} + 16\chi_{k+1,l,m} - \chi_{k+2,l,m}}{12\Delta_{z_e}^2}. \tag{5}$$

Here, the unknown function on the grid, $\chi(z_{e,k}, z_{h,l}, \rho_m)$, is denoted as $\chi_{k,l,m}$. The same FD formula is employed for the second derivative with respect to z_h. We apply the noncentral fourth-order FD formulas for approximation of the first and second partial derivatives of $\chi(z_e, z_h, \rho)$ with respect to ρ. In the calculations, the grid steps over each variable have been taken to be the same, Δ.

Formula (5) defines the theoretical uncertainty of the numerical solution of the order of Δ^4 as $\Delta \to 0$. However, the discontinuity of the square potential at the QW interfaces decreases the convergence rate of the solution over z_e and z_h to order of Δ^2, whereas the convergence rate over variable ρ is kept $\sim \Delta^4$. The precise values of the studied quantities are obtained by the extrapolation of the results of calculations as the grid step goes to zero.

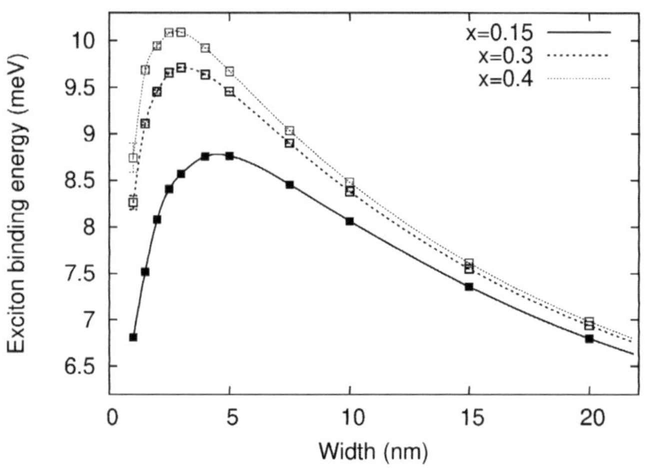

Figure 1. The calculated exciton binding energy for $GaAs/Al_xGa_{1-x}As$ for various aluminium concentrations: $x = 0.15, 0.3, 0.4$.

The nonzero solution of homogeneous equation (3) with trivial boundary conditions can be obtained by the diagonalization of the matrix constructed from the operator of this equation. This square matrix is large, non symmetric and sparse. The typical size of the matrix is of the order of 10^6, so that we keep in the calculations only nontrivial matrix elements of a few diagonals. The full diagonalization of such a matrix is impossible. However, a small part of the spectrum can be easily obtained using the Arnoldi algorithm [20]. As a result, we have calculated the lowest eigenvalue of the matrix and the corresponding eigenfunction. Thus, the ground state energy, E_x, and the corresponding wave function have been obtained for various widths of QW.

We applied the numerical algorithm for solving the SE with parameters for the $GaAs/Al_xGa_{1-x}As$ and $In_xGa_{1-x}As/GaAs$ QWs. The material parameters used for solving the eigenvalue problem (3) are based on the data from Ref. [21] for AlGaAs and from Ref. [8, 22] for InGaAs. In particular, the ratios of potential barriers were taken to be $V_e/V_h = 65/35$. The difference of the electron and hole masses in the QW and in the barrier layers as well as the discontinuity of the dielectric constant at the QW interfaces were ignored.

To check the accuracy of the numerical procedure, we have calculated the exciton binding energy, R_x, which is defined with respect to the quantum confinement energy of the electron, E_e, and the hole, E_h, in QW by the formula $R_x = E_e + E_h - E_x$. Energies E_e and E_h are obtained from the solution of the corresponding one-dimensional SEs for the electron and the hole in QW. The results are shown in Figure 1 for the GaAs/Al_xGa_{1-x}As QW with three barrier heights governed by the aluminium content. The results obtained agree well with other results reported, e.g., in Refs. [18, 21]. We should note here that the usual approach based on the effective mass approximation used in our work may give rise to inaccurate results for very thin (< 10 monolayers) QWs, where one monolayer is ≈ 0.28 nm for GaAs. In the thin QWs, the band mixing at the interfaces becomes important. The interface short-range corrections for this approach can be used in this case to obtain more accurate results [23].

4. Exciton-light coupling

The radiative properties of an exciton are defined by the parameter, called the radiative decay rate or the oscillator strength, see Refs. [1, 4, 5, 24]. The radiative decay rate, Γ_0, characterizes the decay of electromagnetic field emitted by the exciton ensemble after the pulsed excitation: $E(t) = E(0)\exp(-\Gamma_0 t)$. A consistent exciton-light coupling theory is presented in the monograph of Ivchenko [24]. It gives the expression for Γ_0:

$$\Gamma_0 = \frac{2\pi q}{\hbar\epsilon}\left(\frac{e|p_{cv}|}{m_0\omega_0}\right)^2 \left| \int_{-\infty}^{\infty} \Phi(z)\exp(iqz)dz \right|^2, \qquad (6)$$

where $q = \sqrt{\epsilon}\omega/c$ is the light wave vector, ω_0 is the exciton frequency, $|p_{cv}| = \langle u_v|\epsilon \cdot \mathbf{p}|u_c\rangle$

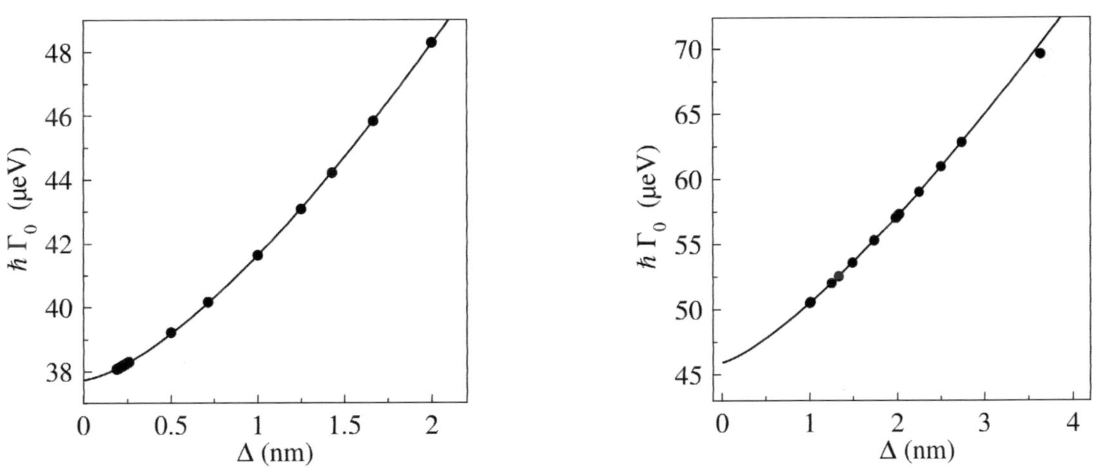

Figure 2. The dependence of the radiative decay rate, $\hbar\Gamma_0$, on the grid step, Δ, for QW width $L = 10$ nm (left panel) and for QW width $L = 200$ nm (right panel). The solid points are the values of the radiative decay rate calculated using the fourth-order FD scheme. The solid line indicates the fit of the calculated values by function $y = a\Delta^b + c$.

is the matrix element of the momentum operator between the single-electron conduction- and valence-band states, and $\Phi(z) \equiv \psi(z, z, 0)$ that is when $z = z_e = z_h$ and $\rho = 0$. The wave function obtained from the microscopic calculation allowed us to calculate the radiative decay rate Γ_0 of the exciton ground state according to Eq. (6). The convergence of $\hbar\Gamma_0$ as the grid step Δ goes to zero is shown in Figure 2. The precise results are obtained by the extrapolation of Δ to 0.

We calculated Γ_0 for GaAs/$Al_{0.3}Ga_{0.7}As$ and $In_{0.02}Ga_{0.98}As$/GaAs QWs of various widths in the ranges of 1 $\div$ 300 nm and 1 $\div$ 200 nm, respectively. In calculations, $|p_{cv}|^2 = m_0 E_p/2$, where $E_p = 28.8$ eV for GaAs and $E_p = 21.5$ eV for InAs are taken from Ref. [22]. The exciton frequency ω_0 is calculated using bandgap $E_g = 1.520$ eV for GaAs [24] and relevant parameters from [21, 22].

Figure 3 shows the radiative decay rate in energy units, $\hbar\Gamma_0$, as a function of the QW width. The radiative decay rate reaches its maximum at the QW width of about 130 nm. It approximately corresponds to the half of the light wavelength in the QW material. So, this maximum of Γ_0 corresponds to the maximal overlap of the exciton wave function $\Phi(z)$ and the light wave, see Eq. (6). For thick QWs, $L > 150$ nm, the calculated decay rate decreases with the well width rise, which can be qualitatively explained by the decrease of the overlap. The applicability of our model for the GaAs/AlGaAs QWs, however, may be limited. The energy distance between the heavy-hole and light-hole exciton states becomes small in the thick QWs so that their coupling by the off-diagonal terms of the Luttinger Hamiltonian omitted in our model becomes important. Heavy-hole-light-hole coupling redistributes oscillator strength between the heavy-hole and light-hole excitons. In the case of InGaAs/GaAs QWs, the strain-induced splitting of the heavy-hole and light-hole excitons suppresses their coupling, therefore our model is still applicable even for thick QWs.

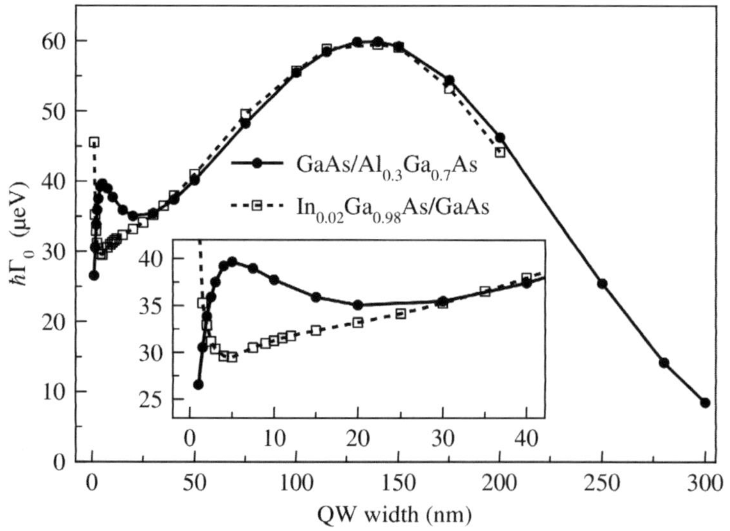

Figure 3. Radiative decay rate for GaAs/$Al_{0.3}Ga_{0.7}As$ (solid curve) and $In_{0.02}Ga_{0.98}As$/GaAs (dashed curve) QWs of various widths from 1 nm to 300 nm and from 1 nm to 200 nm, respectively.

As the QW width decreases, Γ_0 also decreases due to the diminishing overlap integral in Eq. (6). For small QW widths, however, the behavior of Γ_0 for GaAs/$Al_{0.3}Ga_{0.7}As$ and $In_{0.02}Ga_{0.98}As$/GaAs is different. For GaAs/AlGaAs one can see the growth of $\hbar\Gamma_0$ and a local maximum at $L \sim 5$ nm. This result is also confirmed by the experimental data in Ref. [15]. We may suppose that this maximum of Γ_0 is caused by squeezing of the exciton in the narrow QWs by the QW potential. The squeezing increases the probability to find the electron and the hole in the same position ($z_e = z_h$ and $\rho = 0$).

The opposite behavior is observed for $In_{0.02}Ga_{0.98}As$/GaAs as L decreases. One can see that $\hbar\Gamma_0$ diminishes as the QW width L decreases and there is no peak of $\hbar\Gamma_0$ at small QW widths.

We believe that this difference is related to different depths of the QWs. The absence of a peak is the indication of the weak exciton squeezing due to a small QW potential depth.

The rapid growth of $\hbar\Gamma_0$ at $L \to 0$ in the InGaAs/GaAs QWs can be explained by a stronger penetration of an exciton into the barriers as compared to the much deeper GaAs/AlGaAs QWs. Due to the penetration, the overlap of the exciton wave function $\Phi(z)$ and the light wave increases and, in turn, the radiative decay rate increases as $L \to 0$.

5. Conclusions

In summary, we have numerically obtained the radiative decay rate for the exciton ground state. For that, we have solved the three-dimensional equation, which was deduced from the electron-hole Hamiltonian with the Luttinger terms for the valence band. The direct microscopic solution using the fourth-order finite-difference scheme has been carried out for the first time. It allowed us to precisely calculate the exciton ground state energy and the exciton radiative decay rate. We obtained the radiative decay rates for GaAs/AlGaAs and InGaAs/GaAs heterostructures for various QW widths in the ranges of $1 \leq L \leq 300$ nm and $1 \leq L \leq 200$ nm, respectively.

Acknowledgments

Financial support from the Russian Ministry of Science and Education (contract no. 11.G34.31.0067), SPbU (grants No. 11.38.213.2014), RFBR and DFG in the frame of Project ICRC TRR 160 is acknowledged. The authors also thank the SPbU Resource Center "Computational Center of SPbU".

References

[1] Knox R 1963 *Theory of excitons* (New York: Academic press)
[2] Luttinger J M 1956 *Phys. Rev.* **102** 1030
[3] Baldereschi A and Lipari N C 1971 *Phys. Rev. B* **3** 439
[4] Andreani L C 1991 *Solid State Comm.* **77** 641
[5] Citrin D S 1993 *Phys. Rev. B* **47** 3832
[6] Andreani L C 1998 *Physica E* **2** 151
[7] Iotti R C and Andreani L C 1997 *Phys. Rev. B* **56** 3922
[8] D'Andrea A, Tomassini N, Ferrari L, Righini M, Selci S, Bruni M R, Schiumarini D and Simeone M G 1998 *J. Appl. Phys.* **83** 7920
[9] Deveaud B, Clérot F, Roy N, Satzke K, Sermage B and Katzer D S 1991 *Phys. Rev. Lett.* **67** 2355
[10] Zhang B, Kano S S, Shiraki Y and Ito R 1994 *Phys. Rev. B* **50** 7499
[11] Prineas J P, Ell C, Lee E S, Khitrova G, Gibbs H M and Koch S W 2000 *Phys. Rev. B* **61** 13863
[12] Deveaud B, Kappei L, Berney J, Morier-Genoud F, Portella-Oberli M T, Szczytko J and Piermarocchi C 2005 *Chem. Phys.* **318** 104
[13] Trifonov A V, Korotan S N, Kurdyubov A S, Gerlovin I Ya, Ignatiev I V, Efimov Yu P, Eliseev S A, Petrov V V, Dolgikh Yu K, Ovsyankin V V and Kavokin A V 2015 *Phys. Rev. B* **91** 115307
[14] Belykh V V and Kochiev M V 2015 *Phys. Rev. B* **92** 045307
[15] Poltavtsev S V and Stroganov B V 2010 *Phys. Solid State* **52**, 1899 [*Fiz. Tverd. Tela* **52** 1769].
[16] Poltavtsev S V, Efimov Yu P, Dolgikh Yu K, Eliseev S A, Petrov V V and Ovsyankin V V 2014 *Solid State Comm.* **199** 47
[17] Loginov D K, Trifonov A V and Ignatiev I V 2014 *Phys. Rev. B* **90** 075306
[18] Bastard G, Mendez E E, Chang L L and Esaki L 1982 *Phys. Rev. B* **26** 1974
[19] Samarskii A A 1989 *The theory of difference schemes* (Moscow: Nauka)
[20] Sorensen D C, Lehoucq R B and Vu P 1995 *ARPACK: an implementation of the Implicitly Restarted Arnoldi iteration that computes some of the eigenvalues and eigenvectors of a large sparse matrix*
[21] Gerlach B, Wüsthoff J, Dzero M O and Smondyrev M A 1998 *Phys. Rev. B* **58** 10568
[22] Vurgaftman I, Meyer J R and Ram-Mohan L R 2001 *J. Appl. Phys.* **89** 5815
[23] Glinskii G F, Lakisov V A, Dolmatov A G and Kravchenko K O 2000 *Nanotechnology* **11** 233
[24] Ivchenko E L 2005 *Optical spectroscopy of semiconductor nanostructures* (Harrow: Alpha Science Int.)

Si/Zn_xCd_{1-x}Te heterostructures with different Zn contents: growth, electrical and photoelectrical properties

A P Yagunov, T A Gavrikova, V A Zykov and I S Polukhin

Department of Physics of Semiconductors & Nanoelectronics, St.Petersburg Peter the Great Polytechnic University, Polytechnicheskaya 29, St.Petersburg 195251, Russia

E-mail: zykov@rphf.spbstu.ru

Abstract. The Zn_xCd_{1-x}Te films with $x = 0 \div 0.97$ are grown by vacuum deposition in quasi-equilibrium conditions on single crystal Si (111) substrates. The details of the growth process were analyzed. The possibilities of the quasi-closed volume method for producing high quality homogeneous layers of solid solutions were evaluated. It is shown that this method is very effective for the production of films with $x < 0.1$ and may be implemented at condensation temperatures up to ~ 300°C. Studies of electrical and photovoltaic properties showed high quality of Zn_xCd_{1-x}Te layers and structures in general. This is, in particular, evidenced by increasing photosensitivity in the region of absorption of the film material, and of the structure as a whole, increasing integral sensitivity with zinc doping, a significant reduction in non-ideality factor in solid solution films as compared with CdTe films. We have established an increase of mechanical stability of the structure with increasing Zn content, expressed in maintaining the strength during repeated thermal cycling. As a result, the manufactured Si/Zn_xCd_{1-x}Te heterostructures can serve as a basis for fabrication of multijunction solar energy converters.

1. Introduction

Monolithic multijunction structures are promising for optoelectronic applications including the creation of the cascade converters of solar energy. Currently, the greatest value of the conversion efficiency of solar energy was received on multijunction structures based on III-V compounds [1]. However, due to a number of technological and economic reasons, the multijunction structures that combine II-VI compounds with silicon are serious competition for the structures based on III-V compounds [2,3]. Among the possible combinations with silicon, of increased interest is a combination of silicon and cadmium telluride - a compound with an optimal set of physical parameters for photoelectric conversion of solar radiation [4]. The principal limitation in using heterostructures (HS's) of Si/CdTe is a high (~14%) mismatch of lattice parameters of materials. However, the theoretical analysis shows [3] that this problem can be eliminated if, instead of pure CdTe, the Zn_xCd_{1-x}Te solid solutions are used, for which a misalignment is significantly less.

The main goal of the work was to develop a method of growing crystalline Zn_xCd_{1-x}Te films on Si wafers and to create structures which can be used as a platform for further manufacturing tandem two-junction solar cells. The specific goals of the work were as follows:

- the growth of epitaxial Zn_xCd_{1-x}Te films on single-crystal Si substrates with a sufficiently high crystallinity and a controlled content of zinc;

- study of electrical properties of the Si/Zn_xCd_{1-x}Te structures with different content of Zn in solid solutions in order to characterize the obtained films and structures.

2. Growth experiments

The objects of the study are Si/$Zn_xCd_{1-x}Te$ HS's with an anisotype n-Si/p- $Zn_xCd_{1-x}Te$ heterojunction. The samples are planar HS's consisting of a silicon substrate (a single crystal silicon wafer of <111> orientation) coated by a solid solution film of about 1-2 microns thickness. As contacts to silicon and $Zn_xCd_{1-x}Te$ films, we used In-Ga alloys of compositions close to eutectic and a two-layer coating of Ti(~70 Å)/Al(~1 μm), respectively.

The problem of manufacturing high-quality single-crystal films of cadmium telluride has a long history associated primarily with the development of technology of IR devices based on HgCdTe solid solutions [5]. The obvious advantages of vapor-phase techniques for epitaxial growth of CdTe films have been implemented in the framework of different technological solutions from vacuum deposition, including ultrahigh vacuum conditions [6], to vapor-phase reactions techniques using organometallic compounds [7], in conditions close to equilibrium and far from those [8]. All of these methods in the optimum mode allowed us to obtain epitaxial layers, the crystal quality of which is mainly determined by the degree of purity of the precursors and by the ratio of the lattice parameters of CdTe and the substrate material. In this paper, vacuum deposition in a quasi-closed volume has been selected as a method for the films (structures) preparation. This method allowed, for relatively low cost, to carry out the condensation of films in a wide range of condensation conditions, including vapor, the parameters of which in each isothermal section (in the direction of transfer) are as close as possible to the thermodynamic equilibrium ones (the mode of gas-dynamic flow [9]). Moreover, the method allows to obtain uniform films of large area.

The evaporation chamber is a split graphite cylinder, the bottom of which is tightly joined with a graphite block with the evaporated substance (CdTe charge). An upper graphite cover, tightly closing the cylinder, is also a substrate holder. The calculations performed by the method [9] have allowed to design an evaporation chamber, in which the condensation is possible in both a laminar vapor flow (the gas-dynamic mode) and a molecular vapor flow, at temperatures of condensation T_C = 250÷400°C in a wide range of values of relative vapor supersaturation (1÷100) at the crystallization surface.

The films were deposited on monocrystalline silicon wafers, previously cleaned from natural oxide using the modified RCA cleaning method [10]. The method included the removal of organic contaminants from the silicon surface, etching in HF, soaking in deionized water with subsequent mild oxidation. Before loading Si sudstrate into the evaporation chamber, the oxide was removed by soft etching in a dilute HF solution. The final purification step was annealing in vacuum immediately before evaporation. Removal of the oxide is an important procedure because, as pointed out in [11] (and according to our observations), the presence of native oxide on the surface of Si leads to the growth of polycrystalline CdTe films at all substrate temperatures and growth rates.

The composition of the solid solution film was controlled during the manufacturing process by changing the ratio of CdTe and ZnTe in the evaporation source.

According to the method described above, the Si/CdTe heterostructures on Si-substrates with different carrier concentrations were produced. The highest quality films of cadmium telluride have been obtained under quasi-equilibrium conditions with laminar vapor flow and condensation temperatures of ~ 350°C. Analysis of layers morphology showed that they are the textured block structures oriented along the <111> growth direction. The surface is represented by well-faceted crystallites with an average size of more than 1 μm, separated by straight boundaries. The main types of observed structural defects are microtwins and dislocation etch pits. In addition, we observed antiphase domains (evidence of layer-by-layer growth) and micron-sized fragments of hexagonal shape in some samples.

When growing films of $Zn_xCd_{1-x}Te$ solid solutions with a low zinc content (x is of order of a few percent), the growth parameters and morphology of the films practically do not differ from those for the growth of pure cadmium telluride. However, in the experiments on production of solid solution films with $x > 0.1$ it has been found that increasing zinc content in the vapor requires raising the temperature of the evaporation source (above 650°C), and, as a result, changing the geometric parameters of the evaporation chamber and performing the process in the mode of molecular beams. The condensation temperatures do not exceed 400°C, since at higher temperatures the films crystallize on Si (111) substrates predominantly with a hexagonal structure (wurtzite) [12].

The composition of the solid solution films was determined based on the analysis of the transmission spectra in the fundamental absorption edge region for the films of $Zn_xCd_{1-x}Te$ on glass, with a subsequent determination of the band gap Eg and the calculation of x from the composition dependence $E_g(x)$ [13]:

$$E_g(x) = E_g(\mathrm{ZnTe}) - [\, E_g(\mathrm{ZnTe}) - E_g(\mathrm{CdTe})\,]\,(1 - x) + \mathrm{B}x\,(1 - x) \tag{1}$$

In the calculation we used: E_g(ZnTe) = 2.24 eV; E_g(CdTe) = 1.48 eV; B = 0.34 (T = 300 K) [14]. The optical experiments were performed by a Fourier spectrometer Vertex-80x. The light source was a halogen lamp (350-3500 nm).

Considering the fact that:

$$\alpha L = A\frac{\sqrt{h\omega - E_g}}{h\omega}, \tag{2}$$

where α - the absorption coefficient; L - thickness of the film, the value of E_g is determined by the cutoff value on the axis $h\omega$ of the experimental dependences represented in the coordinates $(\alpha L\times h\omega)^2$ - hω, for example, for the samples with the lowest (x = 0.03) and the highest (x = 0.97) concentrations of Zn in solid solutions, presented in figure 1.

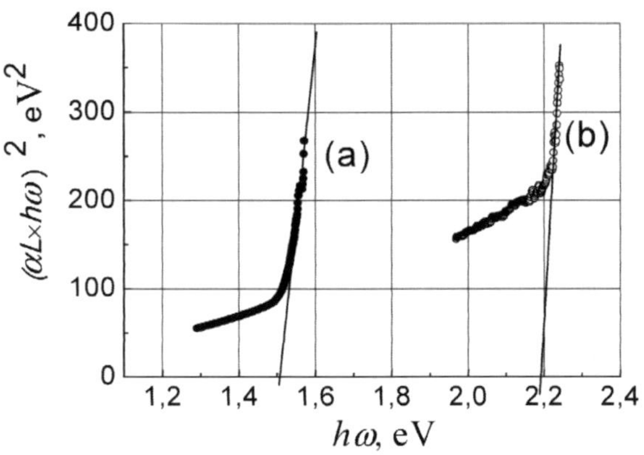

Figure 1. Determination of the band gap of the solid solution films with different Zn contents for the samples with the lowest x = 0.03 (a) and the highest x = 0.97 (b) concentrations of Zn in $Zn_xCd_{1-x}Te$ films

Thus, Si/$Zn_xCd_{1-x}Te$ HS's with the Zn content (x) in solid solution films varying from 0 to 0.97 were fabricated. All the films had p-type conductivity. The hole concentration determined from the capacitance-voltage characteristics was $1\times10^{13}\div1\times10^{15}$ cm^{-3}. The dependence of the carrier concentration on the Zn content is not found in the present work. The carrier concentrations in n-Si substrates were $5\times10^{14}\div1\times10^{18}$ cm^{-3}. The film thickness was 1-2 μm, the size of a structure element for subsequent measurements was 5×5 mm^2.

An important feature of the structure, relevant for subsequent fabrication of two-junction solar energy converters, is the mechanical strength. Indeed, due to a high lattice mismatch between Si and CdTe, as well as a significant divergence between the values of their thermal expansion coefficients, the CdTe layer is in a highly stressed state and can be destroyed by heating resulted from these stresses. At the same time, one of the basic operations during fabrication of multijunction structure components is the formation of a p-n junction in $Zn_xCd_{1-x}Te$ films, which is usually performed by thermal diffusion of impurities from the impurity layer previously deposited on the surface of the wide-gap material. The typical values of the temperature of the diffusion process (for In or Cl) are about 300°C [15]. Therefore, the grown structures must withstand a long-term heating up to these temperatures without destruction. Our experiments show that CdTe films on Si substrates are destroyed (exfoliate) when heated up to these temperatures. However, with the introduction of Zn into the film the mechanical stresses are reduced. Thermocycling of films containing Zn in the range of 20÷300°C does not result in destruction of films on a substrate. This effect was consistently observed at $x > 0.1$. Thus, the introduction of zinc in CdTe films stabilizes the mechanical properties of the structures.

3. Photoelectrical and electrical experiments

All obtained HS's with an anisotype *n*-Si/*p*-$Zn_xCd_{1-x}Te$ heterojunction are photosensitive. At low zinc content in films ($x < 0.1$), the integral photosensitivity increases monotonically with increasing x by about an order of magnitude compared with the structures of Si/CdTe (when $x = 0.1$). This fact apparently testifies to improving the quality of the layers with the introduction of zinc, and, consequently, of the heterostructure as a whole. By improving the quality of the layers, in this case, we mean the reduction in growth defects density (stacking faults, low angle boundaries, regions with wurtzite structure) and, mainly, misfit dislocations with decreasing lattice mismatch of HS components. The increase in photosensitivity was observed in the whole spectral range, both in the long-wavelength region (Si) and in the short-wavelength one ($Zn_xCd_{1-x}Te$) (figure 2(a)).

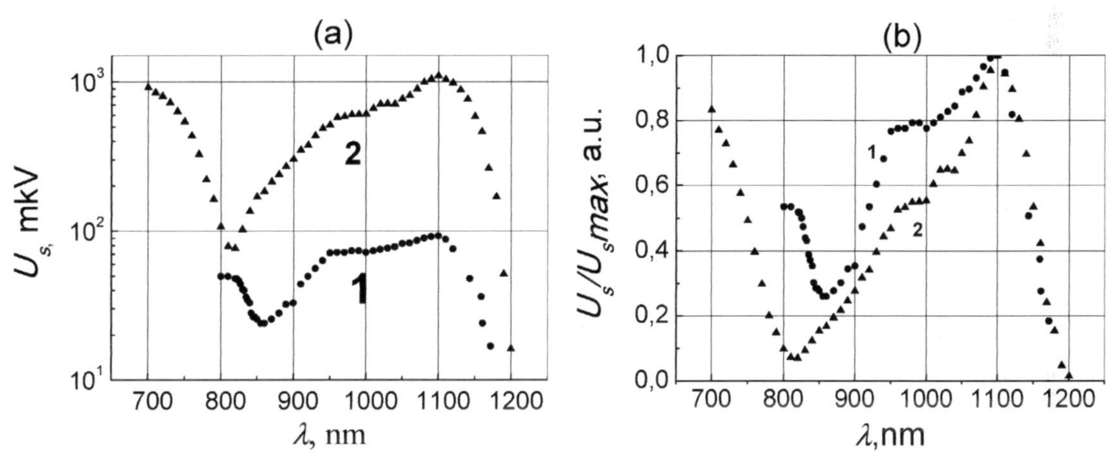

Figure 2. The spectral distribution of the photo-emf (U_s) in Si/CdTe (1) and Si/$Zn_{0.1}Cd_{0.9}Te$ (2) heterostructures. (a) – the absolute value U_s; (b) – the relative units (U_s/U_{smax})

There are two clearly distinguished regions in the spectral characteristics of HS (figure 2(a)) : a long-wave one (with a red edge at about 1100 nm) corresponding to the absorption of Si and a short-wave one (with a red edge at about 840 nm) corresponding to the absorption of the wide-gap material. With increasing Zn content in the film, the photoresponse red edge for wide bandgap semiconductors is shifted to shorter wavelengths (figure 2(b)). The shift value correlates satisfactorily with the results of the fundamental absorption edge studies. The signs of the photo-emf in both regions were the same (negative, minus at Si). The relation between the photo-emf in the long and short wavelength regions depends on the ratio of the concentration of charge carriers in the base region of the HS. For a low content of Zn ($x < 0.1$) we found a tendency to a relative increase in sensitivity in the short-wave region with increasing x.

The current-voltage characteristics of the n-Si/p-$Zn_xCd_{1-x}Te$ HS's are rectifying (figure 3). At low forward bias (U = 0-3 B), the current-voltage characteristics are satisfactorily described by the relation $I \sim \exp(eU / \beta kT)$, where β – the non-ideality factor. The analytical form of the dependence indicates that the conductivity is carried out according to the scheme of thermal emission with significant recombination in the space charge region. The value of β characterizes the contribution of recombination in the space charge region.

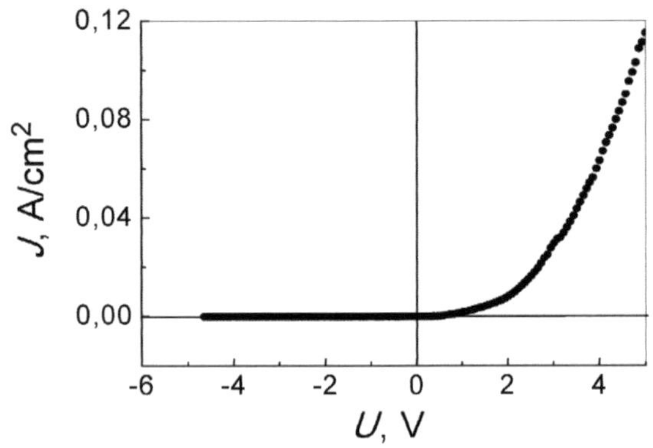

Figure 3. Typical current-voltage characteristic of the *n*-Si/*p*-Zn_xCd_{1-x} Te HS

The main centers of recombination in the space charge of the heterojunction are the states associated with the misfit dislocations. When increasing content of Zn in the film, the mismatch of lattice parameters of Si and the solid solution decreases. As a result, the density of misfit dislocations also decreases, and the contribution of recombination on the boundary states reduces. This is reflected in the decrease in the value of β. This very effect was observed in our study of current-voltage characteristics at low forward bias. For Si/CdTe HS (the greatest mismatch) the experimentally determined β takes a value of 5-7 (figure 4(a)). When we change the film composition by increasing the content of Zn, the value of β decreases and at $x \sim 0.1$ it reaches the minimum value β = 2.1 (figure 4(b)). With a high content of Zn ($x > 0,5$) the value of β increases. We attribute this effect to the change in the conditions of $Zn_xCd_{1-x}Te$ films growth at large values of x, i.e. to the transition from quasi-equilibrium conditions of condensation (small x) to condensation from the molecular flow (large x).

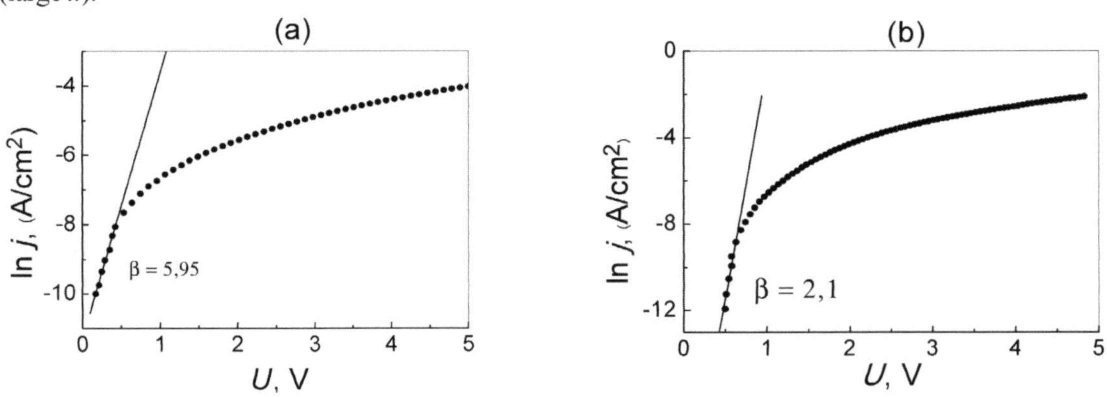

Figure 4. Direct branches of the current-voltage characteristics of the $Zn_xCd_{1-x}Te$ HS's with different contents of Zn in the film: a) $x = 0$, b) $x = 0.1$

At high forward bias the voltage dependence on the current is described by a law $I \sim U^m$, m usually takes values of 1.5 ÷ 2.0. Apparently, this dependence is associated with injection currents in the wide-gap material [16].

4. Results and discussion

As a result of technological experiments, we prepared Si/$Zn_xCd_{1-x}Te$ HS's with various zinc content in the films of solid solutions (x = 0 ÷ 0.97). When manufacturing HS's with $x < 0.1$, the growth of $Zn_xCd_{1-x}Te$ films can be carried out in the gas-dynamic flow mode (quasi-equilibrium conditions of condensation). The films obtained under such conditions have a uniform thickness over an area of about 20 cm^2 and represent large-block texture with a characteristic crystallite size of 1-2 microns. Along with structural uniformity, the films are homogeneous in a number of physical parameters: the variation of the electric and photoelectric parameters of individual elements in different parts of the plate does not exceed a few percent. HS's possess photosensitivity in a wide spectral range (700 – 1100 nm) with the photoresponse signals comparable in the long-wave and short-wave spectral regions. The presence of sensitivity in the absorption region of the solid solution suggests, in particular, the high quality of the film material. With increasing Zn content in the $Zn_XCd_{1-X}Te$ layer we observed increasing in signal in the short-wavelength region and also increasing of the integrated sensitivity of HS. Apparently, this is due to an increase in the structural perfection of the films primarily by reducing the mismatch of the lattice parameters of HS components with increasing *x*. Obviously, this improves the quality of metallurgical boundaries of HS, decreases the density of misfit dislocations, reduces the contribution of recombination at the boundary states. The analysis of current-voltage characteristics showed a significant reduction in the ideality factor β for the HS with a film of the $Zn_XCd_{1-X}Te$ solid solution, which also confirms a significant improvement in the quality of the boundary.

Decreasing mismatch of lattice parameters of the $Si/Zn_XCd_{1-X}Te$ HS components with increasing Zn content has another important consequence. Reduction of mechanical stress is manifested in the stabilization of mechanical properties of the structure and in the ability not to be destroyed at high temperatures. This allows us to use the thermal diffusion technique to form a p-n junction in the layer of the solid solution.

The films of $Zn_XCd_{1-X}Te$ with a high zinc content ($x > 0.1$) were possible to obtain only in the regime of molecular beams. In this case the uniformity and quality of layers are significantly reduced. This, in particular, is reflected in a significant decrease in the integral photosensitivity of HS. However, when designing a solar energy converter it is desirable to use the solid solutions with the Eg value which does not differ much from the optimal value of 1.5 eV, i.e. the materials with a low content of zinc.

Thus, the $Si/Zn_XCd_{1-X}Te$ HS with a zinc content in the film of the solid solution $x \sim 0.1$, obtained by vacuum deposition in quasi-equilibrium conditions have the necessary quality and mechanical properties and can serve as a basis for further manufacturing two-juncnion solar energy converters.

5. Conclusion

The main result of this work is the preparation of $Zn_xCd_{1-x}Te$ HS's with different contents of Zn in the film ($x = 0 \div 0.97$). It was found that the technique of deposition in a quasi-closed volume allows to produce uniform layers of the solid solution on the surface with an area of ~ 20 cm^2. The $Zn_xCd_{1-x}Te$ films make a tight contact with a Si substrate, and are not destroyed during cyclic heating up to 300°C. This quality is manifested only with the introduction of Zn in the film and the structures are stable against destruction even at low zinc content (~ a few percent). Studies of electrical and photovoltaic properties showed a sufficiently high quality of the layers of $Zn_xCd_{1-x}Te$ and structures as a whole. This is, in particular, confirmed by the photosensitivity of the structures in the region of absorption of the film material and structure as a whole, the increase in the integral sensitivity with the introduction of zinc, a significant decrease of the non-ideality factor of the $Zn_xCd_{1-x}Te$ films compared with CdTe films. As a result, we have manufactured the structures that can serve as a basis for further fabrication of multijunction solar energy converters.

References

[1] Green M A, Emery K, Hishikawa Y, Warta W and Dunlop E D 2015 *Progress in Photovoltaics. Research and Applications* **23** 805
[2] Carmody M, Mallick S, Margetis J, Kodama R, Biegala T, Xu D, Bechmann P, Garland J W and Sivananthan S 2010 *Appl. Phys. Lett.* **96** 153502
[3] Dong Xu, Biegala T, Carmody M, Garland J W, Grein C and Sivananthan S 2010 *Appl.Phys. Lett.* **96** 073508

[4] Shockley W and Queisser Y J 1961 *J.Appl.Phys.* **32** 510
[5] Neretina S P, Sochinskii N V and Mascher P 2005 *J. Electron. Materials* **34** 786
[6] Rujirawat S, Almeida A, Chen Y P and Sivananthan S 1997 *Appl.Phys.Lett.* **71** 1810
[7] Ebe H and Nashijima Y 1995 *Appl.Phys.Lett.* **67** 3188
[8] Belyaev A P and Rubets V P 2001 *Semiconductors* **35** 279
[9] Bubnov V Z, Libenson N N, Lurie M S and Filaretov G A 1969 *J.Eng.Phys.and Technol.* **17** 429
[10] Sato S, Yamada S and Suzuki R 1996 *J.Cryst.Growth* **214/215** 5
[11] Yawcheng Lo, Bicknell R N, Myers T H, Schetzina J F 1983 *J.Appl.Phys.* **54** 4238
[12] Sidorov Yu G, Dvoretsky S A, Varavin V S, Mikhailov N N, Yakushev M V and Sabinina I V 2001 *Semiconductors* **35** 1045
[13] Kozlovsky V I, Litvinov V G and Sadofyev Yu G 2000 *Semiconductors* **34** 960
[14] Lozykowski H J and Shastri V K 1991 *J. Appl. Phys.* **69** 3235
[15] Wagah F Mohammad 2011 *J. Opt. and Adv. Mater.* **13** 1289
[16] Lampert M A and Mark P *Current injection in solids* (New-York & London: Acad. Press)

Single-stage fabrication of low-loss dielectric nanoresonators from high-loss material

P A Dmitriev[1], S V Makarov[1], V A Milichko[1], I S Mukhin[1,2], A M Mozharov[2], A A Sitnikova[3], A K Samusev[1,3], A E Krasnok[1] and P A Belov[1]

[1] ITMO University, St. Petersburg 197101, Russia
[2] St. Petersburg Academic University, St. Petersburg 194021, Russia
[3] Ioffe Physical-Technical Institute of the Russian Academy of Sciences, St. Petersburg 194021, Russia

E-mail: p.dmitriev@metalab.ifmo.ru

Abstract. High refractive index subwavelength dielectric nanoresonators, which support both electric and magnetic optical resonances, are a promising platform for many nanophotonic applications. However, the resonant properties of the nanoresonators are highly sensitive to defect concentrations. Therefore, to maintain low defect concentrations, it has generally thought inevitable to use crystalline precursor materials for nanoresonator fabrication. Here, we show that it is possible to fabricate crystalline (low-loss) resonant silicon nanoparticles by femtosecond laser ablation from amorphous (high-loss) silicon thin films. The crystallinity of the fabricated nanoparticles is proven by means of Raman spectroscopy and electron transmission microscopy, and their optical resonant properties are studied by means of dark-field optical spectroscopy and discrete-dipole simulations.

1. Introduction

Dielectric nanoparticles with high refractive indexes and low losses provide excitation of strong magnetic optical resonances [1–9], making them a viable alternative to plasmonic nanoparticles, which have high losses in the visible range, for such applications as electromagnetic field enhancement and sensing [10–13], antireflection [14], perfect reflection [3, 15], light wavefront manipulation [16,17], superdirective scattering [18] and enhancement of nonlinear effects [19,20].

Crystalline silicon (c-Si), being a low-cost CMOS compatible semiconductor and providing the desired optical response because of its large real and low imaginary parts of its refractive index in the visible range, is one of the most popular materials for all-dielectric nanophotonics. The most controllable method of c-Si film structuring — reactive ion etching combined with electron-beam lithography — is appropriated only for microscale structuring. Therefore, various methods of resonant silicon nanoparticle fabrication have been developed during last years, including colloidal chemistry [21], thin film dewetting [22] and laser ablation [6,23]. To achieve the crystalline state of nanoparticles, it was necessary to use initially bulk crystalline samples because amorphous silicon has an up to *two orders of magnitude larger imaginary part* of its refractive index as compared to crystalline silicon in the visible range [24], diminishing the resonant properties of nanoparticles.

Figure 1. A schematic illustration of laser printing of crystalline nanoparticles.

Amorphous thin films are known to crystallize when heated [25], so laser annealing has been used for nanoscale amorphous surface [26] and nanoparticle [23] crystallization. In this paper, we study the fabrication of crystalline nanoparticles from amorphous hydrogenized silicon films, with magnetic Mie-type resonances, by means of an improved method of laser-induced transfer. The technique, so-called laser printing of nanoparticles [23] is shown in figure 1. The main advantage of this method over those reported previously, is that it allows single-stage low cost non-lithographic fabrication of *crystalline* optical nanoresonators on both transparent and opaque materials.

2. c-Si nanoparticle printing

The silicon nanoparticles were printed by a femtosecond laser system (Femtosecond Oscillator TiF-100F, Avesta Poject), which emits laser pulses at a central wavelength of 800 nm, with pulse duration of 100 fs and repetition rate of 80 MHz. Laser pulses selected by a Pockels cell-based pulse picker (Avesta Poject) were tightly focused by an oil immersion microscope objective (Olympus 100×) with a numerical aperture of $NA = 1.4$. According to the relation $d \approx 1.22\lambda/NA$, the estimated diameter of the beam's focal spot size is $d = 0.7$ μm which is close to the value measured by a method based on the dependence of the laser-damaged area on incident laser energy (0.68 μm) [27].

The nanoparticles were fabricated from an 80 nm thick a-Si:H film, which was deposited on a substrate of fused silica by plasma enhanced chemical vapor deposition from a SiH_3 precursor gas. The samples were placed on a three-dimensional air-bearing translating stage driven by brushless servomotors (ABL1000, Aerotech) with sample translation accuracy being at least 100 nm.

The nanoparticles were fabricated by single laser pulses (from a previously undamaged surface) in the forward-transfer geometry, in which the receiving substrate is placed under the film with a spacing of $\sim$ 50 μm (figure 1). This geometry has an advantage over the back-transfer geometry owing to the possibility of nanoparticle printing onto a wide variety of substrates including opaque and structured samples. The silicon nanoparticles were printed at laser energies in the range of $0.5 - 1.2$ nJ providing fluencies in the range of $0.12 - 0.16$ J/cm^2. The nanoparticles are almost spherical in shape (figure 2(b)) and their diameters lie in the range of $50 - 200$ nm depending on the fluence.

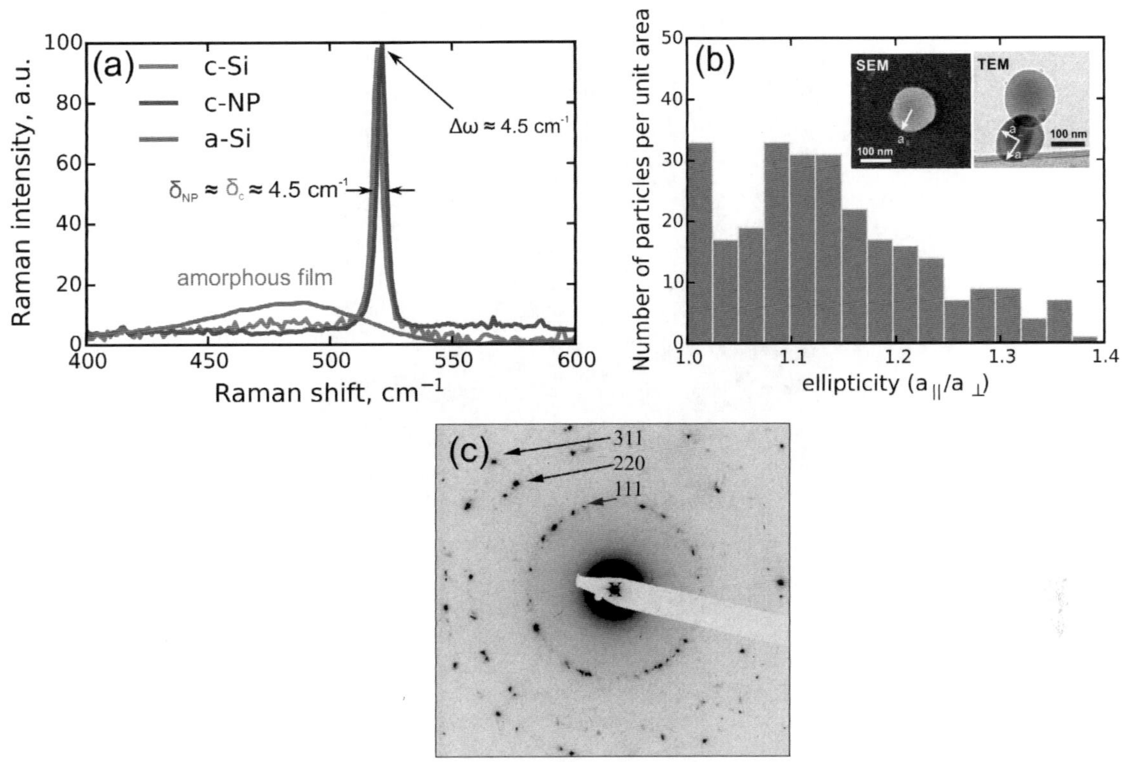

Figure 2. (a) Experimental Raman spectra from an individual nanoparticle with a diameter of about 170 nm (red), initial, donor, film (purple), and reference bulk crystalline silicon (blue). (b) Elipticity distribution of Si nanoparticles; insets: SEM and TEM images of typical Si nanoparticles (c). Electron diffraction picture from the crystalline nanoparticles.

3. Nanoparticle Characterization

3.1. Crystallinity

The crystalline structure of the initial film and the produced individual nanoparticles were characterized by Raman scattering measurements by a micro-Raman setup (Raman spectrometer HORIBA LabRam HR, AIST SmartSPM system). The Raman spectra were recorded under excitation by a 2 mW, 632.8 nm HeNe laser through a 100× microscope objective ($NA = 0.9$) and projected onto a thermoelectrically cooled charge-coupled device (CCD, Andor DU 420A-OE 325) array with a 150 g/mm diffraction grating. Spectra of individual nanoparticles were recorded by precisely positioning the nanoparticle (accuracy down to 100 nm) in the center of the laser beam (1.8 μm in diameter) focused on the substrate.

The initial silicon film was shown to be completely amorphous because of its broad Raman peak centered around 480 cm^{-1}. The Raman spectra of the individual nanoparticles have narrow peaks at 521.5 cm^{-1} indicating a crystalline cubic diamond structure in all the nanoparticles with diameters in the range of $120 - 200$ nm. The reference Raman signal from a bulk crystalline silicon wafer and the literature data suggest that the Raman peak of a pure crystal corresponds to 520 cm^{-1}. The slight positive shift of the peak of nanoparticles $\Delta\nu = 1.5$ cm^{-1} is explained by residual compressive stress [28]. Another important characteristic extracted from the Raman spectra is the size of the nanocrystallite, which is larger than ~ 20 nm because the Raman peaks of the nanoparticles have almost the same halfwidth ($4 - 5$ cm^{-1}) as the peak from a bulk crystalline silicon wafer (4.5 cm^{-1}) [29].

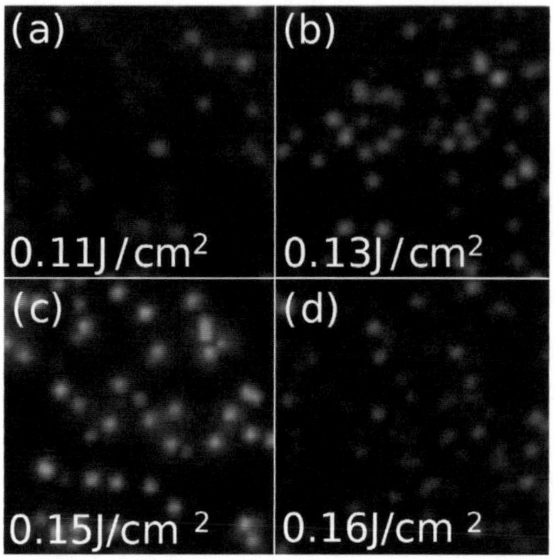

Figure 3. Optical images in the dark-field scheme of silicon nanoparticles fabricated at different peak fluences: 120 (a), 130 (b), 150 (c), 160 mJ/cm^2 (d)

The Raman measurements agree with the characterization of the printed nanoparticles by means of transmission electron microscopy (TEM). We used specimen grids (3 mm-diameter 200-mesh copper grids coated on one side with a 20 nm-thick film of amorphous carbon) to collect nanoparticles ablated from the a-Si film. The size, structure, and composition of the collected nanoparticles were determined using bright and dark field TEM imaging. In our experiments, we observed mainly single-crystal nanoparticles with the mean size in the range of 100 – 200 nm (figure 2(b)). The analysis of the electron diffraction pattern from several nanoparticles shows clear maxima, corresponding to certain crystalline planes (figure 2(c)). TEM imaging also provides information on the oblateness of the particles along the orientation prependicular to the collecting surface, giving the average ellipticity of about $a_{||}/a_{\perp} \approx 1.12$, where $a_{||}(a_{\perp})$ is the particle semi-axis oriented perpendicular (parallel) to the surface of the substrate. Indeed, scanning electron microscopy (SEM, Carl Zeiss, Neon 40) shows that values of $a_{||}$ are almost the same in all directions (figure 2(b)). These results correlate with previously observed oblateness of the printed silicon nanoparticles [8].

As in previous studies [23,30], in our experiments the average size of the printed nanoparticles strongly depends on laser fluence. We observed two different regimes of nanoparticle generation. In the first regime, average nanoparticle sizes grow with an increase of fluence up to 150 mJ/cm^2 manifesting in the change of their colors from blue to red (figure 3(a)-(c)). This growth can be described in terms of the spallation mechanism of laser ablation, where a thin molten layer is spalled due to laser-induced tensile pressure waves [31, 32], breaking into a number of liquid droplets via the Rayleigh-Plateau instability [33]. The photomechanically spalled volume increases as $V \sim \ln(E)$ under the action of a Gaussian beam, owing to the logarithmic dependence of the spalled surface layer area $r_s^2 \sim \ln(E)$ [34], whereas the thickness of the layer remains almost constant [31] or even decreases [32]. In previous studies of nanoparticle printing from crystalline silicon, such an increase of molten volume led to an increase in the number of nanoparticles [30] and their size [23], which agrees with our results (figure 3(a)-(c)). However, using amorphous silicon films it is not possible to reproducibly print one nanoparticle per laser shot, unlike when printing from crystalline films.

The second regime of nanoparticle fabrication corresponds to $F > 150$ mJ/cm^2,

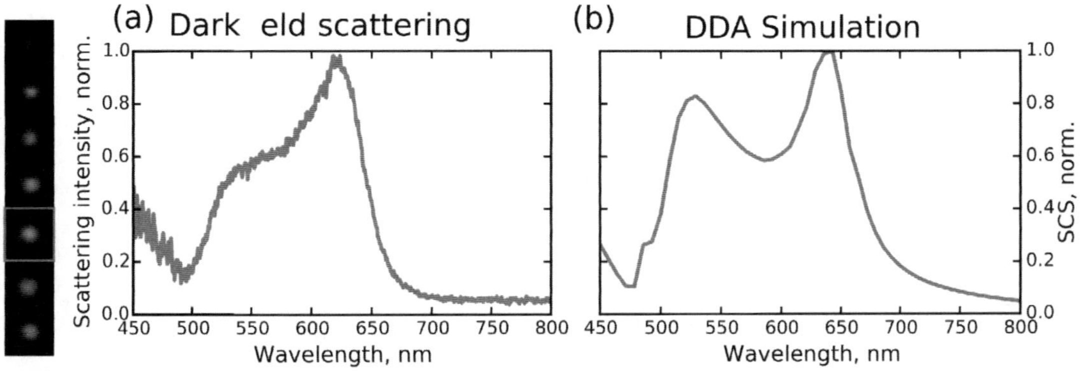

Figure 4. Experimental (a) and simulated (b) spectra for scattered p-polarized incident light (angle of incidence is 70°) from individual nanoparticles with semi-axis parallel to the substrate surface $a_{\perp} = 85$ nm.

where large (red) nanoparticles are followed by small nanoparticles with a much broader size distribution (figure 3(d)). This regime is related to the regime of unstable boiling of superheated silicon [31, 32, 35]. In this regime nanoparticle formation occurs through the explosive decomposition of the material into vapor and small clusters/droplets [32, 36] yielding nanoparticles with average sizes smaller than 100 nm [37].

The almost perfect crystalline phase of the printed silicon nanoparticles (proved by Raman, TEM and SEM characterization) means that it is possible to fabricate crystalline silicon nanoresonators from amorphous films, which is very useful for low-loss all-dielectric nanophotonics.

3.2. Optical properties

The excellent resonant properties of the particles were demonstrated using polarization-resolved scattering spectra measurements in a dark-field scheme, with the nanoparticles illuminated by p-polarized light from a halogen lamp (HL-2000-FHSA) at an angle of 70° to the surface normal and scattered signal collection performed by means of an Mitutoyo M Plan APO NIR 50× objective ($NA = 0.45$) [38]. The scattered light was registered by the same spectrometer as in the Raman measurements.

The excitation of electric dipole (ED) and magnetic dipole (MD) resonances is proven by simulations using the discrete dipole approximation [39, 40]. The scattering geometry is modeled as a c-Si sphere of a given radius R, consisting of 3069 dipoles, on a a-SiO_2 substrate. The sphere is irradiated by a plane wave at an angle of 70 ° in vacuum. Modeling of scattering spectra (figure 4(b)) shows good agreement with the corresponding experimental ones (figure 4(a)). Some differences between the shapes of theoretical and experimental spectra arise from the ellipticity of nanoparticles and the presence of a thin natural oxide layer on the surface of nanoparticles [41]. Such optical and TEM/SEM analysis of average elipticity and crystallinity of nanoparticles will be useful for further applications of the laser printing method.

4. Conclusions

We have experimentally proven that laser fabrication of crystalline (low-loss) silicon nanoparticles, supporting magnetic Mie-type resonance, can be performed from amorphous (high-loss) a-Si:H films, making it unnecessary to use initially crystalline material or any additional annealing. Analysis of the printed nanoparticles shows that they are monocrystalline with similar dimensions at certain fluences. Our method of nanoparticle printing does not limit us to using transparent substrates, meaning that nanoparticles can be printed onto a wide variety of substrates/devices.

Acknowledgments
This work was financially supported by Russian Science Foundation (Grant 15-19-00172). We are also thankful to Sergey Kudryashov, Boris Chichkov and Arseniy Kuznetsov for useful comments and discussion.

References

[1] Popa B I and Cummer S A 2008 *Phys. Rev. Lett.* **100** 207401
[2] Zhao Q, Zhou J, Zhang F and Lippens D 2009 *Mater. Today* **12** 60–69
[3] Evlyukhin A B, Reinhardt C, Seidel A, Lukyanchuk B S and Chichkov B N 2010 *Phys. Rev. B* **82** 045404
[4] García-Etxarri A, Gómez-Medina R, Froufe-Pérez L S, López C, Chantada L, Scheffold F, Aizpurua J, Nieto-Vesperinas M and Sáenz J J 2011 *Optics express* **19** 4815–4826
[5] Krasnok A E, Miroshnichenko A E, Belov P A and Kivshar Y S 2012 *Opt. Express* **20** 20599–20604
[6] Kuznetsov A I, Miroshnichenko A E, Fu Y H, Zhang J and Lukyanchuk B 2012 *Scientific Rep.* **2**
[7] Ginn J C and Brener I 2012 *Phys. Rev. Lett.* **108** 097402
[8] Fu Y H, Kuznetsov A I, Miroshnichenko A E, Yu Y F and Lukyanchuk B 2013 *Nature Comm.* **4** 1527
[9] Krasnok A, Makarov S, Petrov M, Savelev R, Belov P and Kivshar Y 2015 *SPIE Opt. and Optoelectr.* **9502** 950203–950203
[10] Albella P, Poyli M A, Schmidt M K, Maier S A, Moreno F, Saenz J J and Aizpurua J 2013 *The Journal of Physical Chemistry C* **117** 13573–13584
[11] Zambrana-Puyalto X and Bonod N 2015 *Physical Review B* **91** 195422
[12] Bakker R M, Permyakov D, Yu Y F, Markovich D, Paniagua-Dominguez R, Gonzaga L, Samusev A, Kivshar Y, Lukyanchuk B and Kuznetsov A I 2015 *Nano Letters* **15** 2137–2142
[13] Caldarola M, Albella P, Corts E, Rahmani M, Roschuk T, Grinblat G, Oulton R F, Bragas A V and Maier S A 2015 *Nature communications* **6** 7915
[14] Spinelli P, Verschuuren M and Polman A 2012 *Nature communications* **3** 692
[15] Moitra P, Slovick B A, Yu Z G, Krishnamurthy S and Valentine J 2014 *Appl. Phys. Lett.* **104** 171102
[16] Decker M, Staude I, Falkner M, Dominguez J, Neshev D N, Brener I, Pertsch T and Kivshar Y S 2015 *Adv. Opt. Mater.*
[17] Yu Y F, Zhu A Y, Paniagua-Dominguez R, Fu Y H, Lukyanchuk B and Kuznetsov A I 2015 *Laser & Photonics Reviews* **9** 412–418
[18] Krasnok A E, Simovski C R, Belov P A and Kivshar Y S 2014 *Nanoscale* **6** 7354–7361
[19] Shcherbakov M R, Neshev D N, Hopkins B, Shorokhov A S, Staude I, Melik-Gaykazyan E V, Decker M, Ezhov A A, Miroshnichenko A E, Brener I, Fedyanin A A and Kivshar Y S 2014 *Nano Lett.* **14** 6488–6492
[20] Makarov S, Kudryashov S I, Mukhin I, Mozharov A, Milichko V, Krasnok A and Belov P A 2015 *Nano Letters*
[21] Shi L, Tuzer T U, Fenollosa R and Meseguer F 2012 *Adv. Mater.* **24** 5934–5938
[22] Abbarchi M, Naffouti M, Vial B, Benkouider A, Lermusiaux L, Favre L, Ronda A, Bidault S, Berbezier I and Bonod N 2014 *ACS nano* **8** 11181–11190
[23] Zywietz U, Evlyukhin A B, Reinhardt C and Chichkov B N 2014 *Nature Comm.* **5** 3402
[24] Palik E D 1998 *Handbook of optical constants of solids* (Academic press)
[25] Köster U 1978 *Physica status solidi (a)* **48** 313–321
[26] Chimmalgi A, Hwang D J and Grigoropoulos C P 2005 *Nano letters* **5** 1924–1930
[27] Liu J 1982 *Optics letters* **7** 196–198
[28] De Wolf I 1996 *Semiconductor Science and Technology* **11** 139
[29] Campbell I and Fauchet P M 1986 *Solid State Communications* **58** 739–741
[30] Zywietz U, Reinhardt C, Evlyukhin A B, Birr T and Chichkov B N 2014 *Applied Physics A* **114** 45–50
[31] Ionin A, Kudryashov S, Seleznev L, Sinitsyn D, Bunkin A, Lednev V and Pershin S 2013 *Journal of Experimental and Theoretical Physics* **116** 347–362
[32] Wu C and Zhigilei L V 2014 *Applied Physics A* **114** 11–32
[33] Papageorgiou D T 1995 *Physics of Fluids (1994-present)* **7** 1529–1544
[34] Bäuerle D W 2013 *Laser processing and chemistry*
[35] Bulgakova N and Bulgakov A 2001 *Applied Physics A* **73** 199–208
[36] Itina T E 2009 *Applied Surface Science* **255** 5107–5111

[37] Amoruso S, Bruzzese R, Spinelli N, Velotta R, Vitiello M, Wang X, Ausanio G, Iannotti V and Lanotte L 2004 *Applied Physics Letters* **84** 4502–4504
[38] Permyakov D, Sinev I, Markovich D, Ginzburg P, Samusev A, Belov P, Valuckas V, Kuznetsov A I, Luk'yanchuk B S, Miroshnichenko A E *et al.* 2015 *Applied Physics Letters* **106** 171110
[39] Evlyukhin A B, Reinhardt C, Evlyukhin E and Chichkov B N 2013 *JOSA B* **30** 2589–2598
[40] Loke V L and Mengüç M P 2010 *JOSA A* **27** 2293–2303
[41] Evlyukhin A B, Novikov S M, Zywietz U, Eriksen R L, Reinhardt C, Bozhevolnyi S I and Chichkov B N 2012 *Nano letters* **12** 3749–3755

Journal of Physics: Conference Series **690** (2016) 012021 doi:10.1088/1742-6596/690/1/012021

Direct femtosecond laser writing of optical nanoresonators

P A Dmitriev[1], S V Makarov[1], V A Milichko[1], I S Mukhin[1,2], A K Samusev[1,3], A E Krasnok[1] and P A Belov[1]

[1] ITMO University, St. Petersburg 197101, Russia
[2] St. Petersburg Academic University, St. Petersburg 194021, Russia
[3] Ioffe Physical-Technical Institute of the Russian Academy of Sciences, St. Petersburg 194021, Russia

E-mail: p.dmitriev@metalab.ifmo.ru

Abstract. A novel method for single-step, lithography-free, and large-scale laser writing of arrays of metallic (Au) and semiconductor (Si) nanoparticles (nanoresonators) has been developed. Changing energy of femtosecond laser pulses and film thickness it is possible to vary diameter of the nanoparticles, while the distance between them can be varied by laser scanning parameters. This method has an advantage over the most previously demonstrated methods in its simplicity and versatility, while the quality of the structures is good enough for many applications.

1. Introduction

With the advent and rapid progress of fabrication technology in the past few decades, there has been growing interest in the field of plasmonics to explore novel phenomena by manipulating light at nanoscale [1, 2]. Examples of plasmonic devices requiring ordered nanoparticles are plasmonic filters [3, 4], metasurfaces [5], and waveguides [6]. The research area dealing with the interaction of molecules or molecular structures with plasmonic nanostructures is another rapidly growing field, due to potential analytical applications such as surface-enhanced Raman spectroscopy (SERS) and localized surface plasmon resonance spectroscopy [7, 8].

High refractive index dielectric nanoparticles with low dissipative losses provide excitation of strong magnetic optical resonances [9–17] and represent a good alternative to high-loss in visible range plasmonic nanoparticles for such applications as electromagnetic field enhancement and sensing [18–21] and enhancement of nonlinear effects [22, 23]. One of the most popular materials for all-dielectric nanophotonics is crystalline silicon (c-Si), because it is a low-cost CMOS compatible semiconductor with a large real part and a low imaginary part of its refractive index, providing the desirable optical response. Indeed, amorphous silicon has a much larger imaginary part of refractive index (up to two orders of magnitude in visible range) [24] in comparison with the crystalline phase.

To date, the most popular and controllable approaches of nanostructures fabrication are based on direct ion-beam milling or multistage e-beam and nanoimprint lithographies. However, lithography-free and single-step methods are more desirable for large-scale manufacturing. Among single-step techniques, chemical synthesis of monodisperse nanoparticles colloid is a promising method for high-throughput fabrication, however, it requires additional technological steps to order the nanoparticles into a functional nanostructure [25–28]. An alternative cost-effective and versatile approach is to exploit a self-assembly process via dewetting of heated thin supported films of a wide range of materials [29–31]. Laser radiation [29, 32, 33], ion-beam [34] or heating on a hotplate [30, 31] have been utilized as heat sources for launching the dewetting process. Another way is to employ cost-effective laser technologies in order to controllably create nanoparticles via precise ablation of thin films, exploiting single nanoparticle transfer to a receiver substrate after single-shot femtosecond (fs) laser irradiation of a thin film [35–37] or direct two-shots writing of plasmonic nanoantennas coupled with plasmonic lenses [38].

Although many different approaches have been proposed to fabricate nanoparticles, the successful implementation of nanodevices still requires cost-effective and versatile method of controllable large-scale fabrication of ordered metal or dielectric nanoparticles.

In this work we present a conceptually new, single-step, lithography-free and cost-efficient method for large-scale fabrication of nanoparticle-based structures. The combination of simplicity of thin film dewetting process, along with precision and high-productivity of the laser technology enables extreme simplification of the process of nanoparticles fabrication, while

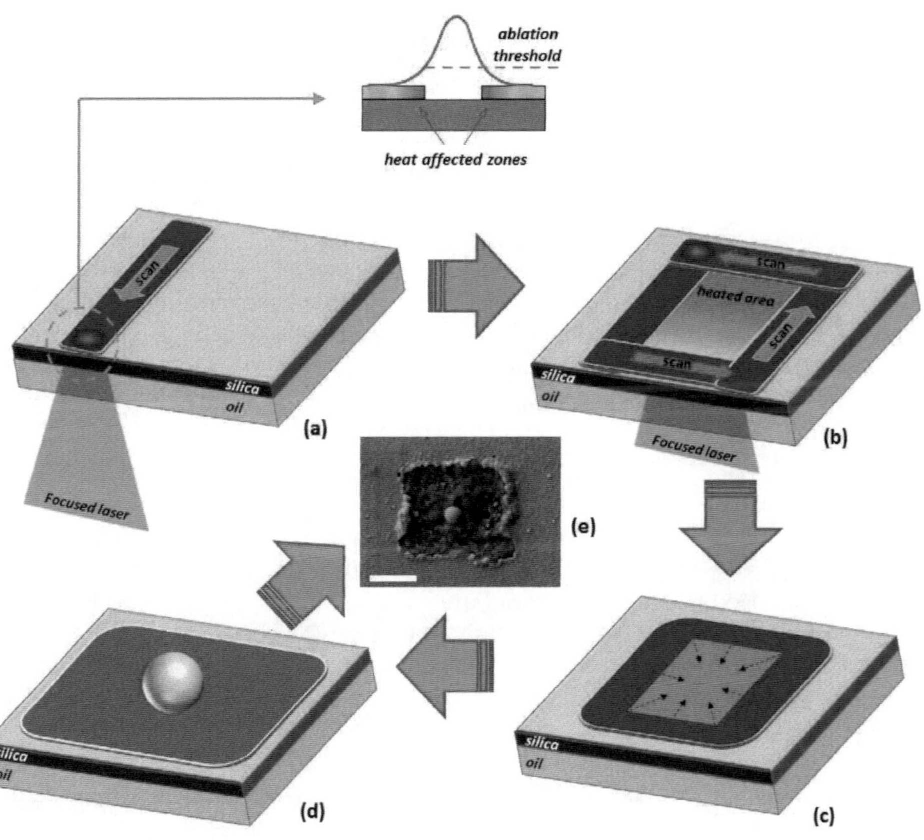

Figure 1. A schematic illustratiion of single nanoparticle formation under femtosecond laser cutting of patch from a thin film on a glass substrate (a-d). SEM images of a fabricated single gold nanoparticle (e) on a fused silica substrate fabricated by fs-laser irradiation of 30nm Au film at a fluence of 40 mJ/cm^2. Scale bars correspond to 1 μm.

the quality of fabricated structure is well-maintained. In other words, we achieve direct single-step "writing" of nanopartices by strongly focused femtosecond laser pulses, where the laser irradiation is used to pattern and heat-up the residual film to the dewetting temperatures. We implement this simple method for writing of 0D, 1D, and 2D structures.

2. Methods

A commercial femtosecond laser system (Femtosecond Oscillator TiF- 100F, Avesta Poject) was used, providing laser pulses at 800 nm central wavelength, with a maximum pulse energy of 5 nJ, and a pulse duration of 100 fs at a repetition rate of 80 MHz. The laser energy was varied and controlled by an acousto-optical modulator (R23080-3-LTD, Gooch and Housego) and a power meter (FielfMax II, Coherent), respectively, while the pulse duration was measured by an autocorrelator (Avesta Poject).

Laser pulses were tightly focused by an oil immersion microscope objective (Carl Zeiss 100×) with a numerical aperture (NA) of 1.4. According to the relation $d \approx 1.22\ \lambda/NA$, the estimated full-width at a half-maximum diameter of the beam focal spot size is $d = 0.68$ μm, which is close to the value measured by the standard method (0.64 μm) based on the dependence of laser-damaged area on incident laser energy [39].

The laser beam was focused on supported Au or Si films with thicknesses in the range of $h = 10 - 100$ nm, thermally evaporated (Bock Edwards Auto 500) on the back side of a 140 μm-thick silica glass without any additional adhesion-improving layers. The samples were then placed on a three-dimensional air-bearing translating stage driven by brushless servomotors (ABL1000, Aerotech), allowing sample translation with various scan speeds up to 300 mm/s. In all experiments, film surface scanning by the laser beam was carried out with velocity of 1 mm/s, providing the number of laser pulses per each point of approximately $N \sim 10^4$.

In order to characterize the crystalline structure of the initial film and the fabricated nanoparticles, we provided Raman scattering measurements. The Raman spectra were recorded by a micro-Raman apparatus (Raman spectrometer HORIBA LabRam HR, AIST-NT SmartSPM system) under excitation by a 632.8 nm HeNe laser through a 100× microscope objective ($NA = 0.9$) and projected onto a thermoelectrically cooled charge-coupled device (Andor DU 420A-OE 325) with a 600 g/mm diffraction grating.

3. Results and discussion

In order to make the method of nanoparticles array fabrication as simple as possible, we replace all lithographical stages of film pattering [40–43] by their direct fs-laser cutting with simultaneous heating of the remain patches up to the dewetting temperatures (figure 1). Although the proposed method is single-step, it is divided into several physical stages.

First, a strongly focused fs-laser beam irradiates a thin film at a fluence slightly higher than the thin film ablation threshold, which increases with growth of the film thickness. Since the scanning conditions correspond to the number of pulses $N \sim 10^4$ and the time delay between the fs-laser pulses in train is 12.5 ns, the temperature in the vicinity of the irradiated area is gradually accumulated pulse-by-pulse and achieves the level of film ablation in a certain region. Importantly, there is almost no ablation debris nearby each groove, being a result of the evaporative mechanism of ablation, rather than the spallative one followed by random dissemination of sub-100 nm nanoparticles in the vicinity of the heated area [44]. In this regime, rather clean and narrow (width of $\Delta \approx 400$ nm $\leq \lambda$) grooves can be written directly on the surface (figure 1).

Second, a single patch can be easily cut (figure 1(b)) with any size larger than the uncertainty of the grooves edges (±50 nm, figure 1(d),(e)). The cut patch is thermally isolated since the thermal conductivities of the silica substrate and air are two and four orders of magnitude smaller than that of gold, respectively. Therefore, the isolated patch can be easier heated up to the temperatures where the film undergoes dewetting process. The dewetting temperature of the films could be much lower than the melting point of bulk material. For instance, a 10 nm Au film on fused silica is dewetted even at 430 K, while 60 nm Au film undergoes dewetting at temperatures lower than 870 K [45]. Moreover, the smaller the patches (with a smaller total volume), the higher the temperatures that can be achieved at a fixed fluence. As a result, the heated cut Au patch transforms into a nanoparticle of the same volume during the dewetting process (figure 1).

Although dewetting is a spontaneous process for homogeneous films [29,32], the reproducible formation of a certain number of nanoparticles has been demonstrated by heating of lithographically cut microscale patches of thin film [31, 34, 40, 41, 46]. Moreover, a single nanoparticle formation from each patch is possible, when the patch width-to-height ratio ($\xi = w/h$, where w is the lateral size of the square patch and h is the film thickness) is smaller than a certain value [40].

3.1. Gold nanoparticle writing

In the case of Au film on SiO_2, this value must fulfill the condition $\xi < 40$, to provide an almost one hundred percent probability of single particle formation with a certain diameter [40]. In our experiments we achieve the most controllable formation of nanoparticles at $\xi \approx 15 - 30$.

This technique enables direct writing of 0D (a single nanoparticle, figure 1(d)), 1D (a line of nanoparticles, figure 1(e)) and 2D (an array of nanoparticles, figure 1(f)) nanoparticle-based structures. The 2D structures are produced by laser beam scanning of the film surface in two orthogonal directions, allowing extremely high production rate of ~ 1 mm^2/min at the available speed of laser scanning of ~ 0.3 m/s. All types of the structures have been successfully fabricated on 20 and 30 nm Au films, while the other thicknesses (10 and 60 nm) provide much less controllable film patterning. Beside the film thickness, another parameter that affects the diameter and quality of the nanoparticles is the width w of the cut patch.

In figure 2, SEM images of the fabricated 2D structures on a 30 nm Au film are shown as a sequence of reduced period between 90°-crossed laser scans at a fluence of 40 mJ/cm^2. The laser fluence is chosen to provide as narrow and reproducible grooves as possible. Evidently, there are four main regimes: (i) square micro-patches formation (figure 2(a)); (ii) unstable transition from square micro-patches to nanoparticles; (iii) formation of ordered and almost identical nanoparticles (figure 2(b),(d)); (iv) formation of disordered different-size nanoparticles (figure 2(c)). The fabricated particles have an almost spherical shape and are partly embedded into the substrate (figure 2(d)), owing to dewetting on a molten substrate.

For the film thickness $h = 30$ nm, we observe reproducible formation of sub-100 nm nanoparticles conglomerates nearby each ordered nanoparticle when periods of scan are smaller than 0.9 μm (see inset of figure 2(c)). Similar behaviour has been observed when some parts of a molten film are not absorbed by larger particles during dewetting process, and interpreted in terms of the multimodal Rayleigh-Plateau instability [43]. Indeed, the conglomerates are formed preferably from the side of last cutting of each patch (see left side in inset of figure 2(c)).

Similar to the scan period (the patch width) dependencies, there are also some ranges of thicknesses where the quality of the 2D nanoparticles array is the most optimal. Non-monotonic behaviour of the size dispersion dependence on the scan period and film thickness can be attributed to two main reasons: (i) the conditions $w/\lambda_m > 1$ and $\xi < 40$ are not fulfilled for very small thicknesses and (ii) the dewetting temperature is much higher for thicker films. Therefore, too thin films can not provide ordered nanoparticles formation, while too thick and large patches can be hardly heated homogeneously during laser cutting to achieve controllable dewetting.

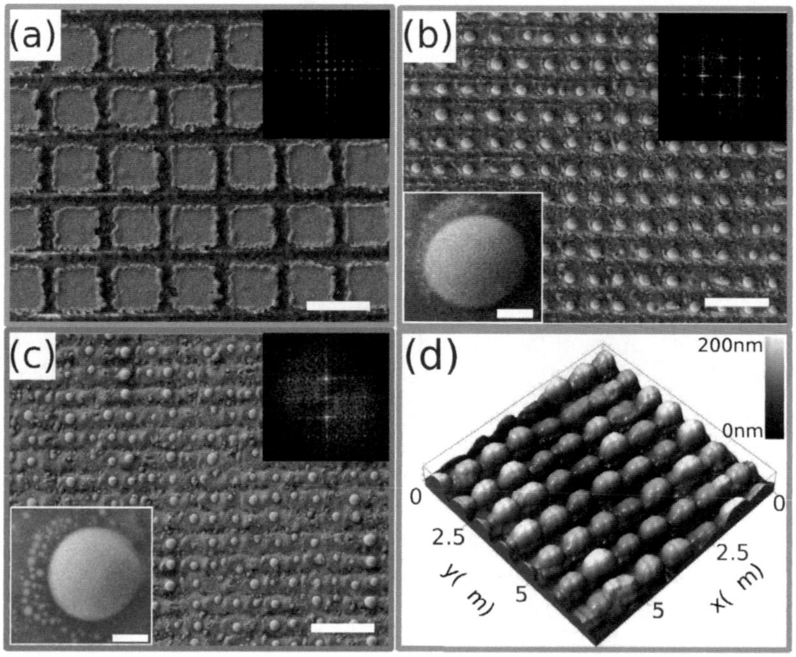

Figure 2. SEM images of the Au nanoparticles arrays fabricated at a fluence of 40 mJ/cm^2 and a periodicity of crossed laser scan of 2.0 (a), 1.0 (b), and 0.8 μm (c). Scale bars are 2.0 μm. The upper-right insets: Fourier spectra of the SEM images. The lower-left insets: enlarged SEM images of typical nanoparticles from the arrays with scale bars of 100 nm. (d) AFM image of the nanoparticles array, corresponding to the picture (b).

3.2. c-Si nanoparticle writing

The novel method of direct laser writing is also applicable for c-Si nanoparticles fabrication from initially amorphous a-Si:H film. Under the optimal conditions of fabrication, the resulting array of nanoparticles has a period of about 0.9 μm, exhibiting bright colors owing to resonant scattering (figure 3(a)) [17].

Although the cutting should produce a rectangular grating, our SEM images show the spherical shape of the nanoparticle. Indeed, the laser fluence $F \approx 100$ mJ/cm^2 provides film heating close to the melting point even in a single shot regime, whereas a 12.5 ns delay between pulses leads to the temperature accumulation and exceeding the ablation threshold. The heat transferring from the ablated area to the surrounding film is accumulated much more efficiently in the cut patches, which are thermally isolated from the rest of the film. These micro-patches are unstable at high temperatures and undergo dewetting to a certain number of similar nanoparticles [47]. In order to provide deeper insight, we model the time dynamics of the cut liquid Si patch with a height of 80 nm and similar widths of 300 nm (Figs. 3B) on a fused silica substrate in the COMSOL software, solving the incompressible Navier-Stokes equations and taking into account the parameters of the used materials. The modeling shows that after ten nanoseconds the patch is transformed into a semisphere with a height of about 130 nm and a width of about 310 nm (Figs. 3B-E), giving qualitative agreement with the experimentally observed shapes.

The corresponding Raman signals from these nanoparticles reveal the crystalline state of the nanoparticles written by the laser, demonstrating a narrow peak at 520 cm^{-1}, with the halfwidth of about 10 cm^{-1} (figure 3), which is larger than the halfwidth of the printed nanoparticles being

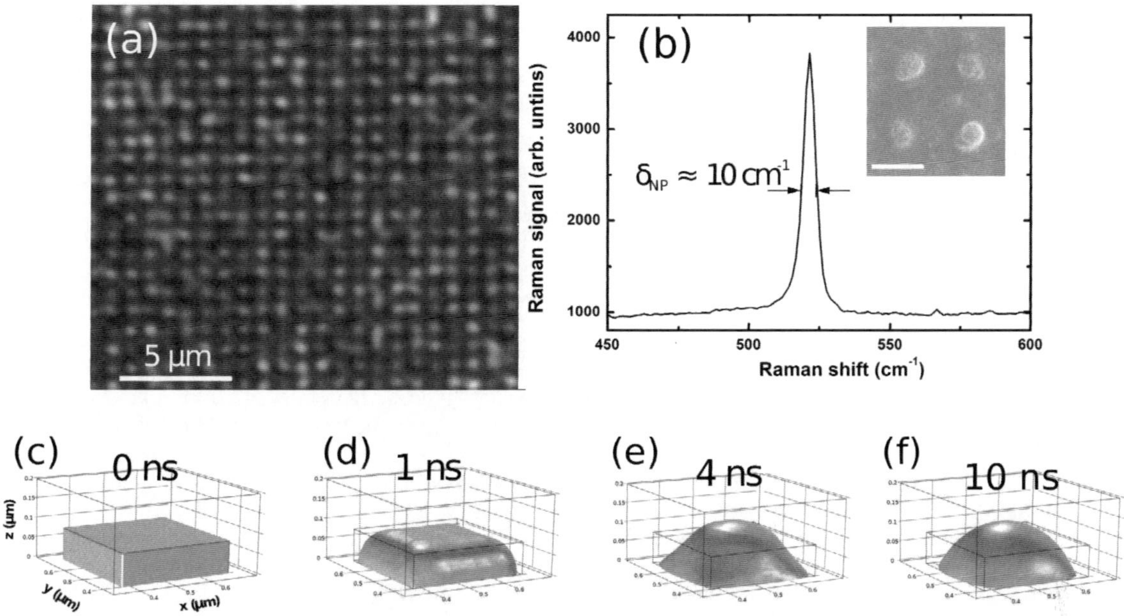

Figure 3. (a) Optical image in the dark-field scheme and array of the silicon nanoparticles fabricated by direct laser writing at a fluence of 100 mJ/cm^2, a repetition rate of 80 MHz and a scan period of 0.9 μm. The inset displays the SEM image of the written nanoparticles covered by a few nm thick gold layer with scale bars of 700 nm. Modelled evolution of the liquid silicon patch with a height of 80 nm and equal sides of 300 nm on the solid fused silica at different time-steps: about 0 ns (b), 1 ns (c), 4 ns (d), and 10 ns (e).

of about $4 - 5$ cm^{-1}. The larger halfwidth corresponds to the smaller mean crystalline size, i.e. less than 10 nm according to the previous studies [48]. The origin of this difference is related to the faster cooling rate for the written nanoparticles on the fused silica substrate as compared with nanoparticles flying in air before deposition in the printing mode, owing to the 20-fold larger value of the glass thermal conductivity in comparison with that for air.

4. Conclusions

In conclusion, the developed novel single-step and lithography-free method of direct nanoparticles fabrication opens new opportunities to fabricate plasmonic and all-dielectric nanoresonator based structures on a large scale. To the best of our knowledge, it is the first demonstration of simultaneous cutting and dewetting of thin film processes that paves the way to extremely simplify the technology of monodisperse and ordered nanoparticles fabrication.

Acknowledgments

This work was financially supported by Russian Science Foundation (Grant 15-19-00172). We are also thankful to Sergey Kudryashov and Arseniy Kuznetsov for useful comments and discussion.

References

[1] Maier S A 2007 *Plasmonics: fundamentals and applications* (New York: Springer)
[2] Luk'yanchuk B, Zheludev N I, Maier S A, Halas N J, Nordlander P, Giessen H and Chong C T 2010 *Nature materials* **9** 707–715
[3] Kravets V, Schedin F and Grigorenko A 2008 *Phys. Rev. Lett.* **101** 087403

[4] Liu Z, Boltasseva A, Pedersen R H, Bakker R, Kildishev A V, Drachev V P and Shalaev V M 2008 *Metamaterials* **2** 45–51
[5] Yu N and Capasso F 2014 *Nature Materials* **13** 139–150
[6] Maier S A, Kik P G, Atwater H A, Meltzer S, Harel E, Koel B E and Requicha A A 2003 *Nature Materials* **2** 229–232
[7] Moskovits M 1985 *Rev. Mod. Phys.* **57** 783
[8] Anker J N, Hall W P, Lyandres O, Shah N C, Zhao J and Van Duyne R P 2008 *Nature Materials* **7** 442–453
[9] Popa B I and Cummer S A 2008 *Physical Review Letters* **100** 207401
[10] Zhao Q, Zhou J, Zhang F and Lippens D 2009 *Mater. Today* **12** 60–69
[11] Evlyukhin A B, Reinhardt C, Seidel A, Lukyanchuk B S and Chichkov B N 2010 *Phys. Rev. B* **82** 045404
[12] García-Etxarri A, Gómez-Medina R, Froufe-Pérez L S, López C, Chantada L, Scheffold F, Aizpurua J, Nieto-Vesperinas M and Sáenz J J 2011 *Optics express* **19** 4815–4826
[13] Krasnok A E, Miroshnichenko A E, Belov P A and Kivshar Y S 2012 *Opt. Express* **20** 20599–20604
[14] Kuznetsov A I, Miroshnichenko A E, Fu Y H, Zhang J and Lukyanchuk B 2012 *Scientific Rep.* **2** 492
[15] Ginn J C and Brener I 2012 *Phys. Rev. Lett.* **108** 097402
[16] Fu Y H, Kuznetsov A I, Miroshnichenko A E, Yu Y F and Lukyanchuk B 2013 *Nature Communication* **4** 1527
[17] Krasnok A, Makarov S, Petrov M, Savelev R, Belov P and Kivshar Y 2015 *SPIE Opt. and Optoelectr.* **9502** 950203–950203
[18] Albella P, Poyli M A, Schmidt M K, Maier S A, Moreno F, Saenz J J and Aizpurua J 2013 *The Journal of Physical Chemistry C* **117** 13573–13584
[19] Zambrana-Puyalto X and Bonod N 2015 *Physical Review B* **91** 195422
[20] Bakker R M, Permyakov D, Yu Y F, Markovich D, Paniagua-Dominguez R, Gonzaga L, Samusev A, Kivshar Y, Lukyanchuk B and Kuznetsov A I 2015 *Nano Letters* **15** 2137–2142
[21] Caldarola M, Albella P, Corts E, Rahmani M, Roschuk T, Grinblat G, Oulton R F, Bragas A V and Maier S A 2015 *Nature communications* **6**
[22] Shcherbakov M R, Neshev D N, Hopkins B, Shorokhov A S, Staude I, Melik-Gaykazyan E V, Decker M, Ezhov A A, Miroshnichenko A E, Brener I, Fedyanin A A and Kivshar Y S 2014 *Nano Letters* **14** 6488–6492
[23] Makarov S, Kudryashov S I, Mukhin I, Mozharov A, Milichko V, Krasnok A and Belov P A 2015 *Nano Letters*
[24] Palik E D 1998 *Handbook of optical constants of solids* (Academic press)
[25] Junno T, Deppert K, Montelius L and Samuelson L 1995 *Appl. Phys. Lett.* **66** 3627–3629
[26] Kraus T, Malaquin L, Schmid H, Riess W, Spencer N D and Wolf H 2007 *Nature Nanotech.* **2** 570–576
[27] Shi L, Harris J T, Fenollosa R, Rodriguez Iand Lu X, Korgel B A and Meseguer F 2013 *Nature Comm.*
[28] Patra P P, Chikkaraddy R, Tripathi R P, Dasgupta A and Kumar G P 2014 *Nature Comm.* **5**
[29] Bischof J, Scherer D, Herminghaus S and Leiderer P 1996 *Phys. Rev. Lett.* **77** 1536
[30] Higgins A M and Jones R A 2000 *Nature* **404** 476–478
[31] Abbarchi M, Naffouti M, Vial B, Benkouider A, Lermusiaux L, Favre L, Ronda A, Bidault S, Berbezier I and Bonod N 2014 *ACS Nano* **8** 11181–11190
[32] Trice J, Thomas D, Favazza C, Sureshkumar R and Kalyanaraman R 2007 *Phys. Rev. B* **75** 235439
[33] Wu Y, Fowlkes J D and Rack P D 2011 *J. Mater. Res.* **26** 277–287
[34] Lian J, Wang L, Sun X, Yu Q and Ewing R C 2006 *Nano Lett.* **6**
[35] Kuznetsov A, Koch J and Chichkov B 2009 *Opt. Express* **17** 18820–18825
[36] Kuznetsov A I, Evlyukhin A B, Gonçalves M R, Reinhardt C, Koroleva A, Arnedillo M L, Kiyan R, Marti O and Chichkov B N 2011 *ACS Nano* **5** 4843–4849
[37] Zywietz U, Evlyukhin A B, Reinhardt C and Chichkov B N 2014 *Nature Comm.* **5**
[38] Ionin A, Kudryashov S, Makarov S, Rudenko A, Kulchin Y N, Vitrik O, Efimov T and Kuchmizhak A 2015 *Opt. Lett.* **40** 224–229
[39] Liu J 1982 *Opt. Lett.* **7** 196–198
[40] Kim D, Giermann A L and Thompson C V 2009 *Appl. Phys. Lett.*
[41] Fowlkes J D, Kondic L, Diez J, Wu Y and Rack P D 2011 *Nano Lett.* **11** 2478–2485
[42] Ye J and Thompson C V 2011 *Adv. Mater.* **23** 1567–1571
[43] Fowlkes J, Roberts N, Wu Y, Diez J, Gonzalez A, Hartnett C, Mahady K, Afkhami S, Kondic L and Rack P 2014 *Nano Lett.* **14** 774–782
[44] Zhigilei L V, Lin Z and Ivanov D S 2009 *J. Phys. Chem. C* **113** 11892–11906
[45] Gadkari P, Warren A, Todi R, Petrova R and Coffey K 2005 *J. Vac. Sci. Tech. A* **23** 1152–1161
[46] Wu Y, Dong N, Fu S, Fowlkes J D, Kondic L, Vincenti M A, de Ceglia D and Rack P D 2014 *ACS Appl. Mater. Interfaces* **6** 5835–5843
[47] Thompson C V 2012 *Annual Review of Materials Research* **42** 399–434

[48] Campbell I and Fauchet P M 1986 *Solid State Communications* **58** 739–741

Misfit stresses in a core-shell nanowire with core in the form of long parallelepiped

S A Krasnitckii[1], A M Smirov[2] and M Yu Gutkin[1-3]

[1]Peter the Great St. Petersburg Polytechnic University, 29 Polytekhnicheskaya str., St. Petersburg, 195251 Russia
[2]ITMO University, 49 Kronverkskii Ave., St. Petersburg, 197101 Russia
[3]Institute of Problems in Mechanical Engineering, Russian Academy of Sciences, 61 Bolshoi Ave., Vasil. Ostrov, St. Petersburg, 199178 Russia

Email: krasnitckii@gmail.com, smirnov.mech@gmail.com, m.y.gutkin@gmail.com

Abstract. We present an analytical solution to the boundary-value problem in the classical theory of elasticity for a core-shell nanowire with a core in the form of a long parallelepiped with a square cross section. The core is placed symmetrically with respect to the cylindrical shell surface. The misfit stresses are found in a concise and transparent form of trigonometric series, which is convenient for practical use in theoretical modeling of misfit relaxation mechanisms. This work was supported by Peter the Great St. Petersburg Polytechnic University and ITMO University.

1. Introduction

Radially inhomogeneous nanowires (NWs) demonstrate excellent electronic and optical properties, which encourage their use in various devices of optoelectronics, nanoscale field effect transistors, storage and data transmission devices, logic devices, sensors, etc. [1-10]. Physical properties of NWs depend on the shape, size, chemical composition, and types of crystalline lattices of NW components as well as on presence of various defects in their structure. In particular, the shape effects can be tightly related with peculiarities in misfit stress distribution over NWs and with the mechanisms of misfit stress relaxation in them. However, in theoretical description of these mechanisms, they commonly use the model of cylindrically symmetric core-shell NWs [11-26] as it is much simpler for analytical modeling. The problem is that this simple approach strictly limits the variety of possible relaxation mechanisms available for theoretical examination. For example, it excludes the glide of straight misfit dislocations along the flat areas of the core-shell interface that is sometimes observed in experiments [27,28]. It is worth noting that flat interface regions often form in radially inhomogeneous NWs [29-32].

It seems that the simplest case of a core with flat faces is the core in the form of a long parallelepiped with a square cross section. To our best knowledge, today there is only one analytical solution which describes the elastically strained state in a core-shell NW with such a core placed symmetrically with respect to the shell surface [33]. The solution was found in the model case of plane strain through the complex potentials method and illustrated by stress maps in Cartesian coordinates. The disadvantage of the work [33] is that the authors did not demonstrate the evidence of the boundary condition fulfillment. Moreover, the case of plane strain is obviously quite far from the case of three-dimensional mismatch of crystalline lattices in real core-shell NWs.

The present work is aimed at the analytical calculation and numerical analysis of the misfit stresses in a core-shell NW with a symmetrically placed core which is characterized by a three-dimensional dilatation eigenstrain (3D misfit strain) with respect to the shell material and has the shape of a long parallelepiped with a square cross section. We show the analytical formulas for the misfit stress components applicable for practical use in theoretical modeling of misfit stress relaxation processes and demonstrate their distribution in the NW cross section, from which the fulfillment of boundary conditions on the shell free surface is evident.

2. Model

Consider a long core-shell NW of radius R, which consists of elastically homogeneous and isotropic shell and a core having the shape of a long parallelepiped (figure 1). The shear modulus G and the Poisson ratio ν are the same for the shell and the core. The core cross section is a square with a side $2a$. The core is centered symmetrically with respect to the shell surface and subjected to the 3D homogeneous dilatation eigenstrain ε^*.

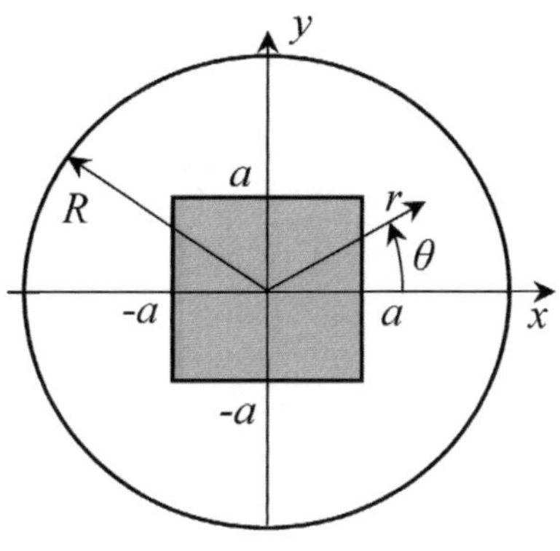

Figure 1. Cross section of a core-shell nanowire of radius R with a core in the form of a long parallelepiped with a square cross section $2a\times 2a$.

The desired stress field σ_{ij} caused by the eigenstrain ε^* can be represented by the sum of a similar stress field σ_{ij}^{∞} created by the core in an infinite medium, and an extra stress field σ_{ij}^* which is needed to satisfy the boundary conditions on the shell free surface:

$$\sigma_{ij} = \sigma_{ij}^{\infty} + \sigma_{ij}^{*}. \tag{1}$$

The boundary conditions are:

$$\sigma_{rr}^{*}(r=R) = -\sigma_{rr}^{\infty}(r=R); \quad \sigma_{r\theta}^{*}(r=R) = -\sigma_{r\theta}^{\infty}(r=R). \tag{2}$$

The non-vanishing stress components σ_{ij}^{∞} are given as follows [34]:

$$\sigma_{xx}^{\infty} = 2C\left[\arctan\frac{x-a}{y+a} + \arctan\frac{x+a}{y-a} - \arctan\frac{x-a}{y-a} - \arctan\frac{x+a}{y+a}\right]; \tag{3a}$$

$$\sigma_{xy}^{\infty} = 2C\ln\frac{[(x-a)^2+(y-a)^2][(x+a)^2+(y+a)^2]}{[(x-a)^2+(y+a)^2][(x+a)^2+(y-a)^2]}; \tag{3b}$$

$$\sigma_{yy}^{\infty} = 2C\left[\arctan\frac{y+a}{x-a} + \arctan\frac{y-a}{x+a} - \arctan\frac{y-a}{x-a} - \arctan\frac{y+a}{x+a}\right]; \tag{3c}$$

$$\sigma_{zz}^{\infty} = \begin{cases} -4\pi C, & |x|\le a,\ |y|\le a \\ 0, & |x|>a,\ |y|>a \end{cases}; \tag{3d}$$

where $C = G\varepsilon^*(1+\nu)/[2\pi(1-\nu)]$.

To satisfy the boundary conditions (2), we should represent Eqs. (3a)-(3c) in the cylindrical coordinate system on the shell surface $r = R$. First, let us consider the following derivative:

$$\frac{d}{d\theta}\left(\arctan\frac{y+a}{x-a}\bigg|_{\substack{x=R\sin\theta \\ y=R\cos\theta}}\right) = \frac{1-\rho\cos(\theta+\pi/4)}{1-2\rho\cos(\theta+\pi/4)+\rho^2}, \tag{4}$$

where $\rho = \sqrt{2}a/R,\ \rho < 1$. The r.h.s. of Eq. (4) can be expanded in a Fourier series [35] as

$$\frac{1-\rho\cos(\theta+\pi/4)}{1-2\rho\cos(\theta+\pi/4)+\rho^2} = \sum_{n=0}^{+\infty}\rho^n\cos(k(\theta+\pi/4)). \tag{5}$$

Integrating series (5) over θ, we find the trigonometric series for the desired function:

$$\arctan\frac{y+a}{x-a}\bigg| = \Theta(\theta) + \sum_{n=1}^{+\infty}\frac{1}{n}\rho^n\sin(k(\theta+\pi/4)), \tag{6}$$

where $\Theta(\theta) = \begin{cases}\theta+\pi, & -\pi < \theta \le -\arccos a/R \\ \theta, & -\arccos a/R < \theta \le \arccos a/R. \\ \theta-\pi, & \arccos a/R < \theta < \pi\end{cases}$

Applying the same procedure to every function figuring in Eqs. (3a)-(3c), after some algebra we find the following compact formulas for the stress fields σ_{ij}^{∞} on the shell surface $r = R$:

$$\sigma_{xy}^{\infty}(r=R) = 4C\sum_{k=1}^{+\infty}\frac{(-1)^k}{2k-1}\rho^{4k-2}\cos[(4k-2)\theta]; \tag{7a}$$

$$\sigma_{xy}^{\infty}(r=R) = 4C\sum_{k=1}^{+\infty}\frac{(-1)^k}{2k-1}\rho^{4k-2}\sin[(4k-2)\theta]; \tag{7b}$$

$$\sigma_{yy}^{\infty}(r=R) = -4C\sum_{k=1}^{+\infty}\frac{(-1)^k}{2k-1}\rho^{4k-2}\cos[(4k-2)\theta]. \tag{7c}$$

We can now rewrite the stress tensor σ_{ij}^{∞} in the cylindrical coordinate system. Using the well-known formulas:

$$\sigma_{rr}^{\infty} = \sigma_{xx}^{\infty}\cos^2\theta + 2\sigma_{xy}^{\infty}\cos\theta\sin\theta + \sigma_{yy}^{\infty}\sin^2\theta; \tag{8a}$$

$$\sigma_{r\theta}^{\infty} = (\sigma_{yy}^{\infty} - \sigma_{xx}^{\infty})\cos\theta\sin\theta + \sigma_{xy}^{\infty}(\cos^2\theta - \sin^2\theta); \tag{8b}$$

$$\sigma_{\theta\theta}^{\infty} = \sigma_{xx}^{\infty}\sin^2\theta - 2\sigma_{xy}^{\infty}\cos\theta\sin\theta + \sigma_{yy}^{\infty}\cos^2\theta, \tag{8c}$$

with Eqs. (7) we come to the final formulas for the stress components σ_{ij}^{∞} in the cylindrical coordinate system:

$$\sigma_{rr}^{\infty}(r=R) = -4C\rho^2\sum_{n=0}^{+\infty}\frac{(-1)^n}{2n+1}(\rho)^{4n}\cos 4n\theta; \tag{9a}$$

$$\sigma_{r\theta}^{\infty}(r=R) = -4C\rho^2\sum_{n=0}^{+\infty}\frac{(-1)^n}{2n+1}(\rho)^{4n}\sin 4n\theta; \tag{9b}$$

$$\sigma_{\theta\theta}^{\infty}(r=R) = 4C\rho^2\sum_{n=0}^{+\infty}\frac{(-1)^n}{2n+1}(\rho)^{4n}\cos 4n\theta. \tag{9c}$$

Let us now introduce the complex variable $\xi = x + \mathrm{i}y = re^{\mathrm{i}\theta}$, $\mathrm{i} = \sqrt{-1}$. The components of the extra stress tensor σ_{ij}^{*} in the cylindrical coordinate system can be calculated through the complex potentials [36] as:

$$\sigma^*_{rr} + \sigma^*_{\theta\theta} = 2[F'(\xi) + \overline{F'(\xi)}]; \tag{10a}$$

$$\sigma^*_{\theta\theta} - \sigma^*_{rr} + 2\mathrm{i}\sigma^*_{r\theta} = 2[\overline{\xi}F'(\xi) + \chi''(\xi)]\mathrm{e}^{2\mathrm{i}\theta}, \tag{10b}$$

where $F'(\xi)$ and $\chi''(\xi)$ are the complex potentials that are unknown analytical functions of the complex variable ξ, and $\overline{F'(\xi)}$ and $\overline{\chi''(\xi)}$ are the function conjugate with $F'(\xi)$ and $\chi''(\xi)$, respectively.

Subtracting Eq. (10b) from Eq. (10a), we obtain the following formula which is convenient for the fulfillment of the boundary conditions:

$$\sigma^*_{rr} - \mathrm{i}\sigma^*_{r\theta} = F'(\xi) + \overline{F'}(\overline{\xi}) - [\overline{\xi}F'(\xi) + \chi''(\xi)]e^{2\mathrm{i}\theta}. \tag{10c}$$

Taking into account that the stress must be finite at $r \to 0$, we will search the functions $F'(\xi)$ and $\chi''(\xi)$ in the form of power series:

$$F'(\xi) = \sum_{n=1}^{+\infty} A_n \xi^n, \qquad \chi''(\xi) = \sum_{n=1}^{+\infty} B_n \xi^n, \tag{11}$$

where A_n and B_n are complex constants in the general case.

Substituting series (9a,b) and (11) into Eq. (10c) and taking into account that $r = R$ on the outer surface, after some algebra we find that:

$$-4C\rho^2 \sum_{n=1}^{+\infty} \frac{(-1)^n}{2n+1} (\rho)^{4n} \mathrm{e}^{-4in\theta} = A_0 + \overline{A_0} + \sum_{n=1}^{+\infty} \overline{A_{4n}} R^{4n} \mathrm{e}^{-4in\theta} -$$
$$- \sum_{n=1}^{+\infty} (4n-1) A_{4n} R^{4n} \mathrm{e}^{4ni\theta} - \frac{1}{R^2} \sum_{n=1}^{+\infty} B_{4n-2} R^{4n} \mathrm{e}^{4in\theta}. \tag{12}$$

Comparing the coefficients at $\mathrm{e}^{4in\theta}$ in both parts of this equation, we have:

$$A_0 = \overline{A_0} = 2C\rho^2; \quad A_{4n} = \overline{A_{4n}} = 4C\rho^2 \frac{(-1)^n}{2n+1}\left(\frac{\rho}{R}\right)^{4n}; \quad B_{4n-2} = -4C\rho^2(-1)^n \frac{4n-1}{2n+1} \frac{\rho^{4n}}{R^{4n-2}}. \tag{13}$$

Substitution of Eqs. (11) and (13) to Eqs. (10a,b) gives the final components of the extra stress tensor:

$$\sigma^*_{rr} = 4C\rho^2 \left[1 - \sum_{n=1}^{+\infty} \frac{(-1)^n}{2n+1} \left(2(2n-1) - (4n-1)\left(\frac{R}{r}\right)^2\right) \left(\frac{\rho r}{R}\right)^{4n} \cos 4n\theta\right]; \tag{14a}$$

$$\sigma^*_{r\theta} = 4C\rho^2 \sum_{n=1}^{+\infty} \frac{(-1)^n}{2n+1} \left(4n - (4n-1)\left(\frac{R}{r}\right)^2\right) \left(\frac{\rho r}{R}\right)^{4n} \sin 4n\theta; \tag{14b}$$

$$\sigma^*_{\theta\theta} = 4C\rho^2 \left[1 + \sum_{n=1}^{+\infty} \frac{(-1)^n}{2n+1} \left(2(2n-1) - (4n-1)\left(\frac{R}{r}\right)^2\right) \left(\frac{\rho r}{R}\right)^{4n} \cos 4n\theta\right]. \tag{14c}$$

Thus, the problem under consideration (figure 1) is solved.

3. Results

To illustrate our solution, we show in figures 2(a) and 2(b) the distribution of stress components $\sigma^\infty_{rr}, \sigma_{rr}$ and $\sigma^\infty_{r\theta}, \sigma_{r\theta}$ over the shell surface $r = R$. It is evident that the σ_{rr} and $\sigma_{r\theta}$ components satisfy the boundary conditions (2) with account for Eq. (1). Figures 2(c) and 2(d) demonstrate the maps of the stress components σ^∞_{xy} and σ_{xy}, respectively, in the cross section of the NW. As is seen, the shell free surface noticeably screens the misfit shear stress.

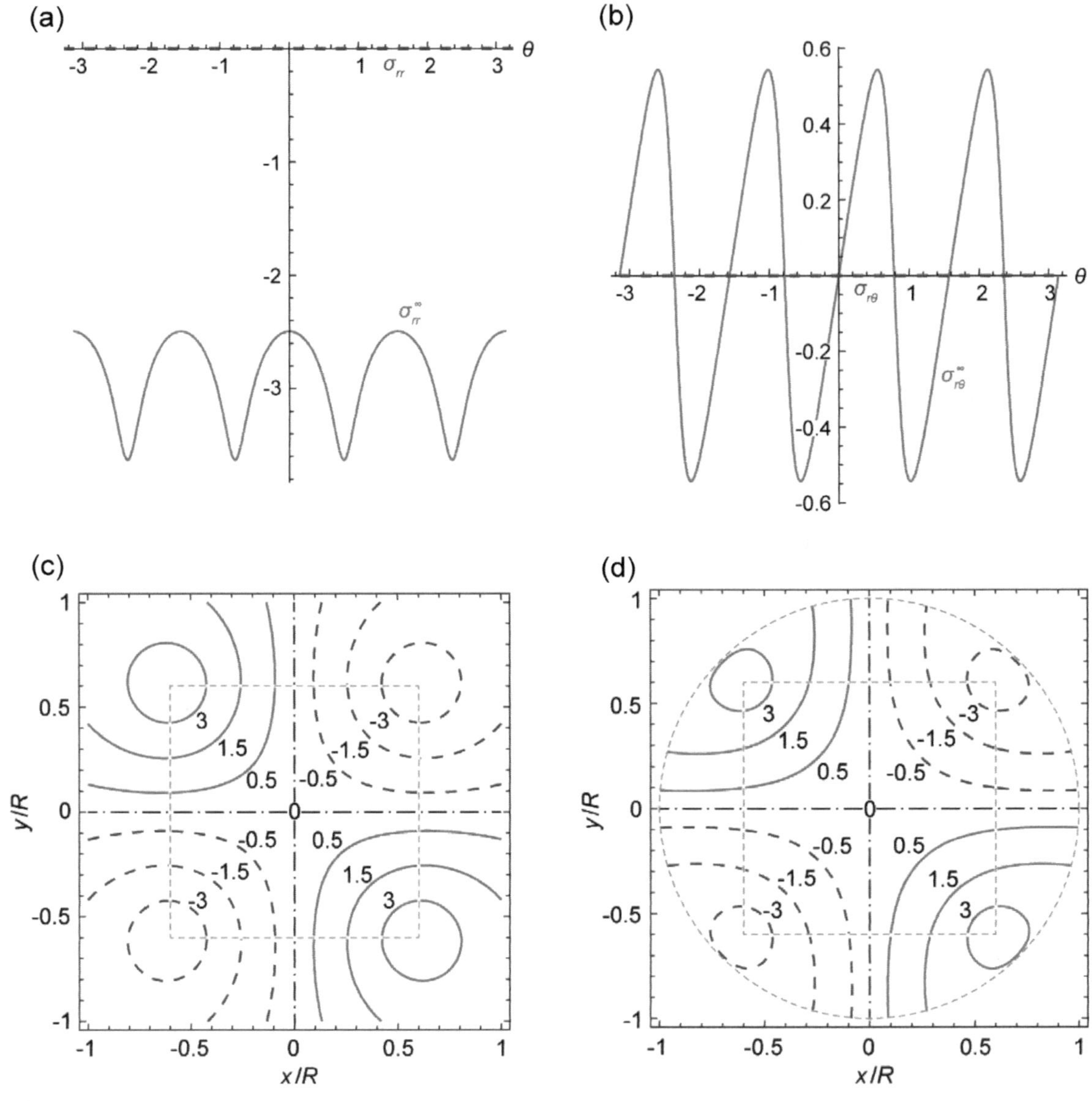

Figure 2. Distribution of the misfit stress components (a,b) $\sigma_{rr}^{\infty}, \sigma_{rr}$ and $\sigma_{r\theta}^{\infty}, \sigma_{r\theta}$ over the shell surface $(r = R, -\pi < \theta < \pi)$, and (c,d) σ_{xy}^{∞} and σ_{xy} in the cross section of a core-shell nanowire in the case of $a = 0.6R$. The stress values are given in units of C.

4. Conclusions

We can conclude that our analytical solution (1) satisfies the boundary conditions of the problem (2), gives an opportunity to analyze the misfit stress distribution in detail, and (3) is represented in a concise and transparent form which is applicable for theoretical modeling of misfit stress relaxation processes.

References

[1] Lauhon L J, Gudiksen M S and Lieber C M 2004 *Phil. Trans. R. Soc. Lond.* A **362** 1247
[2] Mieszawska A J, Jalilian R, Sumanasekera G U and Zamborini F P 2007 *Small* **3** 722

[3] Agarwal R 2008, *Small* **4** 1872
[4] Comini E, Baratto C, Faglia G, Ferroni M, Vomiero A and Sberveglieri G 2009 *Prog. Mater. Sci.* **54** 1
[5] Dubrovskii V G, Cirlin G E and Ustinov V M 2009 *Semiconductors* **43** 1539
[6] Barth S, Hernandez-Ramirez F, Holmes J D and Romano-Rodriguez A 2010 *Prog. Mater. Sci.* **55** 563
[7] Gao Q, Tan H H, Jackson H E, Smith L M, Yarrison-Rice J M, Zou J and Jagadish C 2011 *Semicond. Sci. Technol.* **26** 014035
[8] Garnett E C, Brongersma M L, Cui Y and McGehee M D 2011 *Annu. Rev. Mater. Res.* **41** 269
[9] Kenry C T 2013 *Prog. Mater. Sci.* **58** 705
[10] Amato M, Palummo M, Rurali R and Ossicini S 2014 *Chem. Rev.* **114** 1371
[11] Gutkin M Yu, Ovid'ko I A and Sheinerman A G 2000 *J. Phys. Condens. Matter* **12** 5391
[12] Sheinerman A G and Gutkin M Yu 2001 *Phys. Stat. Solidi* A **184** 485
[13] Ovid'ko I A and Sheinerman A G 2004 *Phil. Mag.* **84** 2103
[14] Liang Y, Nix W D, Griffin P B and Plummer J D 2005 *J. Appl. Phys*. **97** 043519
[15] Raychaudhuri S and Yu E T 2006 *J. Appl. Phys*. **99** 114308
[16] Aifantis K E, Kolesnikova A L and Romanov A E 2007 *Phil. Mag.* **87** 4731
[17] Colin J 2010 *Phys. Rev.* B **82** 054118
[18] Wang X, Pan E and Chung P W 2010 *Int. J. Plasticity* **26** 1415
[19] Chu H J, Wang J, Zhou C Z and Beyerlein I J 2011 *Acta Mater.* **59** 7114
[20] Gutkin M Yu, Kuzmin K V and Sheinerman A G 2011 *Phys. Stat. Solidi* B **248** 1651
[21] Haapamaki C M, Baugh J and LaPierre R R 2012 *J. Appl. Phys.* **112** 124305
[22] Chu H J, Zhou C Z, Wang J and Beyerlein I J 2013 *J. Mech. Phys. Solids* **61** 2147
[23] Dayeh S A, Tang W, Boioli F, Kavanagh K L, Zheng H, Wang J, Mack N H, Swadener G, Huang J Y, Miglio L, Tu K N and Picraux S T 2013 *Nano Lett.* **13** 1869
[24] Salehzadeh O, Kavanagh K L and Watkins S P 2013 *J. Appl. Phys.* **114** 054301
[25] Enzevaee C, Gutkin M Yu and Shodja H M 2014, *Phil. Mag.* **94** 492
[26] Gutkin M Yu and Smirnov A M 2015 *Acta Mater*. **88** 91
[27] Ding Y, Fan F, Tian Z and Wang Z L 2010, *J. Am. Chem. Soc.* **132** 12480
[28] Bhattarai N, Casillas G, Ponce A and Jose-Yacaman M 2013 *Surf. Sci.* **609** 161
[29] Wang Z L, Dai Z R, Gao R P, Bai Z G and Gole J L 2000 *Appl. Phys. Lett.* **77** 3349
[30] Hu J, Bando Y, Liu Z, Sekiguchi T, Golberg D and Zhan J 2003 *J. Am. Chem. Soc.* **125** 11306
[31] Teo B K, Li C P, Sun X H, Wong N B and Lee S T 2003 *Inorg. Chem*. **42** 6723
[32] Qian F, Li Y, Gradečak S, Park H G, Dong Y, Ding Y, Wang Z L and Lieber C M 2008 *Nature Materials* **7** 701
[33] Zou W-N, He Q-C and Zheng Q-S 2012 *Int. J. Solids Struct.* **49** 1627
[34] Malyshev K L, Gutkin M Yu, Romanov A E, Sitnikova A A and Sorokin L M 1988 *Sov. Phys.-Solid State* **30** 1176
[35] Prudnikov A P Brychkov Yu A Marichev O I 1986 Integrals and Series **1** Elementary Functions *Gordon & Breach Sci. Publ* (New York)
[36] Muskhelishvili N I 1977 Some Basic Problems of the Mathematical Theory of Elasticity *Noordhoff Int. Publ.*(Leyden, The Netherlands)

Lasing of InP/AlInAs quantum dots in AlInAs microdisk cavity

D V Lebedev[1], A M Mintairov[1, 2], M M Kulagina[1], S I Troshkov[1], J Kapaldo[2], J L Merz[2], S Rouvimov[2], G Juska[3], A Gocalinska[3], S T Moroni[3] and E Pelucchi[3]

[1] Ioffe Institute, 26 Polytechnicheskaya st, 194021, St. Petersburg, Russia
[2] University of Notre Dame, EE Department, Notre Dame, IN 46556, USA
[3] Tyndall National Institute, University College Cork, "Lee Maltings", Dyke Parade, Cork, Ireland

E-mail: Lebedev.dmitri@mail.ioffe.ru

Abstract. AlInAs microdisk cavities having quality factor $Q \sim 15\,000$ were fabricated from lattice-matched InP/AlInAs (interface/aggregation) quantum dot (QD) structures using wet chemical etching. The QD emission coupled to whispering gallery modes was observed at spectral range 920 - 1000 nm at temperatures of 10 - 160 K. The laser generation with threshold power density of 50 W/cm^2 at $T = 10$ K was observed under optical pumping. It was found that the spontaneous emission coupling factor β equals 0.23 for these microdisks. The low temperature lasing operation and the small coupling factors observed suggest the existence of small QDs formed at the InP/InAlAs interface as the main active elements.

1. Introduction

Whispering gallery mode (WGM) semiconductor microdisk (MD) cavities with embedded quantum dots (QDs) represent modern nano-photonic platform for the realization of low-threshold lasers [1], single photon sources [2] and experiments on cavity quantum electrodynamics [3] and quantum optics [4]. For these cavities which have sizes of few micrometers and quality factors Q from 10^3 - 10^5, lasing under optical pumping from cryogenic to room temperatures was demonstrated in the spectral range 0.32 - 1.3 μm using InGaAs/GaAs [5], InP/GaInP [1], CdSe/ZnSe [6] and GaN/AlN [7] QD systems having a type-I band alignment. Recently MD lasers with GaAs/GaSb [8] QDs having a band alignment type-II were also reported. The lowest, subnanowatt, threshold was achieved at $T = 10$ K for InP/GaInP QDs having a large lateral size (~ 100 nm) providing strong coupling of the WGM with the QD emission, and low density (~ 10^9 cm^{-2}) suggesting single dot lasing. Accounting for the low Q values used (~ 5000) one can suggest that further increase in QD size and quality factor will lead to lower thresholds which is very important for understanding the fundamental limitations on efficiency for the generation of coherent light in the solid state and for the creation of the nanophotonic devices having low energy consumption.

Here we investigated the novel type-II InP/AlInAs QDs which can show a large size (~ 250 nm) and a reduced density (~ 10^8 cm^{-2} showing nevertheless a broad phenomenology with growth conditions as discussed in Ref. [9]), as an active layer of MD lasers having a diameter in the 2.5 - 4 μm range. The QD emission coupled to whispering gallery modes was indeed observed at the spectral range

920 - 1000 nm at temperatures of 10 - 160 K. The laser generation with threshold power density of ~50 W/cm^2 at T = 10 K was observed under optical pumping. For these microdisks it was found that the spontaneous emission coupling factor β equals 0.23. Indeed, the low temperature lasing operation and the small coupling factors observed suggest the existence of small QDs formed at InP/InAlAs interface as the main active elements instead of their larger size relatives reported in [9].

2. Experimental

The structures were grown on InP substrate using metal-organic vapour phase epitaxy. The structures consist of a 100 nm thick InP buffer layer, 250 nm thick AlInAs waveguide with InP QD embedded at the waveguide centre. The aggregation dots were formed following a deposition of 0.75 nm InP [9]. We expect a type-II band alignment of these QDs with conduction/valence band off sets 300/220 meV [10]. Thus, an electron is confined in a QD with activation energy of ~ 100 meV. Accounting for the type-II band structure we used modulation doping of QDs which was achieved by the insertion of a 20 nm n-type AlInAs ($n = 5\cdot 10^{17}$ cm^{-3}) layer at a distance of 20 nm from the QDs. From this structure MDs having diameters of 2.5 - 4 μm were fabricated by optical photolithography and wet chemical etching using HCl, H_3PO_4 and CH_3COOH.

Photoluminescence (PL) experiments were conducted under CW (532 nm) and pulsed (635 nm, frequency 10 – 80 MHz) excitation in a standard microphotoluminescence set-up with photon correlation capability, as described elsewhere [11]. The system enabled exciting and probing regions as small as a few μm^2 using an objective with 50 times magnification and numerical aperture of 0.5. Spectra were registered by Horyba Jobin Yvon THR1000 monochromator coupled with a Symphony 2 CCD camera. Experiments were conducted at T = 10 – 300 K. The optical power on the sample was regulated by a set of neutral filters with different transmissions.

3. Results and discussion

Figure 1 shows the μPL spectrum of the MD having the diameter of 3.2 μm at T = 77 K under CW excitation. This spectrum contains a broad band corresponding to the emission of QDs located near the centre of the MD which are weakly coupled to WGM, and sharp emission lines corresponding to WGMs generated by QDs located near the circumference. The intensities of the WGMs show strong temperature quenching (figure 2) which may be approximated by an exponential decay: $I(T) \sim I_0/(1+A\exp(-E_a/kT))$, where k is the Boltzman constant and E_a = 40 meV. The quenching is normally caused by thermal escape of holes.

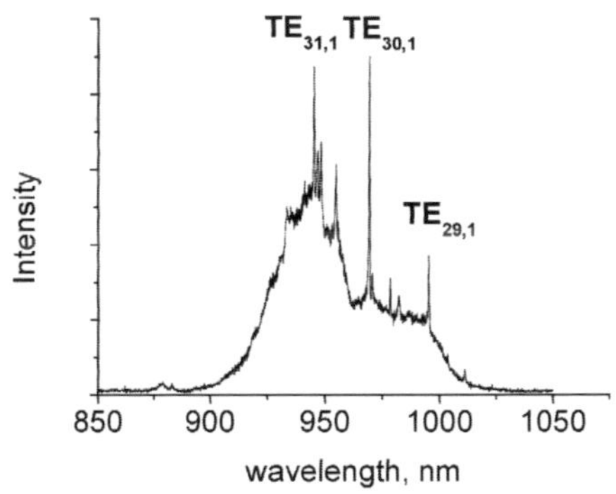

Figure 1. μ-PL spectra of 3.2 μm diameter InP/AlInAs MD at T=77 K at CW 532 nm excitation.

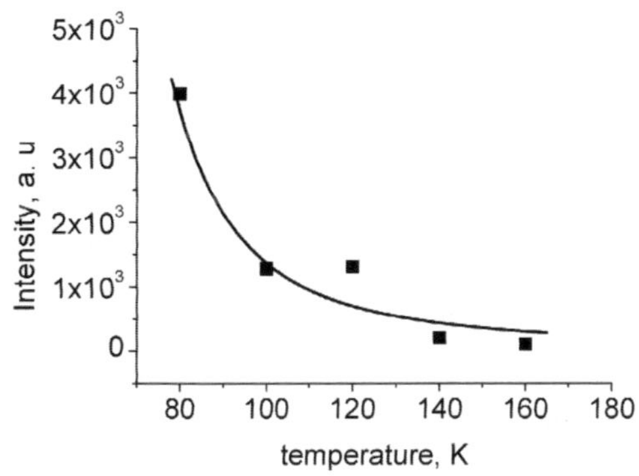

Figure 2. Intensity of WGM versus temperature.

We observed a lasing behaviour of these MDs at temperatures from 10 - 90 K. The lasing spectra at $T = 10$ K for three MDs denoted 1, 2 and 3 having diameter of 2.5 μm are shown in figure 3. Single mode lasing at $\lambda = 952$ and $\lambda = 947$ nm was observed for MD 1 and 2, respectively, and two mode lasing was observed for MD 3 at $\lambda = 926$ nm and 936 nm. A width of the lasing mode $\Delta\lambda$ of 0.06 nm was observed which corresponds to a quality factor $Q = \Delta\lambda/\lambda$ of ~15000. From Input-Output (I/O) power dependences shown in figure 4 the lasing threshold was estimated to be 50 W/cm^2.

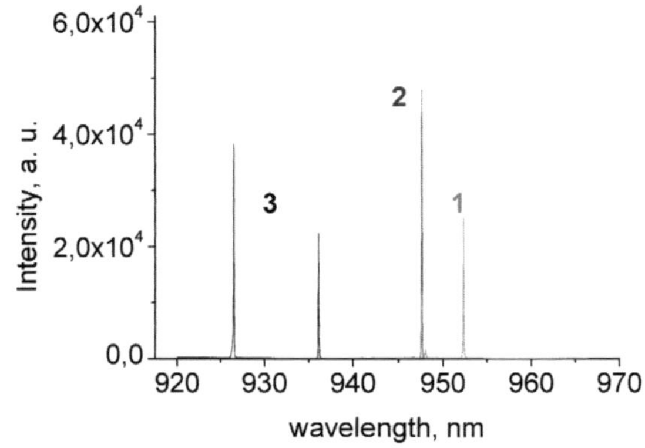

Figure 3. Lasing spectra of three InP/AlInAs QD MDs having diameter of 2.5 μm at 10 K and pump power of 100 μW (300 W/cm^2).

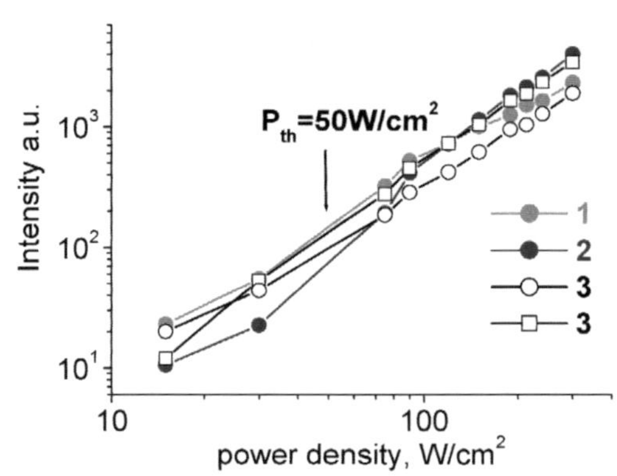

Figure 4. Input-Output power dependence of InP/AlInAs QD MDs having diameter of 2.5 μm at 10 K.

These dependences do not show a strong "knee" at the laser threshold region which is typical for micro-lasers. An absence of strong knee indicates a large spontaneous emission coupling factor β. To investigate Input-Output behavior in the threshold region we fitted the experimental data with the expression introduced by Bjork and Yamamoto for microcavities which relates the photon number p to the excitation intensity I [12]:

$$I(p) = A\left[\frac{p}{1+p}(1+\xi)(1+\beta p) - \xi\beta p\right]$$

Here, A is the scale factor $A = h\omega\gamma/\delta\beta$, where ω is the frequency of the mode, γ is the cavity decay rate, and δ is the photon conversion efficiency. For an optical excitation I corresponds to the pumping power and p is proportional to the output intensity. The dimensionless parameter $\xi = N_0\beta V/\gamma\tau_{sp}$, where N_0 is the transparency carrier concentration of the gain material, V is the volume of the active material and τ_{sp} is the spontaneous emission lifetime, may be considered as a number of photons in the cavity at transparency condition.

Fitting the data under variation of A and ξ yields $\beta = 0.23$ for MD 1 (see figure 5). For comparison, the theoretical curves with $\beta = 1$ and $\beta = 0.1$ are also introduced. It was discussed that the power threshold is defined when the mean number of photons in the laser mode is equal to unity. From the rate equation analysis [13] it follows that the threshold power corresponds to the "knee" position.

We carried out time resolved experiments which showed that the emission decays of WGMs τ_{res} are 1.4 times shorter than that for the background τ_0 (figure 6). Decay times were extracted from monoexponential fit of the PL decay. It is known that spontaneous emission coupling ratio may be estimated from [14]:

$$\beta \approx 1 - \frac{\tau_{cav}}{\tau_{free}}$$

where τ_{free} is the lifetime for the emission in free space, τ_{cav} is the lifetime for the emission in cavity mode. Assuming that τ_{cav}/τ_{free} is close to τ_0/τ_{res}, the spontaneous emission coupling factor may be estimated. It was estimated that $\beta \approx 0.29$ which is in good agreement with the β value obtained from the fitted I-O curves.

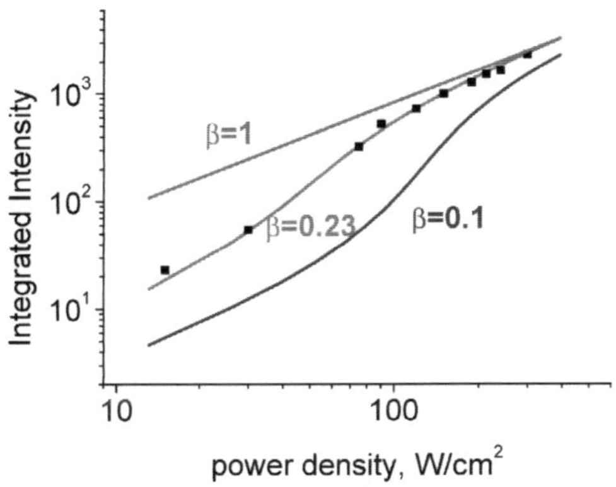

Figure 5. Input-Output power dependence for MD 1 obtained at 15K. Theoretical curves with different β values are included.

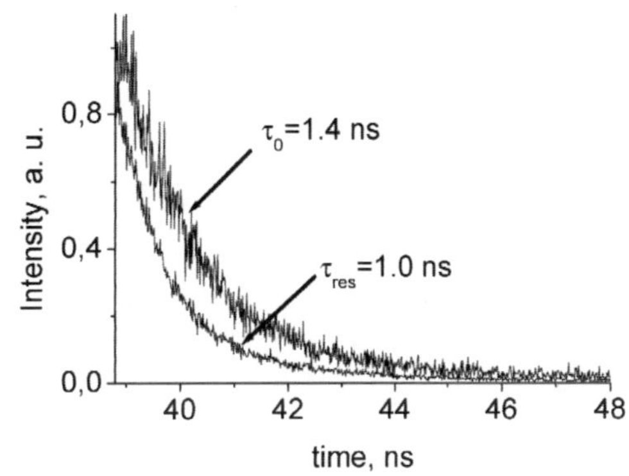

Figure 6. Emission decays of WGM (τ_{res}) and background (τ_0).

The β factor obtained in this work (0.23) is similar to that (0.195) obtained for InP/GaInP QD MDs prepared by wet etching [15] but much lower than that (0.8) for InP/GaInP QD MDs prepared by ICP-RIE etching [16]. Since wet etch InP/GaInP QD have lateral size ~20 nm we suggest that in our structures the emission occurs through "small" QDs formed at InP/InAlAs interface [17]. This could explain the low-temperature operation of the lasers and the relatively low thresholds.

4. Conclusion

Here we used the novel type-II InP/AlInAs nanostructure [9] as an active layer of MD lasers having the diameter of 2.5 - 4 μm. The QD emission coupled to whispering gallery modes was observed at spectral range 920 - 1000 nm at temperatures of 10 - 160 K. The laser generation with threshold power density of 50 W/cm^2 at $T = 10$ K was observed under optical pumping. For these microdisks it was found that the spontaneous emission coupling factor β equals 0.23. The low temperature lasing operation and the small coupling factors observed suggest the existence of small QDs formed at InP/InAlAs interface.

Acknowledgements
This work has been supported by Ministry of the Education and Science of the Russian Federation (contract No.14.Z50.31.0021, 7th April 2014) and by the Irish Higher Education Authority Program for Research in Third Level Institutions via the INSPIRE programs and by SFI under the IPIC award 12/RC/2276 and grant 10/IN.1/I3000.

References

[1] Chu Y, Mintairov A M, He Y, Merz J L, Kalugnyy N A, Lantratov V M and Mintairov S A 2011 *Phys. Stat. Sol.* C **8** 325
[2] Michler P, Kiraz A, Becher C, Schoenfeld W V, Petroff P M, Zhang L, Hu E and Imamoglu A 2000 *Science* **290** 2282
[3] Peter E, Senellart P, Martrou D, Lemaître A, Hours J, Gérard J M and Bloch J 2005 *Phys. Rev. Lett.* **95** 067401
[4] Berger C, Huttner U, Mootz M, Kira M, Koch S W, Tempel J S, Aßmann M, Bayer M, Mintairov A M and Merz J L 2014 *Phys. Rev. Lett.* **113** 093902
[5] Srinivasan K, Borselli M, Painter O, Stintz A and Krishna S 2006 *Opt. Soc. Am.* **14** 1094
[6] Renner J, Worschech L, Forchel A, Mahapatra S and Brunner K 2006 *Appl. Phys. Lett.* **89** 231104
[7] Bürger M, Callsen G, Kure T, Hoffmann A, Pawlis A, Reuter D and As D J 2013 *Appl. Phys. Lett.* **103** 021107
[8] Hsu K S, Chiu T T, Lin W, Chen K L, Shih M H, Lin S and Chang Y 2011 *Appl. Phys. Lett.* **98** 051105
[9] Gocalinska A, Manganaro M, Juska G, Dimastrodonato V, Thomas K, Joyce B, Zhang J, Vvedensky D and Pelucchi E 2014 *Appl. Phys. Lett.* **104** 141606
[10] Pocas L C, Duarte J L, Dias I F L, Laureto E, Lourenco S A, Toginho Filho D O, Meneses E A, Mazzaro I and Harmand J C 2002 *J. Appl. Phys.* **91** 8999
[11] Juska G, Murray E, Dimastrodonato V, Chung T H, Moroni S T, Gocalinska A and Pelucchi E 2015 *J. Appl. Phys.* **117** 134302
[12] Björk G, Karlsson A and Yamamoto Y 1994 *Phys. Rev* A **50** 1675
[13] Björk G and Yamamoto Y 1991 *IEEE J. Of quantum electronics* **27** 2386
[14] Solomon G, Pelton M and Yamamoto Y 2001 *Phys. Rev. Lett.* **86** 3903
[15] Witzany M, Rossbach R, Schulz W M, Jetter M, Michler P, Liu T L, Hu E, Wiersig J and Jahnke F2011 *Phys. Rev.* B **83** 205305
[16] Chu Y, Mintairov A M, He Y, Merz J L, Kalyuzhnyy N A, Lantratov V M and Mintairov S A 2009 *Phys. Lett.* A **373** 1185
[17] Duez V, Vanbe´sien O, Lippens D, Vignaud D, Wallart X and Mollot F 1999, *J. Appl. Phys.* **85** 2022

Thermoelectric ionization of $D^{(-)}$-centers in a semiconductor quantum wire with a parabolic confinement potential in a strong longitudinal electric field

A B Grunin and I A Kirillov
Penza State University, 40 Krasnaya str., Penza, 440026, Russia

E-mail: kirillov.ia1990@yandex.ru

Abstract. Conductivity of a semiconductor quantum wire with a parabolic confinement potential was studied within the framework of the zero-range potential model, containing donor impurities, in the longitudinal with respect to its axis electric field at low temperatures. A dispersion equation determining the dependence of the binding energy of the impurity on the electric field strength was obtained. It has been shown that the bound state of an electron on an impurity is destroyed by a strong electric field, i.e. the thermoelectric phenomenon of ionization occurs. It was found that under conditions of low temperatures the dependence of the current density in a quantum wire in the longitudinal electric field has a superlinear character.

1. Introduction

At the present time the range of investigated nanostructures is being significantly increased: from quantum wells, quantum dots, quantum cylinders, quantum wires, quantum channels, nanospheres, fullerenes to carbon based nanotubes, graphene, thready nanocrystals [1]. The expansion and complication of subjects of theoretical and experimental works in this area is associated with the desire to understand the depth of quantum processes in nanostructures, as well as with the ability to create new semiconductor optoelectronic and nanoelectronic devices.

The purpose of this work is to theoretically study the phenomenon of the ionization of donor impurity centers ($D^{(-)}$-centers) in a quantum wire (QWire) with a parabolic confinement potential and conductivity of QWire in a strong longitudinal (along its axis) electric field.

There are a lot of studies in which scientists investigate transport and optical properties of semiconductor nanostructures in low electric fields [1,2,3], but the influence of impurities and strong electric fields has been researched to a much lesser extent. There are many simultaneous effects in bulk semiconductors in a strong electric field, such as the Frenkel effect, the Zener effect, impact ionization, etc. [4]. However, these effects have not been sufficiently studied in nanostructures. This fact explains the relevance of the present study.

2. Models

In this paper, we consider cylindrical QWire with a length L_z and a diameter L. The electric field strength $\vec{F}$ is directed along the axis of QWire. To describe one-electron states in QWire we use the confinement potential in the form of a spherical oscillator well:

$$V\left(\vec{r}\right)=\frac{m^{*}\omega_{x}^{2}x^{2}}{2}+\frac{m^{*}\omega_{y}^{2}y^{2}}{2}+\frac{m^{*}\omega_{z}^{2}z^{2}}{2}, \tag{1}$$

where $\vec{r}=(x,y,z)$ are Cartesian coordinates ($-L/2\leq x\leq L/2$; $-L/2\leq y\leq L/2$; $-L_z/2\leq z\leq L_z/2$); m^* is the electron effective mass in QWire; ω_x, ω_y, ω_z are the binding potential characteristic frequencies of QWire along the axis *OX*, *OY* and *OZ*, respectively. Moreover, $\omega_x=\omega_y=\omega>>\omega_z$.

For a theoretical description of single-electron states in a quantum wire, various models of the confinement potential are used in the literature: an infinite waveguide with a constant cross section [3, 5, 6], the potential of the saddle point for the restrictions in quantum channels [3, 7, 8], as well as a symmetrical quadratic potential [3, 9, 10]. In a closer approximation of the confining potential shape it is necessary to find self-consistent solutions of the Poisson equation and the Schrödinger equation. The numerical solution of these equations gives results close to the parabolic potential for the channels in a two-dimensional gas but with a flat lower boundary (with the lower part cut off [10]). Since this form of the potential is very close to parabolic, the harmonic potential corresponds to the experimental results [9].

Practical implementation of the transversal parabolic potential can be performed by introducing impurities into QWire and barrier layers, similar to the case of multiwell quantum structures. Also, the restriction can be implemented by the influence of an external magnetic field [11].

As is well known [4], the contact potential difference occurs upon the metal-semiconductor contact and is proportional to the square of the width of the barrier layer. Therefore, the confining potential along the axis of QWire is also similar to a parabolic one, but with the lower part cut off.

Wave functions and energy spectrum of an electron in QWire with an external longitudinal electric field, respectively, have the form:

$$\psi_{n_1,n_2,n_3}(x,y,z)=\frac{1}{\pi^{3/4}a\sqrt{2^{n_1+n_2+n_3}n_1!n_2!n_3!a_z}}H_{n_1}\left(\frac{x}{a}\right)H_{n_2}\left(\frac{y}{a}\right)H_{n_3}\left(\frac{z-z_0}{a_z}\right)\exp\left(-\frac{x^2+y^2}{2a^2}-\frac{(z-z_0)^2}{2a_z^2}\right), \tag{2}$$

$$E_{n_1,n_2,n_3}=\hbar\omega\left(n_1+\frac{1}{2}\right)+\hbar\omega\left(n_2+\frac{1}{2}\right)+\hbar\omega_z\left(n_3+\frac{1}{2}\right)-\frac{e^2F^2}{2m^*\omega_z^2}, \tag{3}$$

where $n_1,n_2,n_3=0,1,2,...$ are oscillator quantum numbers; $a=\sqrt{\hbar/(m^*\omega)}$ is the characteristic length of the oscillator along the axis *OX* (or *OY*); $a_z=\sqrt{\hbar/(m^*\omega_z)}$ is the characteristic length of the oscillator along the axis *OZ*; $H_n(x)$ are Hermite polynomials [12]; $z_0=|e|F/(m^*\omega_z^2)$; e is the electron charge; F is the electric field strength.

The confining potential of QWire must have a finite depth. Thus, the amplitude of the confining potential (1) is an empirical parameter. The value of the confining potential amplitude along the axis *OX* (or *OY*) is equal to $V_{0x(0y)}=m^*\omega^2L^2/8$; along the axis *OZ* – $V_{0z}=m^*\omega_z^2L_z^2/8$. Therefore, expressions (2) and (3) are valid when $V_{0x(0y)}/(\hbar\omega)>>1$ and $V_{0z}/(\hbar\omega_z)>>1$.

Let us consider $D^{(-)}$-center localized at the point $\vec{R}_a=(x_a,y_a,z_a)$. The impurity potential is modeled by the zero-range potential [13] with the intensity $\gamma=2\pi\hbar^2/(\alpha m^*)$. In the Cartesian system of reference, this potential can be written as:

$$V_\delta\left(\vec{r},\vec{R}_a\right)=\gamma\delta(x-x_a)\delta(y-y_a)\delta(z-z_a)\left[1+(x-x_a)\frac{\partial}{\partial x}+(y-y_a)\frac{\partial}{\partial y}+(z-z_a)\frac{\partial}{\partial z}\right], \tag{4}$$

where α is determined by the binding energy $E_i = -\hbar^2\alpha^2/(2m^*)$ for the same $D^{(-)}$-center in a massive semiconductor; $\delta(x)$ is the Dirac delta-function.

3. Calculations

The wave function $\Psi_\lambda^{QWire}(\vec{r},\vec{R_a})$ for the electron which is localized at the $D^{(-)}$-center, in the effective mass approximation, satisfies the Schrödinger equation:

$$\left(E_\lambda - \hat{H}\right)\Psi_\lambda^{QWire}(\vec{r},\vec{R_a}) = V_\delta(\vec{r},\vec{R_a})\Psi_\lambda^{QWire}(\vec{r},\vec{R_a}), \tag{5}$$

where $\hat{H}$ is the Hamiltonian operator; $|E_\lambda| = \hbar^2\lambda^2/2m^*$ are eigenvalues of the Hamiltonian $\hat{H}_\delta = \hat{H} + V_\delta(\vec{r},\vec{R_a})$.

Equation (5) is solved by the method of Green-functions. The one-electron Green-function $G(\vec{r},\vec{r_1},E_\lambda)$ associated to the Schrödinger equation (5) corresponding to the energy E_λ and to the source at the point $\vec{r_1} = (x_1, y_1, z_1)$, can be written as:

$$G(\vec{r},\vec{r_1},E_\lambda) = \sum_{n_1,n_2,n_3} \frac{\Psi^*_{n_1,n_2,n_3}(\vec{r_1})\Psi_{n_1,n_2,n_3}(\vec{r_1})}{E_\lambda - E_{n_1,n_2,n_3}}. \tag{6}$$

The Lippman-Schwinger equation for $D^{(-)}$-state in QWire in an external electric field is:

$$\Psi_\lambda^{QWire}(\vec{r},\vec{R_a}) = \int_{-\infty}^{+\infty}\int_{-\infty}^{+\infty}\int_{-\infty}^{+\infty} dx_1dy_1dz_1 G(\vec{r},\vec{r_1},E_\lambda)V_\delta(\vec{r_1},\vec{R_a})\Psi_\lambda^{QWire}(\vec{r_1},\vec{R_a}). \tag{7}$$

By substituting the expression for zero-range potential (4) into equation (7) we obtain:

$$\Psi_\lambda^{QWire}(\vec{r},\vec{R_a}) = \gamma G(\vec{r},\vec{r_1},E_\lambda)\left(\hat{T}\Psi_\lambda^{QWire}\right)(\vec{R_a},\vec{R_a}), \tag{8}$$

where:

$$\left(\hat{T}\Psi_\lambda^{QWire}\right)(\vec{R_a},\vec{R_a}) == \lim_{\substack{x\to x_a\\ y\to y_a\\ z\to z_a}}\left(1+(x-x_a)\frac{\partial}{\partial x}+(y-y_a)\frac{\partial}{\partial y}+(z-z_a)\frac{\partial}{\partial z}\right)\Psi_\lambda^{QWire}(\vec{r},\vec{R_a}). \tag{9}$$

The equation which determines the dependence of the binding energy of $D^{(-)}$-center on the parameters of QWire, the position $\vec{R_a} = (x_a, y_a, z_a)$ of the impurity, and the electric field strength $\vec{F}$ can be obtained by applying the operator $\hat{T}$ to the left and right hand sides of equation (8). This gives us:

$$\alpha = \frac{2\pi\hbar^2}{m^*}\left(\hat{T}G\right)(x_a, y_a, z_a, x_a, y_a, z_a, E_\lambda). \tag{10}$$

By summing the terms of the right side of equation (6) over the quantum numbers n_1, n_2, n_3, we obtain the one-electron Green-function with a distinguished divergent part [12]:

$$G(\vec{r},\vec{R_a},\beta,\beta_z,\eta) = C(\vec{r},\vec{R_a})\int_0^{+\infty}\exp(-bt)\frac{1}{(1-\exp(-2\beta t))\sqrt{1-\exp(-2\beta_z t)}}\times$$

$$\times\exp\left(\frac{2\exp(-\beta_z t)(x_a x + y_a y) - \exp(-2\beta_z t)(x_a^2 + x^2 + y_a^2 + y^2)}{a^2(1-\exp(-2\beta_z t))}\right)\times$$

$$\times\exp\left(\frac{2\exp(-\beta t)(z_a-z_0)(z-z_0)-\exp(-2\beta t)\left((z_a-z_0)^2+(z-z_0)^2\right)}{a_z^2\left(1-\exp(-2\beta t)\right)}\right)dt-$$

$$-\frac{1}{2\beta_z\sqrt{2\beta}}\int_0^\infty \exp(-bt)\exp\left(-\frac{1}{t}\left(\frac{(x-x_a)^2+(y-y_a)^2}{2a^2}+\frac{(z-z_a)^2}{2a_z^2}\right)\right)t^{-3/2}dt+$$

$$+\frac{\sqrt{\pi}}{2\beta_z\sqrt{\beta}}\frac{\exp\left(-\sqrt{2b}\sqrt{\frac{(x-x_a)^2+(y-y_a)^2}{a^2}+\frac{(z-z_a)^2}{a_z^2}}\right)}{\sqrt{\frac{(x-x_a)^2+(y-y_a)^2}{a^2}+\frac{(z-z_a)^2}{a_z^2}}}. \tag{11}$$

where $C\left(\vec{r},\vec{R_a}\right)=-\frac{\beta\beta_z}{\pi^{3/2}E_d a^2 a_z}\exp\left(-\frac{x_a^2+x^2+y_a^2+y^2}{2a^2}-\frac{(z_a-z_0)^2+(z-z_0)^2}{2a_z^2}\right)$; $\beta=\hbar\omega/E_d$; $\beta_z=\hbar\omega_z/E_d$; E_d – the effective Bohr energy; $b=(E_D/E_d)\beta\beta_z=\eta^2\beta\beta_z+\beta_z+\beta/2$; $E_D=E_{0,0,0}-E_\lambda$ – the binding energy of the $D^{(-)}$-state; $\eta^2=|E_\lambda|/E_d$.

By substituting (11) from (10) we obtain the dispersion equation for the $D^{(-)}$-state of the electron in QWire with the influence of a strong longitudinal electric field:

$$\sqrt{|E_i|}=\frac{\sqrt{2}\pi\hbar^3}{m^{*3/2}}\left(-\frac{\sqrt{\pi b}}{\beta_z\sqrt{2\beta}}+C\left(\vec{R}_a,\vec{R}_a\right)\int_0^\infty \exp(-bt)\left(\frac{1}{\left(1-\exp(-2\beta t)\right)\sqrt{1-\exp\left(1-\exp(-2\beta_z t)\right)}}\times\right.\right.$$

$$\left.\left.\times\exp\left(\frac{2\left(x_a^2+y_a^2\right)\exp(-\beta_z t)}{a^2\left(1+\exp(-\beta_z t)\right)}+\frac{2(z_a-z_0)^2\exp(-\beta t)}{a_z^2\left(1+\exp(-\beta t)\right)}\right)-\frac{t^{-3/2}}{2\beta_z\sqrt{2\beta}}\right)dt\right). \tag{12}$$

4. Results

Figures 1 and 2 show the results of numerical analysis of the dispersion equation (12) which determines the dependence of the binding energy E_D of $D^{(-)}$-state on the value of the strong longitudinal electric field strength $\vec{F}$, the parameters of confining potential of QWire, the characteristics of impurity (the power of zero-radius potential γ, coordinates $\vec{R_a}=(x_a,y_a,z_a)$ of the impurity center) in GaAs.

Figure 1 shows that the increase of the electric field strength $\vec{F}$ leads to a strong decrease of the binding energy E_D of the impurity state (see curves 1 and 2, figure 1) due to the quantum-confined Stark effect for one-electron states in QWire [14]. Since the wave function of the electron on an impurity is formed from electron states of the conduction band, the displacement of impurity levels in the electric field is similar to the displacement of the energy levels of the electron in the conduction band. However, displacement of the impurity level is much lower in the same electric field than that for the Stark effect for the conductivity band. Therefore, the dynamics of size-quantization levels and impurity level leads to the destruction of the bound state of the electron on an impurity in the electric field, i.e. its ionization at a certain value of the electric field strength $\vec{F}$ (see curve 1, figure 1). The impurity level is deeper at a higher power of the zero-potential radius γ, and, therefore, thermoelectric ionization occurs at higher field strength $\vec{F}$ (see curve 2, figure 1).

Figure 2 shows the dependence of the binding energy E_D of $D^{(-)}$-state on the coordinates x_a (or y_a) of an impurity in QWire of GaAs. The decrease of the binding energy E_D when approaching the boundary of QWire is due to the influence of the confinement potential. This effect is known as a "position disorder effect" [15]. A comparison of curves 1 and 3 in figure 2 shows that the region of existence of impurity states decreases with increase of the electric field strength $\vec{F}$ "pulling" to the axis of the QWire, and at a certain value of the electric field strength the bound states do not occur.

Thus, the thermoelectric ionization effect, which occurs in QWire with impurities in the strong longitudinal electric field at low temperatures, results in increase of the concentration of free electrons in the conduction band. This effect is called the Frenkel effect in bulk semiconductors [4].

The range of values of the electric field strength $\vec{F}$, which was considered by the thermoelectric ionization effect, is from $2\cdot10^6$ V/m to $2{,}3\cdot10^6$ V/m. In general, the considered range of values of the electric field strength $\vec{F}$ is determined by the characteristics of impurity centers (γ, $\vec{R_a}$) and the parameters of QWire (L, L_z, V_0).

The numerical solution of the equation of electrical neutrality for the impurity conductivity at low temperatures makes it possible to estimate the dependence of the current density $\vec{j}$ on the electric field strength $\vec{F}$ in QWire, provided that the dependence of the electron mobility μ on the electric field strength $\vec{F}$ is not too strong [16].

Figure 3 shows the dependence of the current density $\vec{j}$ on the electric field strength $\vec{F}$, which has a superlinear character. The sharp increase of current density occurs at a certain value of the electric field strength, which is the closest to the value of the ionization electric field strength, when practically all impurities are ionized. The increase of temperature accelerates the process of ionization (see curves 1 and 3, figure 3). Therefore, the current density increases due to the corresponding dynamics of the Fermi level. Increasing the concentration of impurity centers N_D also leads to an increasing current density (see the curves 2 and 3, figure 3).

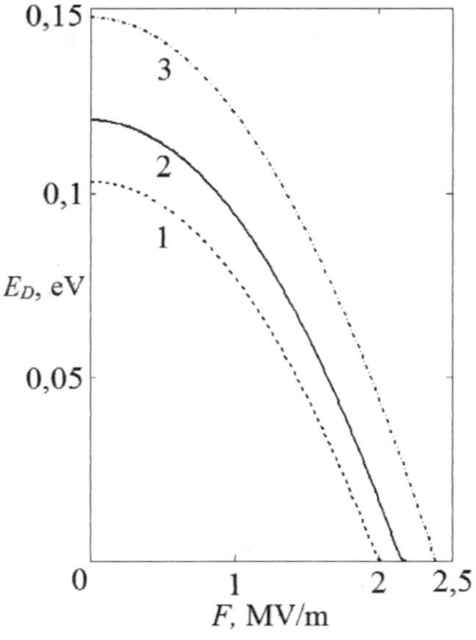

Figure 1. The binding energy E_D dependence (for $D^{(-)}$-center in QWire, based on GaAs) on the value of the longitudinal electric field strength F, (curve 3 shows the dependence of ground state energy $E_{0,0,0}$ on F); $L = 9{,}1$ nm; $L_z = 91$ nm; $V_{0x} = V_{0y} = 0{,}2$ eV; $V_{0z} = 0{,}02$ eV; $x_a = 0$; $y_a = 0$; $z_a = 0$. Curve 1 – $E_i = 2\times10^{-3}$ eV; 2 – $E_i = 9\times10^{-3}$ eV.

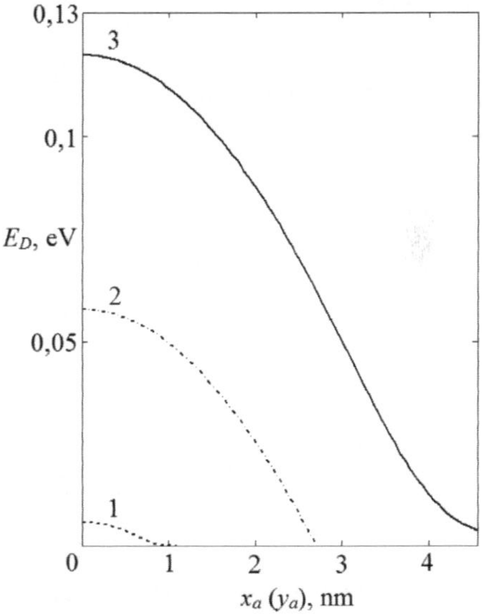

Figure 2. The binding energy E_D dependence (for $D^{(-)}$-center in QWire, based on GaAs), on the coordinate x_a (or y_a); $L = 9{,}1$ nm; $L_z = 91$ nm; $V_{0x} = V_{0y} = 0{,}2$ eV; $V_{0z} = 0{,}02$ eV; $z_a = 0$; $E_i = 9\times10^{-3}$ eV. Curve 1 – $F = 2{,}1\times10^6$ V/m; 2 – $F = 1{,}6\times10^6$ V/m; 3 – $F = 10^3$ V/m.

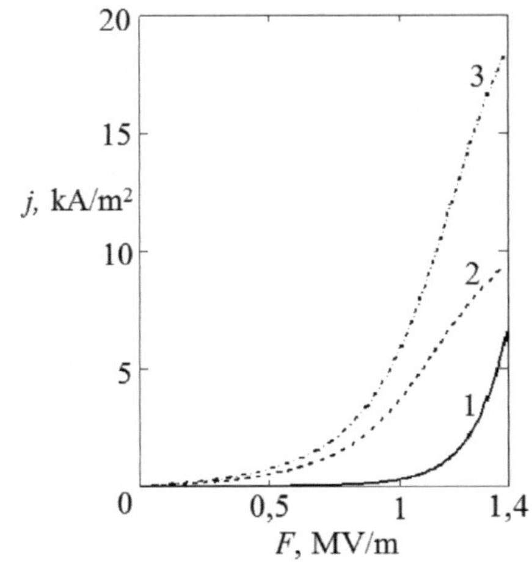

Figure 3. The current density j dependence (for $D^{(-)}$-center in QWire, based on GaAs), on the value of the longitudinal electric field strength F; $L = 9,1$ nm; $L_z = 91$ nm; $V_{0x} = V_{0y} = 0,2$ eV; $V_{0z} = 0,02$ eV; $E_i = 9\times10^{-3}$ eV; $\mu = 0,86$ m^2/(V·s). Curve 1 – N_D = 10^{17} m^{-3}, T = 50 K; 2 – $N_D = 0,5\times 10^{17}$ m^{-3}, T = 70 K; 3 – $N_D = 10^{17}$ m^{-3}, T = 70 K.

5. Conclusion
From a fundamental point of view, the considered effect of the thermoelectric ionization in QWire allows us to estimate the binding energy and the concentration of impurities. From a practical point of view, this effect makes it possible to control the concentration of free charge carriers and, hence, electrical conductivity of QWire, that is important to develop new semiconductor nanoelectronic devices [17, 18].

References
[1] Buravlev A D *et al* 2014 *Phys. and tech. of semicond.* **48** 358
[2] Geiler V A, Margulis V A, Filina L I 1998 *Jour. of exp. and theor. phys.* **113** 1377
[3] Bagraev N T *et al* 2005 *Physics and technics of semiconductors* **39** 716
[4] Fistul V I 1975 Introduction in physics of semiconductors *Moscow Nauka* 250
[5] Chu C S, Sorbello R S 1989 *Phys. Rev. B* **40** 5941
[6] Bagwell P F 1990 *Phys. Rev. B* **41** 10354
[7] Fertig H A, Halperin B I 1987 *Phys. Rev. B* **36** 7969
[8] Levinson Y B, Lubin M I, Sukhorukov E V 1992 *Phys. Rev. B* **45** 11936
[9] Martin T, Feng S 1990 *Phys. Rev. Lett.* **64** 1971
[10] Beenakker C W J, Van Houten H ed. By Ehrenreich H and Turnbull D 1991 *in Solid State Physics New York Academ. Press* **44** 83
[11] Huant S, Najda S P 1990 *Phys. Rev. Lett.* **65** 1486
[12] Nikiforov A F, Uvarov V B 1978 Special functions for mathematical physics *Moscow Nauka* 320
[13] Pahomov A A, Halipov K V, Yassievich I N 1996 *Phys. and tech. of semicond.* **30** 1387
[14] Krevchik V D, Zaytsev R V, Evstifeev V V 2000 *Phys. and tech. of semicond.* **34** 1244
[15] Pokutniy S I 2000 *Phys. and tech. of semicond.* **34** 1120
[16] Sinyavskiy E P, Karapetyan 2014 *Phys. and tech. of semicond.* **48** 229
[17] Bagraev *et al* 2005 *Phys. and tech. of semicond.* **39** 716
[18] Dubrovskiy V G, Cirlin G E 2009 *Phys. and tech. of semicond.* **43** 1585

Electrical characterization of organic monolayers/silicon hybrid structures

I V Malyar[1], V O Lukyanova[1], E G Glukhovskoy[1], S B Venig[1] and D A Gorin[1]

[1] Department of Nano- and Biomedical Technologies, Saratov State University, Astrakhanskaya, 83, Saratov, 410012, Russia

E-mail: imalyar@yandex.ru

Abstract. Hybrid structures consisting of an *n*-Si substrate covered with a cationic polyethylenimine layer and an anionic dye layer were fabricated. Their electrical properties were characterized using tunnel current measurements by a tungsten probe. The obtained I-V curves were analyzed using a modified diode equation. We found out that adsorption of the cationic polyelectrolyte decreases the barrier height of *n*-Si/organic monolayer/W structure, while adsorption of the low molecular weight anionic dye increases it. The results demonstrate that along with thermionic mechanism other ones are present which prevail at low voltage. In particular, electron tunneling dominates for a single monolayer on the silicon substrate, while, apparently, the Pool–Frenkel mechanism is typical for two monolayers.

1. Introduction

Hybrid structures consisting of organic and inorganic components are a subject of interest in various research fields [1] as they combine well-known and developed inorganic materials with a huge amount of organic ones properties of which can be varied in wide ranges. In particular, such structures based on inorganic semiconductors can be used in microelectronics [2], sensors [3] and photovoltaic applications [4].

There are several models describing electron transfer in hybrid structures [2] which defines their electrical properties. Moreover, models of charge transport through organic molecules were also elaborated [5] to estimate electrical properties of hybrid structures with thick organic layers. Therefore, the conduction mechanism for a certain hybrid structure should be considered separately.

In the present work we examine the electrical properties of hybrid structures consisting of a silicon substrate covered with a cationic polyelectrolyte monolayer and an anionic dye layer using tunnel current measurements.

2. Experimental details

Wafers of *n*-Si (100) with native oxide and specific resistivity of $\rho = 3\text{-}6\ \Omega\cdot\text{cm}$ were used as substrates. They were cut into 13 mm × 13 mm squares and treated in a peroxide-ammonia solution to remove organic contaminations and to activate negative OH-groups on the surface [6]. Then a cationic polyelectrolyte polyethylenimine (PEI) was adsorbed under different conditions as it was described in [7], i.e. some silicon substrates were illuminated during PEI adsorption. However, a halogen lamp (Philips 13186 EPX/EPV) was used as a light source to maximize effect of illumination instead of a He-Cd laser. According to [7], illumination increases the negative surface charge of silicon substrates due to tunneling of photogenerated electrons into the native oxide. Thus, it enhances interaction between the negatively-charged substrate and cationic polyelectrolyte molecules decreasing both thickness and roughness of the adsorbed monolayer.

Similarly, an anionic dye Photosens consisting of sulfonated hydroxialuminum phthalocyanines $AlPcS_x$ with x = 2, 3 or 4 (the mean x = 3.1) [8] was adsorbed. The deposition time and $AlPcS_x$ concentration were 10 min and 0.2 mg/mL, respectively.

Thus, 7 different types of samples were prepared: *n*-Si, *n*-Si / PEI, *n*-Si / PEI^{ill}, *n*-Si / PEI / $AlPcS_x$; *n*-Si / PEI / $AlPcS_x^{ill}$; *n*-Si / PEI^{ill} / $AlPcS_x$; *n*-Si / PEI^{ill} / $AlPcS_x^{ill}$, where superscript "ill" indicates illumination during adsorption.

Electrical properties of the samples were characterized using tunnel current-voltage (I-V) curves measured with a tungsten probe by means of NTEGRA Spectra (NT-MDT, Russia) under ambient conditions. A conducting carbon tape was used as a bottom contact. Initially, the sample surface was scanned in the constant current regime. Then ten I-V curves were measured in each of six selected points on the scan by applying a linear bias from -5 to +5 V to sample, while probe was grounded.

3. Results and discussion

Figure 1 demonstrates typical I-V curves for different samples. All dependences reveal rectification, which is typical for the Schottky barrier. Therefore, these curves were analyzed using a diode equation with ideality factor *n* indicating different conductance mechanisms:

$$I = I_s \exp\left(\frac{qV}{nk_BT}\right)\left(1-\exp\left(-\frac{qV}{k_BT}\right)\right) \tag{1}$$

where $I_s = \sigma A^* T^2 \exp\left(-\frac{q\varphi_{Bn}}{k_BT}\right)$ is the saturation current, q is the elementary charge, A^* – the effective Richardson constant, σ – contact area, $q\varphi_{Bn}$ – the Schottky barrier height for electrons, T – temperature, k_B is the Boltzmann constant.

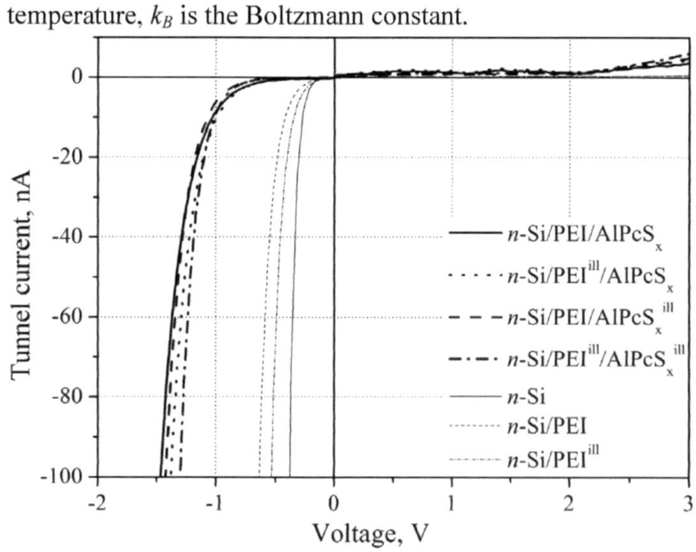

Figure 1. Typical I-V characteristics of an *n*-Si substrate and ones covered with polyethylenimine (PEI) and sulphonated hydroxylaluminum phthalocyanine ($AlPcS_x$) monolayers. Superscript "ill" indicates illumination during adsorption of organic monolayer.

In report [9, 10] the equation (1) was used to analyze I-V curves for semiconductor/organic monolayer/metal hybrid structures, where along with thermionic current mechanism other ones occur. Therefore, the ideality factor exceeds 1. The direct branches of experimental I-V curves were plotted in ln*I* vs. *V* scale (figure 2) using the following equation:

$$\ln\left[\frac{I}{1-\exp(-qV/k_BT)}\right] = \ln\left(\sigma A^* T^2\right) + \frac{q}{k_BT}\left(-\varphi_{Bn} + \frac{V}{n}\right). \tag{2}$$

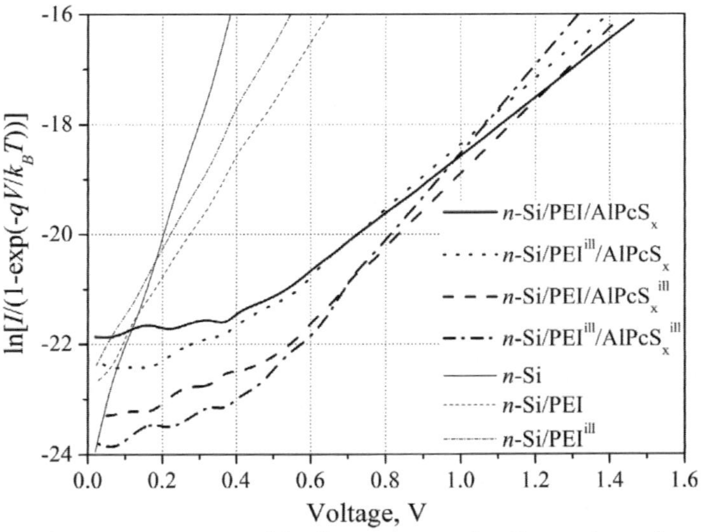

Figure 2. Typical I-V characteristics of an *n*-Si substrate and ones covered with polyethylenimine (PEI) and sulphonated hydroxyaluminum phthalocyanine ($AlPcS_x$) monolayers. Superscript "ill" indicates illumination during adsorption of organic monolayer.

The curves presented in figure 2 reveal a linear dependence at high voltage and nonlinear one at low voltage. The linear region corresponds to thermionic (Schottky) emission and, thereby, the ideality factor can be extracted from its slope, while the Schottky barrier height can be estimated from its interception of ln*I* axe. However, the contact area is necessary to define. According to [11] a radius of freshly etched tungsten probes is about 40 nm, then, suggesting that electron tunneling occurs only from a surface layer of δ = 1 nm, the effective contact area, σ, for electrons can be calculated as [12]: $\sigma = \pi d^2/4$, where $d = 2\sqrt{R^2 - (R-\delta)^2}$ – is an effective diameter. Thus, the effective contact area is $\sigma = 248.7$ nm^2 and term $\ln\left(\sigma A^* T^2\right)$ for *n*-Si (100) is equal to -9.82. Table 1 presents calculated averaged parameters.

The results in table 1 demonstrate that deposition of the organic monolayer changes both the barrier height and the ideality factor. The barrier height drastically depends on the surface state charge which is negative for *n*-Si substrates [13]. Therefore, deposition of positively-charged PEI decreases the barrier height and, on the contrary, negatively-charged dye increases it. Both organic layers significantly increase the ideality factor. Its high values for all samples indicate non-thermionic current mechanisms which, apparently, prevail at low voltage and correspond to nonlinear regions in figure 2. Moreover, deposition conditions, i.e., photoassistance during adsorption, also influence the barrier height and the ideality factor. Since illumination increases electrostatic interaction between a silicon substrate and adsorbed molecules, then it enhances the effect of organic layer on hybrid structure parameters. In particular, photoassisted adsorption of a polyelectrolyte enhances electrical passivation of silicon surface [14].

There are several current mechanisms in dielectrics as tunneling, thermionic emission, Frenkel-Poole emission, etc. [2, 13]. Therefore, we plotted I-V curves in a double logarithmic scale (figure 3) to find out the current mechanisms at low voltage from the power law relation $I \propto V^{\alpha}$.

All curves in figure 3 were approximated by linear dependences, and an exponent α was calculated from their slopes (Table 1). We observe a reciprocal proportion between the ideality factor *n* and the exponent α for hybrid samples, i.e., $\alpha \cdot n \approx const$. Constant is equal to 3.55±0.15 for Si/PEI samples and 2.35±0.13 for Si/PEI/$AlPcS_x$ samples. Thus, these parameters are complementary.

A linear dependence (α ≈ 1) between tunnel current and voltage for Si/PEI samples indicates electron tunneling through a rectangular barrier [15]. According to [16] the slope value of about 0.5 indicates the Pool–Frenkel mechanism in Si/PEI/$AlPcS_x$ samples. However, we suppose a combination of mechanisms mentioned above for these samples. In addition, we suppose a complex conductivity mechanism consisting of thermionic (Schottky) emission and electron tunneling for a silicon substrate without any organic coating.

Table 1. The calculated parameters of an *n*-Si substrate and ones with poly(ethylenimine) (PEI) and sulphonated hydroxyaluminum phthalocyanine ($AlPcS_x$) monolayers. *n* is the ideality factor, φ_b, is the Schottky barrier height, α is an exponent in the power law connecting tunnel current and applied voltage. Superscript "ill" indicates illumination during polyelectrolyte adsorption.

Parameters	Samples						
	Si	Si/PEI	Si/PEI^{ill}	*n*-Si / PEI / $AlPcS_x$	*n*-Si / PEI^{ill} / $AlPcS_x$	*n*-Si / PEI / $AlPcS_x^{ill}$	*n*-Si / PEI^{ill} / $AlPcS_x^{ill}$
n	1.81	3.65	3.16	7.37	6.57	5.93	4.88
φ_b, mV	373±5	336±5	331±5	360±6	371±7	401±5	428±5
α	1.58	1.01	1.07	0.31	0.35	0.43	0.45

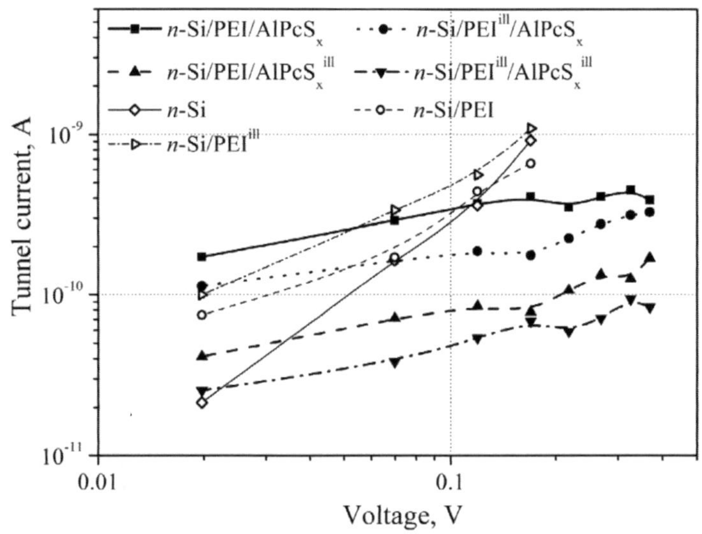

Figure 3. Typical I-V characteristics of an *n*-Si substrate and ones with polyethylenimine (PEI) and sulphonated hydroxyaluminum phthalocyanine ($AlPcS_x$) monolayers. Superscript "ill" indicates illumination during adsorption.

4. Conclusions

We fabricated hybrid structures based on the *n*-Si substrates with organic monolayers of the cationic polyelectrolyte (PEI) and the low molecular weight anionic dye (Photosens). Electrical properties of the samples were characterized using tunnel current measurements. We observed rectification in I-V curves; therefore, the modified diode equation was used to calculate the barrier height and to monitor the current mechanism. We found out that adsorption of PEI decreases the barrier height for *n*-Si, while adsorption of Photosens increases it. The results demonstrate that along with thermionic mechanism other ones occur which prevail at low voltage. In particular, electron tunneling dominates for a single monolayer on a silicon substrate, while, apparently, the Pool–Frenkel mechanism is typical for two monolayers.

Acknowledgements
This work was supported by the Russian Science Foundation (grant 14-12-00275) and Russian Federation President Grant for young scientists.

References

[1] Sanchez C, Belleville P, Popalld M and Nicole L 2011 *Chem. Soc. Rev.* **40** 696
[2] Ortiz R P, Facchetti A and Marks T J 2010 *Chem. Rev.* **110** 205
[3] Wang S, Kang Y, Wang L, Zhang H, Wang Y, Wang Y 2013 *Sensors and Actuators B: Chemical* **182** 467

[4] Wright M and Uddin A 2012 *Solar Energy Materials & Solar Cells* **107** 87
[5] Ulgut B and Abruna H D 2008 *Chem. ReV.* **108** 2721
[6] Aswal D K, Lenfant S; Guerin D; Yakhmi J V, Vuillaume D 2006 *Anal. Chim. Acta* **568** 84
[7] Malyar I V, Gorin D A, Stetsyura S V, Santer S 2013 *Langmuir.* **29** 16058
[8] Svenskaya Y, Parakhonskiy B, Haase A, Atkin V, Lukyanets E, Gorin D, Antolini R 2013 *Biophysical Chemistry* **182** 11
[9] Vilan A, Shanzer A and Cahen D 2000 *Nature* **404** 166
[10] Hiremath R K, Rabinal M K, Mulimani B G and Khazi I M 2008 *Langmuir* **24** 11300
[11] Malyar I V, Gorin D A , Stetsyura S V 2012 *Proc. SPIE* **8700** 870009
[12] Feenstra R M, Stroscio J A and Fein A P 1987 *Surface science* **181** 295
[13] Sze S 1985 *Semiconductor Devices: Physics and Technology, 2nd ed* (Wiley: New York)
[14] Stetsyura S V, Kozlowski A V, Malyar I V 2015 *Technical Physics Letters* **4** 168
[15] Simmons J G 1963 *Journal of Applied Physics* **34** 1793
[16] Nabok A V, Hassan A K, Ray A K and Toldi G N 2002 *Materials Science and Engineering* **22** 387

Journal of Physics: Conference Series **690** (2016) 012026 doi:10.1088/1742-6596/690/1/012026

Spin properties of the ultra–short capped single-walled carbon nanotubes (0, 9)

A V Tuchin, G I Glushkov and L A Bityutskaya

Voronezh State University, 1 Universitetskaya sq., Voronezh 394006, Russia

E-mail: a.tuchin@bk.ru

Abstract. Results of the numerical simulation of the electronic structure of ultra–short single-walled carbon nanotubes (0, 9) of D_{3h}, D_{3d} and D_3 symmetries at the singlet and triplet spin states were presented. The dependencies of the energy gap, ionization potential, electron affinity and work function on the length of nanotubes at the singlet and triplet spin states were obtained. The analysis of the dimensionally dependent change of the electron density of states and energy gap for electrons with spin up and down allowed to establish the dependence of the spin polarization value on the length and symmetry of nanotubes. It was revealed that the energy required to turn from singlet to triplet is in the IR range that makes ultra-short single-walled carbon nanotubes (0, 9) a promising material for the design of the element base for spintronics and optoelectronics.

1. Introduction

In recent years a signigificant success was achieved in the field of theoretical and experimental studies aimed at the nanosystems characterization. The magnetic properties can be obtained at the nanoscale in materials which are nonmagnetic at the macroscale [1]. The present-day thin metal film ferromagnetic structures are characterized by a high Curie temperature and radiation tolerance. Otherwise, the principal limitation of melal spintronics is the impossibility to design amplifiers. That is why there is a great number of studies devoted to the semiconductor spintronics which is able to combine non-volatility, low leakage currents and rather high operation rates. The main actual task of semiconductor spintronics is a search for new materials which demonstrate a fairly high spin polarization of charge carriers and the construction of devices with a pretty easy registration of the magnetoresistance difference [2]. The relatively low spin-orbit interaction and ballistic transport of charge carriers in carbon nanotubes (CNTs) cause a long spin coherence time [3]. Catalyst particles which are used in CVD- synthesis can act as spin polarizers, however, the technology of doped CNTs controlled synthesis is rather complicated. The theoretical and experimental study of undoped CNTs observes the probability of the intrinsic magnetic moment occurrence [4–7].

The particular emphasis is deserved by the magnetic properties of ultra-short single-walled carbon nanotubes (us-SWCNTs) [5]. The distinctive consideration of us-SWCNTs is caused by a significant achievement in the field of synthesis methods with a narrow distribution in the chirality and length control up to angstroms [8, 9]. Lots of theoretical studies of carbon nanotubes demonstrate that the decrease in their length less than 10 nm causes great changes in the electronic structure and fundamental parameters such as the energy gap (E_{LH}) between the lowest unoccupied (LUMO) and the highest occupied (HOMO) molecular orbitals, ionization potential (IP), electron affinity (EA), work function (W) [10–17]. It is expected that the properties and fundamental parameters of us-SWCNTs strongly depend on their spin states. Thus, the intrinsic magnetic moment in pure carbon nanotubes makes possible their application as functional nanomaterials for spintronics and optoelectronics.

Literature data describing the origin of intrinsic magnetism in nanotubes are not complete and do not contain the dependencies of main spin-dependent parameters on the tube length. The aim of this work is the theoretical study of the electronic structure of us-SWCNTs (0, 9) at the singlet and triplet states.

2. Computational details

The numerical simulations of the electronic structure of the capped zigzag us-SWCNT (0, 9) were carried out using the DFT (density functional theory) method in the local spin density approximation (LSDA) with 3-21G basis set by means of the Gaussian 09 program package in the Supercomputing center of Voronezh State University.

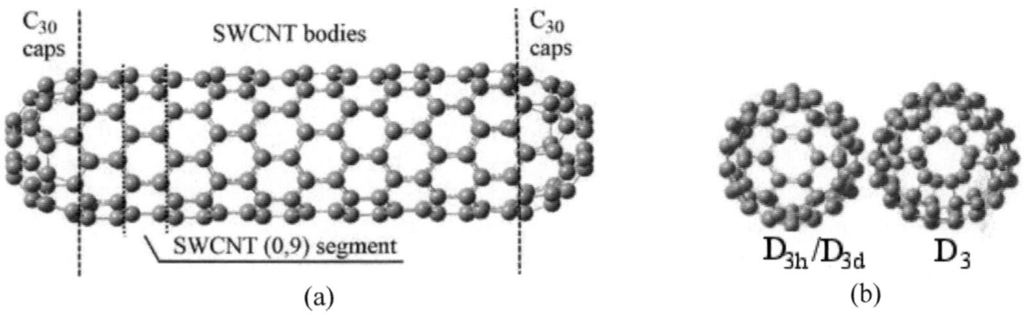

(a) (b)

Figure 1. Structural images of capped us–SWCNTs (0, 9) (a). The mutual orientation of two C_{60} hemisphere caps of D_{3h}/D_{3d} and D_3 symmetries (b).

A stoichiometric formula of capped us–SWCNT (0, 9) is C_{60+18i}. Caps are obtained by the dissection of the fullerene molecule perpendicularly to the C_{3v} axis (figure 1). Regardless of number i there may be two orientations of the caps, corresponding to: 1) D_{3h} (when $i = 2p + 1$) and D_{3d} (when $i = 2p$), (where p is an integer) and 2) D_3 symmetries. The nanotube length varied from 1 to 3 nm. According to the Coopmans theorem, the ionization potential *IP* and electron affinity *EA* equal to the HOMO and LUMO energy with opposite sign. In case of zero number of segments, the capped us-SWCNT (0, 9) represents fullerene C_{60}. Therefore, fullerene is used as a test object which has been well studied both theoretically and experimentally [18–20]. Our calculated values for fullerene C_{60} $IP = 6.69$ eV and $EA = 4.86$ eV differ from the average experimental values $IP = 7.56 \pm 0.06$ eV [21–24] and $EA = 2.68 \pm 0.02$ eV [25, 26]. Using the assumption of Cioslowski et al. [11], the differences between average experimental data and calculated values for fullerene C_{60} and us-SWCNTs (0, 9) are equal. In our research we used the following corrections: $\Delta IP = 0.87$ eV и $\Delta EA = -2.18$ eV.

3. Results and discussion

3.1. *The electronic structure of the capped us-SWCNT (0, 9) at the singlet state*

The analysis of the all-electron total energy showed that us-SWCNTs with D_3 symmetry are more stable than D_{3h} and D_{3d} ones in all the investigated length range. The exception is the nanotube C_{78} that is in agreement with the calculations of the standard enthalpy formation of us-SWCNT (0, 9), which were carried out in paper [11]. It was found out that the length increase causes the monotonic decrease of the energy gap. The range of the gap E_{LH} energies is from 1.58 eV to 0.59 eV and from 0.81 eV to 0.29 eV for nanotubes of D_{3h}/D_{3d} and D_3 symmetries, respectively. The energy gap is non-zero just in rather close length ranges and seeks to zero at the sizes more than 3 nm. This non-zero gap energy is caused by the restriction of the electrons motion along the nanotube axis and by the impact of two caps.

This fact together with thermodynamical stability of us-SWCNTs (0, 9) makes it possible to classify them as a whole family of semiconductor nanomaterials. The dependencies of the ionization potential, electron affinity and work function on the tube length are monotonically decreasing. The values belong to the ranges IP = 6.49 – 7.30 eV, EA = 4.67 – 4.97 eV, W = 2.65 – 2.84 eV for us-SWCNTs with D_{3h}/D_{3d} symmetry and IP = 6.37 – 7.29 eV, EA=4.69–5.37 eV, W = 3.02 – 3.44 eV for nanotubes with D_3 symmetry. It is remarkably that in the approximation of the infinite length and flat elementary cell zigzag, CNT (0, 9) has a zero energy gap or Eg = 100 – 140 meV if curvature effects are taken into account [27–30].

3.2. The electronic structure of the capped us-SWCNT (0, 9) at the triplet state

The dependencies of the energy gap between the frontier orbitals on the number of segments for us-SWCNTs (0, 9) at the singlet and triplet states are shown in figure 2. It was found that at the singlet-triplet transition the conductivity of us-SWCNTs (0, 9) of D_3 symmetry decreases that is determined by the increase of the energy gap between frontier orbitals, so $E_{LH}(\uparrow)$ и $E_{LH}(\downarrow)$ changes to 0.1 – 0.5 eV in relation to the energy E_{LH} at the singlet state. In its turn, the decrease of spin-dependent gaps between frontier orbitals to 0.4 – 1.5 eV with respect to the energy E_{LH} of the singlet state of us-SWCNTs of D_{3h}/D_{3d} symmetry points out to the significant increase of their conductivity. The relations between spin-dependent transport properties and the length are qualitatively different for nanotubes of D_{3h}/D_{3d} and D_3 symmetries. The value of the electron spin polarization of the nanotubes D_3 harshly decreases with the increase of the tube length, that is due to the decrease of the difference of energy gaps $E_{LH}(\uparrow) - E_{LH}(\downarrow)$ from 0.231 to 0.011 eV in the range of segments number i = 2 – 8. For us-SWCNTs (0, 9) of D_{3h}/D_{3d} symmetry with the number of segments $i \geq 4$, the value of the spin polarization increases almost linearly with the rise of the nanotubes length. Thus, the difference of gaps between frontier orbitals $E_{LH}(\uparrow) - E_{LH}(\downarrow)$ is 0.056 eV and 0.107 eV at i = 4 and i = 11. The wide length range of the spin polarization existence and the increase of the difference of electron conductivities for two spin channels at the length increase of us-SWCNTs of D_{3h}/D_{3d} symmetry determine the interest in their application in spintronic devices.

At the triplet state of us-SWCNTs (0, 9) the number of electrons with spin up is two more than the number of electrons with spin down. The energies of molecular orbitals with the majority and minority spin for the triplet state of us-SWCNT (0, 9) are shifted up and down relative to energies of the singlet state. This shift of orbital energies is determined by the us-SWCNT symmetry.

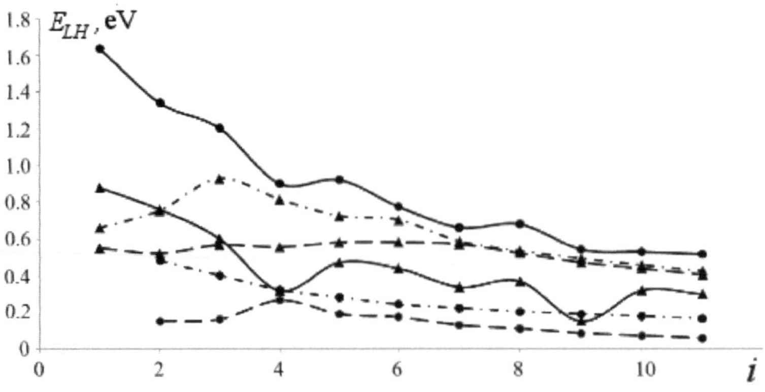

Figure 2. Dependencies of energy gaps E_{LH} at the singlet state (solid lines) and spin-dependent $E_{LH}(\uparrow)$ (dotted line), $E_{LH}(\downarrow)$ (dash-dot line) at the triplet state on the number of segments of the capped us–SWCNTs (0, 9) of D_{3h}/D_{3d} (circles) and D_3 (triangles) symmetries.

In the triplet state of the nanotube of D_3 symmetry, the energies of the LUMO and HOMO for electrons with majority and minority spins are quantitatively shifted almost symmetrically relative to the energies of frontier orbitals at the singlet state. Thus, the dependencies of energy gaps $E_{LH}(\uparrow)$ and $E_{LH}(\downarrow)$ converge to similar values (see figure 2). It follows from the analysis of the electron density of states, that the symmetrical shift of energies of molecular orbitals leads to the intersection of the DOS(↑) and DOS(↓) at the Fermi level. That results in a loss of spin polarization for long us-SWCNTs (0, 9) of D_3 symmetry.

The gaps between frontier orbitals of us-SWCNTs of D_{3h}/D_{3d} symmetries decrease during the singlet-triplet transition that is caused by the unsymmetrical changes of orbital energies. As a result, the energies of the HOMO(↓) and LUMO(↑) are close to HOMO and LUMO levels of the singlet state. That points out to the destabilization of the HOMO and stabilization of the LUMO. As follows from the DOS-spectrum, the asymmetric shift of the electron energies at the singlet-triplet transition defines the spin noncompensation at the Fermi level and, hence, leads to a possibility of spin-dependent transport of charge carriers.

According to the aim of the investigation, the difference of total energies at the singlet and triplet states was analyzed in the relation to the nanotube length. It is remarkably that the singlet-triplet transition energy (the difference $\Delta = E_{tot}(S = 0) - E_{tot}(S = 1)$, where E_{tot} and S are the total energy and total electron spin of the nanotube) is positive in value and inversely proportional to the number of segments (figure 3). The ground state corresponds to the singlet state. The energy of the transition to the triplet state corresponds to the IR – range of spectrum. Since the probability of a direct transition from the singlet to the triplet state is low, switching can be practically realized by a laser radiation through intermediaries or an additional influence on the system, for example, by an external electric field.

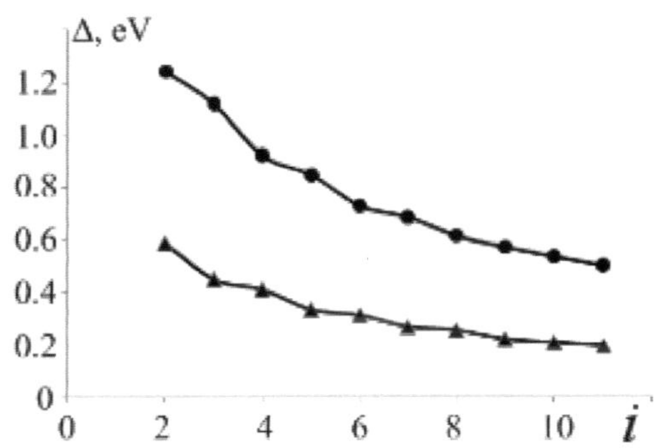

(a)

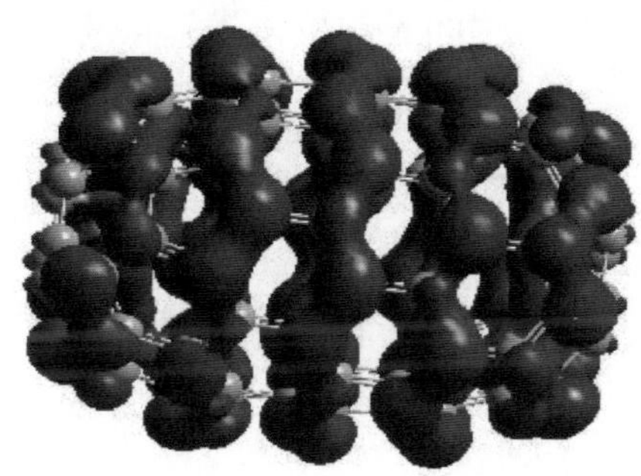
(b)

Figure 3. The singlet-triplet transition energy versus the number of segments of the capped us–SWCNTs (0, 9) of D_{3h}/D_{3d} (circles) and D_3 (triangles) symmetries (a) and the visualization of the electron spin density distribution (b)

The electron spin density is localized on the carbon atom segments which form the us-SWCNTs (0, 9) body (figure 1 (a), figure 3 (b)). The segment addition does not lead to the qualitative change of the electron density distribution. Therefore, the ionization potential, electron affinity, energy gap, work function, energy of the singlet-triplet transition and value of the spin polarization depend monotonically on the length of us-SWCNTs (0, 9).

4. Conclusions

The size confinement in the capped zigzag us-SWCNTs (0, 9) causes the size-dependent restructuring of the electronic structure. All main us-SWCNT parameters at the singlet and triplet spin states, such as the energy gap, ionization potential, electron affinity and work function, monotonically decrease with the increase of the nanotube length. The thermodynamic stability, non-zero energy gap, the energy of the singlet-triplet transition energy lying in the IR-range, the wide length range of the spin polarization existence and the increase of the difference of electroconductivities for two spin channels with the increase of the length of us-SWCNTs (0, 9) of D_{3h}/D_{3d} symmetry determine the promise for their application as materials for design of the element base for spintronics and optoelectronics.

Acknowledgments
The work was supported by the Federal Target Program "Research and development on priority area of scientific-technological complex of Russia for 2014 - 2020 years", the agreement № 14.574.21.0112 from 10.21.2014, the unique identifier of the project is RFMEFI57414X0112.

References

[1] Bhaattacharya S, Akande A and Sanvito S 2014 *Chemical Communications* **50** 6626
[2] Awschalom D D, Flatte M E 2007 *Nature Communications* **3** 153
[3] Murat A, Rungger I, Jin C, Sanvito S and Schwingenschlögl U 2014 *Journal of Physical Chemistry C* **118** 3319
[4] Sun L, Weil P, Wei J, Sanvito S and Houl S 2011 *Journal of Physics: Condensed Matter* **23** 425301
[5] Wu J, and Hagelberg F 2010 *Physical Review B* **81** 155407
[6] Kamil L, Ritschel M, Albrecht L, Krupskaya Y, Buchner B, Klingeler R 2010 *Journal of Physics: Conference Series* **200** 072061
[7] Minot E D , Yaish Y, Sazonova V, McEuen P 2004 *Nature* **428** 536
[8] Sanchez-Valencia J R, Dienel T, Gröning O, Shorubalko I, Mueller A, Jansen M, Amsharov K, Ruffieux P, Fasel R 2014 *Nature* **512** 61
[9] Kato T, Hatakeyama R 2010 *ACS Nano* **4** 7395
[10] Tuchin A V, Nestrugina A V, Bityutskaya L A, Bormontov E N 2014 *Journal of Physics: Conference Series* **541** 012008
[11] Cioslowski J, Rao N and Moncrieff D 2002 *Journal of the American Chemical Society* **124** 8485
[12] Rocherfort A, Salahub D R and Avouris 1999 *Journal of Physical Chemistry B* **103** 641
[13] Buonocore F, Trani F, Ninno D, Matteo A Di, Cantele G and Iadonisi G 2008 *Nanotechnology* **19** 025711
[14] Wang B-C, Wang H-W, Lin I-C, Lin Y-S, Chou Y-M and Chiu H-L 2002 *Journal of the Chinese Chemical Society* **50** 939
[15] Tuchin A V, Ganin A A, Zhukalin D A, Bitytskaya L A and Bormontov E N 2014 *Recent Advances in Biomedical & Chemical Engineering and Materials Science* **1** 40
[16] Lu D, Li Y, Rotkin S V, Ravaioli U, Schulten K 2004 *Nano Letters* **4** 2383
[17] Yumura T, Hirahara K, Bandow S, Yoshizava, Iijima S 2004 *Chemical Physics Letters* **386** 38
[18] Parker S F, Bennington S M, Taylor J W, Herman H, Silverwood I, Albers P, Refson K 2011 *Physical Chemistry Chemical Physics* **13** 11192
[19] Tuchin A V, Bityutskaya L A, Bormontov E N 2015 *European Physical Journal D* **69** 1
[20] Schettino V, Pagliai M., and Cardini G. 2002 *Journal of Physical Chemistry A* **106** 1815
[21] Hertel I V, Steger H, de Vries J, Weisser B, Menzel C, Kamke W 1992 *Physical Review Letters* **68** 784
[22] Yoo R K, Ruscic B, Berkowitz J 1992 *Journal of Chemical Physics* **96** 911
[23] de Vries J, Steger H, Kamke B, Menzel C, Weisser B, Kamke W, Hertel I V 1992 *Chemical Physics Letters* **188** 159
[24] H. Steger, J. Holzapfel, A. Hielscher, W. Kamke, I. V. Hertel 1995 *Chemical Physics Letters* **234** 455
[25] Brink C, Andersen L H, Hvelplund P, Mathur D, Voldstad J D 1995 *Chemical Physics Letters* **233** 52

[26] Wang X B, Ding C F and Wang L S 1999 *Journal of Chemical Physics*. **110** 8217
[27] Dresselhaus M S, Dresselhaus G and Saito R 1995 *Carbon* **33** 883
[28] Saito R, Fujita M, Dresselhaus G and Dresselhaus M 1992 *Physical Review B* **46** 1804
[29] Odom T W, Huang J L, Kim P and Lieber C M 1998 *Nature* **391** 62
[30] Ouyang M, Huang J L and Lieber C M 2002 *Accounts of Chemical Research* **35** 1018

Optical properties of quazi-zero dimensional structures with two-electron impurity centers

V D Krevchik, A V Razumov, P S Budyansky and I M Moiko

Faculty of Engineering, IT and Electronics, Department of Physics, Penza State University, Krasnaya 40, Penza 440026, Russia

E-mail: physics@pnzgu.ru

Abstract. A generalization of the zero-radius potential (ZRP) method in case of two-electron impurity centers in quantum dots with zero nuclear charge was made in this work. As a part of the semi-empirical model of the variational method, there was obtained an analytical equation for the first ionization potential of the two-electron impurity center. In the dipole approximation, the coefficient of impurity absorption of light at the two-electron double ionization of impurity centers in case of quantum dots synthesized in a transparent dielectric matrix is calculated.

1. Introduction

A large number of impurities in semiconductors in bound state may have not one, but two electrons, therefore behave like helium-like impurity centers. The difficulty of theoretical study of two-electron impurity states are due to the complex nature of the interaction of electrons in the outer shell of the impurity center with the valence electrons of the nearest lattice atoms [1]. As a result, it is impossible to predict not only the position of multicharge levels, but also their possible charge states. Of particular interest is the process of double photoionization of two-electron impurity centers with a single photon. This reaction is one of the fundamental reactions involving multiple particles. Theoretical approaches to the study of such reactions are developed in works [2, 3]. The purpose of this work is to calculate the first ionization potential of a two-electron impurity center in a quantum dot (QD) by using the variation method, and to study theoretically the features of double photoionization of two-electron impurity centers in a quazi-zero dimensional structure which is a dielectric matrix with synthesized QDs.

2. The energy spectrum of a two-electron impurity center in a semiconductor QD

Let's consider a semi-empirical model of two-electron impurity centers in QDs. Short-range potential in this model is approximated by a potential well of depth V_0, radius of which d is much smaller than the radius of the localized state. As an empirical parameter, we take the energy of doubly ionized impurity, i.e., the second ionization potential, E_2. Thus, the task is to calculate the first ionization potential, E_1, the value of which can be calculated by the Ritz variation method. The chosen units of length and energy are, respectively, $a_d = \varepsilon\hbar^2/m^* e^2$ – the effective Bohr radius and $E_d = \hbar^2/2m^* a_d^2$ – the effective Bohr energy. At the same time, it is recognized that the power of short-range potential,

$V_0 d^2$ in case of $d \to 0$ remains finite. The two-electron wave function $\Psi(\rho_1,\rho_2)$ satisfies the Schrödinger equation

$$\mathbf{H}(\rho_1,\rho_2)\Psi(\rho_1,\rho_2)=E\Psi(\rho_1,\rho_2), \tag{1}$$

where $\mathbf{H}(\rho_1,\rho_2)=\mathbf{H}(\rho_1)+\mathbf{H}(\rho_2)+\frac{\hbar^2}{m^* a_d^2 |\vec{\rho}_1-\vec{\rho}_2|}$, $\mathbf{H}(\rho_i)=-\frac{\hbar^2}{2m^* a_d^2}\Delta_i-\frac{\hbar^2}{2m^* a_d^2}V(\rho_i)$, $\rho_d=d/a_d$,

$V(\rho_i)=\begin{cases} V_0^*, \ \rho_i \le \rho_d \\ 0, \quad \rho_i > \rho_d \end{cases}$, $V_0^*=V_0/E_d$, $\rho_i=r_i/a_d$, r_i – coordinates of electrons in case of $i=1,2$, m^* – the effective mass of electron.

Usually, the simplest form of the two-electron wave function $\Psi(\rho_1,\rho_2)$ leading to a satisfactory approximation in helium-like centers is the product of one-electron wave functions.

$$\Psi(\rho_1,\rho_2)=\Psi(\rho_1)\Psi(\rho_2), \tag{2}$$

Using the results of [4], in the model of a spherically symmetrical potential well (the "hard walls" model), an equation for the wave function of an electron localized on the short-range potential in a QD $\Psi(\rho_i)$ is:

$$\Psi(\rho_i)=\frac{B}{\rho_i}\begin{cases} \dfrac{\operatorname{sh}\left(R_0^*\eta^{-1}-\rho_d\eta^{-1}\right)}{\operatorname{sh}\left(R_0^*\eta^{-1}\right)}\dfrac{\sin(\chi_0\rho_i)}{\sin(\chi_0\rho_d)}, \ \rho_i \le \rho_d \\ \dfrac{\operatorname{sh}\left(R_0^*\eta^{-1}-\rho_i\eta^{-1}\right)}{\operatorname{sh}\left(R_0^*\eta^{-1}\right)}, \ \rho_i \ge \rho_d \end{cases}, \tag{3}$$

where $\chi_0=\sqrt{V_0^*-\eta^{-2}}$, $V_0^*=V_0/E_d$, $\eta=\sqrt{E_d/|E_2|}$, $R_0^*=R_0/a_d$, R_0 – the QD radius, $B=\sqrt{2\eta^{-1}/\left(\operatorname{th}\left(R_0^*\eta^{-1}\right)-R_0^*\eta^{-1}\operatorname{cosec}\left(R_0^*\eta^{-1}\right)\right)}$.

In view of (3), the trial two-electron wave function can be written as:

$$\Psi(\rho_1,\rho_2)=\frac{B^2}{\rho_1\rho_2}\begin{cases} \dfrac{\operatorname{sh}^2\left(R_0^*\eta^{-1}-\rho_d\eta^{-1}\right)}{\operatorname{sh}^2\left(R_0^*\eta^{-1}\right)}\dfrac{\sin(\chi_0\rho_1)\sin(\chi_0\rho_2)}{\sin^2(\chi_0\rho_d)}, \ \rho_i \le \rho_d \\ \dfrac{\operatorname{sh}\left(R_0^*\eta^{-1}-\rho_1\eta^{-1}\right)\operatorname{sh}\left(R_0^*\eta^{-1}-\rho_2\eta^{-1}\right)}{\operatorname{sh}^2\left(R_0^*\eta^{-1}\right)}, \ \rho_i \ge \rho_d. \end{cases} \tag{4}$$

The energy of a two-electron impurity center is determined by the minimum average value of the Hamiltonian $\varepsilon\left(R_0^*,\eta\right)$:

$$\varepsilon\left(R_0^*,\eta\right)=E_d\left\langle\Psi(\rho_1,\rho_2)\middle|\mathbf{H}(\rho_1,\rho_2)\middle|\Psi(\rho_1,\rho_2)\right\rangle\Big/\left\langle\left|\Psi(\rho_1,\rho_2)\right|^2\right\rangle. \tag{5}$$

Making a limit pass ($d \to 0$) for $\varepsilon\left(R_0^*,\eta\right)$, the result is:

$$\varepsilon\left(R_0^*,\eta\right)=-\frac{2^6\eta^{-2}\pi^2E_d}{\operatorname{sh}^4\left(R_0^*\eta^{-1}\right)\left(\operatorname{th}\left(R_0^*\eta^{-1}\right)-R_0^*\eta^{-1}\operatorname{cosec}\left(R_0^*\eta^{-1}\right)\right)^2}\times$$
$$\times\left\{2^{-4}\left(2\eta^{-1}R_0^*-\operatorname{sh}\left(2\eta^{-1}R_0^*\right)\right)^2+\frac{\eta}{2^3}\times\right.$$
$$\times\left[-4\eta^{-1}R_0^*-8\operatorname{ch}^3\left(\eta^{-1}R_0^*\right)\operatorname{sh}\left(\eta^{-1}R_0^*\right)\operatorname{Chi}\left(2\eta^{-1}R_0^*\right)+2\ln\left(2\exp(1+C)\eta^{-1}R_0^*\right)\times\right.$$
$$\times\operatorname{sh}\left(2\eta^{-1}R_0^*\right)+\left(\operatorname{Chi}\left(4\eta^{-1}R_0^*\right)-\ln(2)\right)\operatorname{sh}\left(4\eta^{-1}R_0^*\right)-\operatorname{Shi}\left(2\eta^{-1}R_0^*\right)\left(1-2\operatorname{ch}\left(2\eta^{-1}R_0^*\right)-\right.$$
$$\left.\left.\left.-\operatorname{ch}\left(4\eta^{-1}R_0^*\right)\right)+\operatorname{ch}\left(4\eta^{-1}R_0^*\right)\operatorname{Shi}\left(4\eta^{-1}R_0^*\right)\right]\right\}, \qquad (6)$$

where C – the Euler constant.

Minimization of the parameter η leads to a rather cumbersome transcendental equation for finding extreme values η_{extr}, that is not shown in this work.

In figure 1, the relation between the first and the second ionization potentials of a two-electron impurity center with zero nuclear charge in semiconductor QDs in units of Bohr energy E_d, obtained by numerical calculations (curves 1,2,3) is shown.

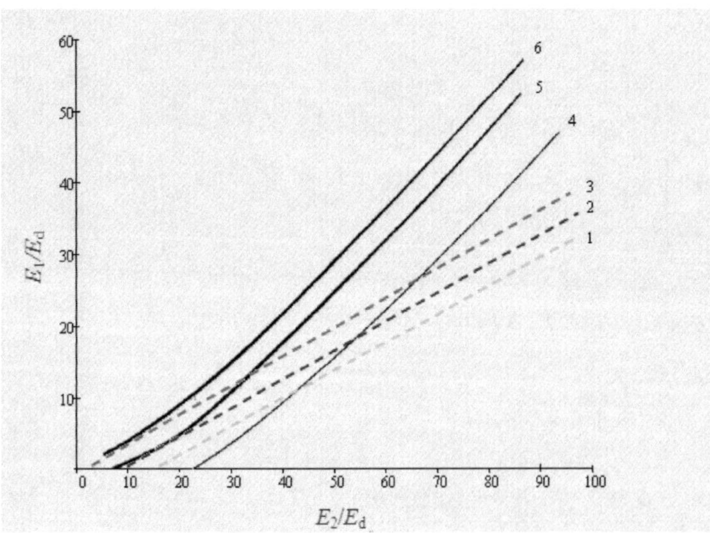

Figure 1. Relation between the first and the second ionization potentials of a two-electron impurity center with zero nuclear charge in semiconductor QDs (curves 1, 2, 3), for various values of the QD radius (in Bohr units) R_0^*: $1-R_0^*=3$; $2-R_0^*=1$; $3-R_0^*=0,5$ and in the semiconductor bulk [1] (curves 4, 5, 6) for core Z of a two-electron impurity center: $4-Z=0$; $5-Z=1$; $6-Z=2$.

As shown in figure 1, the energy of mutual attraction is not compensated by the corresponding Coulomb attraction to the nucleus (Z = 0) because the threshold value of E_2/E_d, from which the existence of a two-electron bound state is possible, is very large. In case of QDs (curves 1, 2, 3 in figure 1), dimensional quantization energy growth and reduction in the QD radius causes reduction in the threshold value of E_2/E_d.

3. The light absorption coefficient at the double photoionization of a two-electron impurity center in a quazi-zero dimensional structure

The atoms of helium and helium-like ions in the ground state are the simple systems, in which it is possible to observe the double photoionization by a single photon. In the study of semiconductor nanostructures, such a reaction is of a particular interest due to a new situation caused by size quantization. In this section, we will look at the photoionization of a two-electron impurity center in the ground state in a semiconductor QD with the wave function of the form:

$$\Psi(\rho_1,\rho_2)=\frac{2^2\eta^{-2}}{\left(\mathrm{th}\left(R_0^*\eta^{-1}\right)-R_0^*\eta^{-1}\,\mathrm{cosec}\left(R_0^*\eta^{-1}\right)\right)^2}\frac{\mathrm{sh}\left(R_0^*\eta^{-1}-\rho_1\eta^{-1}\right)\mathrm{sh}\left(R_0^*\eta^{-1}-\rho_2\eta^{-1}\right)}{\rho_1\rho_2\,\mathrm{sh}^2\left(R_0^*\eta^{-1}\right)}. \quad (7)$$

The wave function of the final state is determined by the product of one-electron wave functions in a spherical QD:

$$\Phi(\rho_1,\rho_2)=\Psi_{\mathrm{n,l,m}}(\rho_1,\varphi_1,\theta_1)\Psi_{\mathrm{n,l,m}}(\rho_2,\varphi_2,\theta_2), \quad (8)$$

where:

$$\Psi_{\mathrm{n,l,m}}(\rho_i,\varphi_i,\theta_i)=\frac{\sqrt{2}}{R_0\sqrt{\rho_i}}\frac{J_{l+1/2}\left(k_{nl}\rho_i\right)}{J_{l+3/2}\left(\xi_{nl}\right)}Y_{lm}(\varphi_i,\theta_i), \quad (9)$$

here $i = 1,2$; $Y_{l,m}$ – normalized spherical functions; l, m – orbital and magnetic quantum numbers; $J_\nu(x)$ – the Bessel functions of the first kind of order ν; $k_{nl}=\xi_{nl}/R_0$; ξ_{nl} – n-th root of the Bessel function of the l-th order.

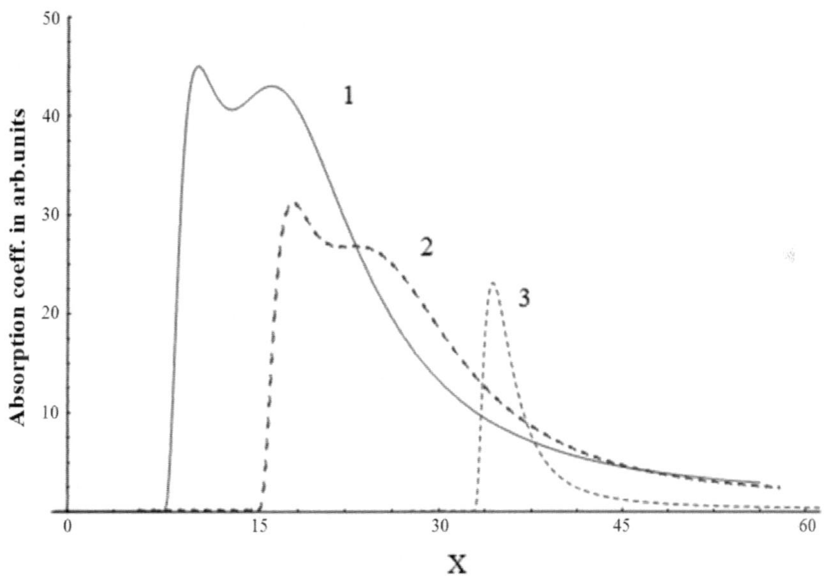

Figure 2. The spectral dependence of the absorption coefficient of light in photoionization of two-electron impurity centers in case of $\overline{R}_0^*=1$: 1 – $E_2 = 0,04$ eV ($E_1 = 0,012$ eV); 2 – $E_2 = 0,05$ eV ($E_1 = 0,015$ eV); 3 – $E_2 = 0,08$ eV ($E_1 = 0,027$ eV); $X=\hbar\omega/E_d$.

The energy of one-electron states undisturbed by impurities in a spherical QD is defined as:

$$E_{n,l}=\frac{\hbar^2\xi_{nl}^2}{2m^*R_0^2}. \quad (10)$$

The effective Hamiltonian of the interaction with the field of the light wave with a single

Journal of Physics: Conference Series **690** (2016) 012027 doi:10.1088/1742-6596/690/1/012027

polarization vector $\mathbf{e}_\lambda$ and wave vector $\mathbf{q}$ is determined by the equation:

$$\mathbf{H}_{\text{int}} = -i\lambda_0 \hbar \left(\frac{2\pi\hbar^2 \alpha^*}{\varepsilon\omega m^{*2}} I_0 \right)^{1/2} \exp(i\mathbf{qr})(\mathbf{e}_\lambda \nabla_{\mathbf{r}}), \tag{11}$$

where λ_0 – the local field coefficient that takes into account the difference in amplitudes of the local and average macroscopic fields; I_0 – the light intensity; ω – the frequency of absorbed light; ε – the static permittivity of the QD material; α^* – the fine structure constant based on the dialectical constant.

The matrix element which determines the amount of oscillator strength of dipole optical transitions of electrons from the ground state of the two-electron impurity center (6) in the state $\Psi_{n,m,l}(\rho,\varphi,\theta)$ of discrete spectrum of a QD, can be written as:

$$M = i\lambda_0 \sqrt{\frac{2\pi\alpha^*}{\omega} I_0} \Big[\left(E_{n,l,m} - E_1\right) \left\langle \Psi^*_{n,l,m}(\rho_1,\theta_1,\varphi_1)\psi^*(\rho_2) \right| (\mathbf{e}_\lambda, \mathbf{r}_1) \left| \Psi(\rho_1,\rho_2) \right\rangle + \\ + \left(E_{n,l,m} - E_2\right) \left\langle \Psi^*_{n,l,m}(\rho_1,\theta_1,\varphi_1)\psi^*(\rho_2) \right| (\mathbf{e}_\lambda, \mathbf{r}_2) \left| \Psi(\rho_1,\rho_2) \right\rangle \Big]. \tag{12}$$

We assume that the variance u of the QD size occurs during the decay phase of the supersaturated solid solution and is satisfactorily described by the Lifshitz-Slezov equation [5]:

$$P(u) = \begin{cases} \dfrac{3^4 e u^2 \exp\left[-1/(1-2u/3)\right]}{2^{\frac{5}{3}} (u+3)^{\frac{7}{3}} (3/2-u)^{\frac{11}{3}}}, & u < \dfrac{3}{2}, \\ 0, & u > \dfrac{3}{2}, \end{cases} \tag{13}$$

where $u = R_0 / \overline{R_0}$, R_0 и $\overline{R_0}$ – the QD radius and its average value, respectively ; e – the base of the natural logarithm.

Taking into account the QD size dispersion, the coefficient of impurity light absorption, $K(\omega)$, can be represented as:

$$K(X) = \sum_{n=1}^{N} P(u_n) \frac{2^3 \lambda_0^2 \pi \alpha^*}{X} \times \frac{E_d \left(\eta^{-2} + \xi_{n1}^2 \overline{R}_0^{*-2} u_n^{-2}\right)^{-6}}{a_d^2 \overline{R}_0^{*2} u_n^2 \operatorname{sh}^6\left(\overline{R}_0^* u_n \eta^{-1}\right) \pi \xi_{n1} \overline{R}_0^{*-1} u_n^{-1} \left| J_{l+3/2}(\xi_{nl}) \right|^2} \times \\ \times \frac{2^3 \eta^{-3}}{\left(\operatorname{th}\left(\overline{R}_0^* u_n \eta^{-1}\right) - \overline{R}_0^* u_n \eta^{-1} \operatorname{cosec}\left(\overline{R}_0^* u_n \eta^{-1}\right)\right)^3} \left(2\xi_{nl}^2 \overline{R}_0^{*-2} u_n^{-2} + |E_1|/E_d + |E_2|/E_d\right)^2 \times \\ \times \xi_{n1}^3 \overline{R}_0^{*-3} u_n^{-3} \left| \sqrt{\xi_{n1} \overline{R}_0^* + i\eta^{-1}} \left(\xi_{n1} \overline{R}_0^* - i2\eta^{-1}\right) \operatorname{ch}\left(\overline{R}_0^* u_n \eta^{-1}\right) \times \right. \\ \times \operatorname{Si}\left(\sqrt{\frac{2}{\pi} \overline{R}_0^* u_n \left(\xi_{n1} \overline{R}_0^{*-1} u_n^{-1} - i\eta^{-1}\right)} \right) + i\sqrt{\xi_{n1} \overline{R}_0^{*-1} u_n^{-1} - i\eta^{-1}} \left(\xi_{n1} \overline{R}_0^{*-1} u_n^{-1} + i2\eta^{-1}\right) \operatorname{ch}\left(\overline{R}_0^* u_n \eta^{-1}\right) \times \\ \times \operatorname{Si}\left(\sqrt{\frac{2}{\pi} \overline{R}_0^* u_n \left(\xi_{n1} \overline{R}_0^{*-1} u_n^{-1} + i\eta^{-1}\right)} \right) + \sqrt{\xi_{n1} \overline{R}_0^{*-1} u_n^{-1} + i\eta^{-1}} \left(\xi_{n1} \overline{R}_0^{*-1} u_n^{-1} - i2\eta^{-1}\right) \operatorname{sh}\left(\overline{R}_0^* u_n \eta^{-1}\right) \times$$

$$
\times \mathrm{Ci}\left(\sqrt{\frac{2}{\pi}\bar{R}_0^* u_n\left(\xi_{n1}\bar{R}_0^{*-1}u_n^{-1}-i\eta^{-1}\right)}\right)+\sqrt{\xi_{n1}\bar{R}_0^{*-1}u_n^{-1}-i\eta^{-1}}\left(\xi_{n1}\bar{R}_0^{*-1}u_n^{-1}+i2\eta^{-1}\right)\mathrm{sh}\left(\bar{R}_0^* u_n\eta^{-1}\right)\times
$$

$$
\times \mathrm{Ci}\left(\sqrt{\frac{2}{\pi}\bar{R}_0^* u_n\left(\xi_{n1}\bar{R}_0^{*-1}u_n^{-1}+i\eta^{-1}\right)}\right)\Bigg]\Bigg|^2 \Big|\eta^{-1}\Big[\left(\eta^{-2}+\xi_{n1}^2\bar{R}_0^{*-2}u_n^{-2}\right)\bar{R}_0^* u_n\cos\left(\xi_{n1}\right)-\bar{R}_0^* u_n\sin\left(\xi_{n1}\right)\times
$$

$$
\times\left(\eta^{-2}+2\xi_{n1}\bar{R}_0^{*-1}u_n^{-1}+\xi_{n1}^2\bar{R}_0^{*-2}u_n^{-2}\right)\Big]+\left(\eta^{-2}\left(\xi_{n1}\bar{R}_0^{*-1}u_n^{-1}-1\right)+k_{nl}^2\left(\xi_{n1}\bar{R}_0^{*-1}u_n^{-1}+1\right)\right)\mathrm{sh}\left(\bar{R}_0^* u_n\eta^{-1}\right)\Big|^2, \quad (14)
$$

where $u_n=\xi_{n1}^2\big/\left(\bar{R}_0^{*2}\left(X-|E_1|/E_d-|E_2|/E_d\right)\right)$; N – is an integral part of the solution of the transcendent equation $\xi_{n1}^2=3\bar{R}_0^{*2}\left(X-|E_1|/E_d-|E_2|/E_d\right)/2$.

The spectral absorption coefficient of the light associated with the double photoionization of two-electron impurity centers is shown in figure 2. As can be seen, the characteristic of double photoionization of helium atoms is the emergence of a "double-hump" profile on the photo-absorption curve, the distance between the peaks in the tops of the spectral curve corresponds to the difference between the ionization potentials. In figure 2, dynamics of the second peak with growth of E_2/E_d is also shown: the second peak transforms into a step (curve 2 in figure 2) and then disappears (curve 3 in figure 2). Indeed, with growth of E_2/E_d the bond energy of two-electron impurity state grows as well, and at a certain value of E_2/E_d the double photoionization becomes impossible.

3. Conclusion

In this work, generalization of the ZRP in case of two-electron impurity centers in QDs was made. In the semi-empirical model of a two-electron impurity center in a spherical QD, the first ionization potential was calculated. It is shown that in contrast to the bulk semiconductor, because of the quantum size effect, the existence of two-electron bound impurity states with zero nuclear charge in QDs is possible only at lower threshold values of E_2/E_d.

In the dipole approximation in the frame of the effective mass method, the impurity light absorption coefficient for a quazi-zero dimensional structure at the double ionization of a two-electron impurity center by a single photon is calculated. It is shown that a characteristic feature of the spectrum of double photoionization is a "double-humped" profile of the spectral curve. The distance between peaks of the spectral curve is determined by the difference between the first and the second ionization potentials of the two-electron impurity center.

References

[1] Belorusets E D and Grinberg A A On the energy spectrum of multiply charged impurity centers in semiconductors 1978 *Solid State Physics* **12** 1970

[2] Knyr V A and Zaitsev S A The reaction of the double photoionization of atoms in the J-matrix approach (in Russian) 2010 *Bulletin of PNU* **1(16)**, 13

[3] Mikhailov A I and Mikhailov I A Double ionization and excitation ionization in the Compton scattering of high-energy photons for metastable states of heliumlike ions 1999 *JETP* **89** 1023.

[4] Krevchik V D and Levashov A V Energy spectrum and optical properties of quantum dot–impurity center complex 2002 *Semiconductors* **36** 208

[5] Lifshitz L M and Slezov V V Kinetics of diffusive decomposition of supersaturated solid solutions 1958 *JETP* **8** 331

Journal of Physics: Conference Series **690** (2016) 012028 doi:10.1088/1742-6596/690/1/012028

Quantum phase transition and conductivity of parallel quantum dots with a moderate Coulomb interaction

V S Protsenko[1,2], **A A Katanin**[1,2]

[1] Institute of Metal Physics, Kovalevskaya str. 18, Ekaterinburg 620990, Russia
[2] Ural Federal University, Mira str. 19, Ekaterinburg 620002, Russia

E-mail: protsenko.vladimir15@mail.ru, katanin@mail.ru

Abstract. We study the effects of electronic interactions on transport through parallel quantum dots connected symmetrically to leads, focusing on the case of an intermediate value of the on–site Coulomb interaction at each quantum dot. We apply both the mean-field (MF) approximation and the functional renormalization group (fRG) approach with using the Litim regularization scheme in frequency space to treat the effect of interaction and calculate the energy levels, the magnetization, the square of local spin as well as the linear conductance as a function of gate voltage. Already for intermediate values of the Coulomb interaction, the system exhibits a quantum phase transition from the parallel aligned (high spin state) to the paramagnetic (low spin) regime. This phase transition is accompanied by the appearance of the step discontinuities in the conductance, which can be observed in the experiment.

1. Introduction

A quantum dot is formed by a nanostructure that restricts the motion of electrons in all three directions. The Coulomb interaction can play a crucial role in the electron transport through quantum dots. Even in the simplest system consisting of a single quantum dot at sufficiently low temperatures, the Coulomb interaction can lead to the Kondo plateau in the total conductance [1, 2].

Especially interesting cases are those in which electrons can pass through the mesoscopic system following different paths, which can result in quantum interference effects [3, 4]. In the article [5] within the numerical renormalization group [6] it has been shown that the strong Coulomb interaction in such geometries leads to the parallel spin alignment state as the ground state of the systems in some range of gate voltages. Moreover, it has been found that this phase transition is accompanied by the appearance of the resonance in conductance. On the other hand, in the non-interacting case the non-magnetic state is realized. Therefore, the problem of influence of the intermediate Coulomb interaction on the spin alignment and electron transport is not trivial.

In this paper, we consider the simplest ring configuration, namely parallel quantum dots connected to two common leads. We focus on the case of an intermediate value of the on-site Coulomb interaction U in each quantum dot, and introduce a weak magnetic field H. The local Coulomb interaction is treated within the functional renormalization group (fRG) approach [2, 7–9] with using the Litim regularization scheme [10]. As a result, via fRG we calculate the energy levels, the magnetization, the square of local spin as well as the linear conductance as function of the gate voltage.

2. The model

The system of two quantum dots coupled in parallel, which are connected to common leads, can be described by the following Hamiltonian:

$$\mathcal{H} = \mathcal{H}_l + \mathcal{H}_d + \mathcal{H}_c. \tag{1}$$

Here $\mathcal{H}_l$ is the part of the Hamiltonian that describes the leads. We assume that the left and the right leads are equivalent and described by a non-interacting Hamiltonian. The next term in (1) is the Hamiltonian of the quantum dots:

$$\mathcal{H}_d = \sum_{\sigma}\sum_{j=1}^{2} \epsilon_{j,\sigma} d^{\dagger}_{j,\sigma} d_{j,\sigma} + \sum_{\sigma}\sum_{j=1}^{2} \frac{U_j}{2}\left(d^{\dagger}_{j,\sigma} d_{j,\sigma} - \frac{1}{2}\right)\left(d^{\dagger}_{j,\bar{\sigma}} d_{j,\bar{\sigma}} - \frac{1}{2}\right), \tag{2}$$

where $d^{\dagger}_{j,\sigma}(d_{j,\sigma})$ denote creation (annihilation) operators for an electron with spin $\sigma \in \{\uparrow,\downarrow\}$ localized on the j-th ($j \in \{1,2\}$) quantum dot. The index $\bar{\sigma}$ is the complement of σ ($\bar{\sigma} =\uparrow$ if $\sigma =\downarrow$ and vice versa). In general, each quantum dot level $\epsilon_{j,\sigma}$ consists of a constant part ϵ_j^0 and the part which includes the gate voltage V_g, as well as the magnetic field $\mathcal{H}$:

$$\epsilon_{j,\sigma} = \epsilon_j^0 + V_g - \sigma\frac{\mathcal{H}}{2}. \tag{3}$$

Therefore, the level positions can be shifted by applying a gate voltage or a magnetic field. The second term in $\mathcal{H}_d$ accounts for the two–particle interaction between electrons, where the parameter U_j is the on–site Coulomb repulsion of spin up and down electrons. Finally, the last terms in $\mathcal{H}$ take into account the coupling between dots and leads. It is given by:

$$\mathcal{H}_c = -\sum_{\sigma}\sum_{l=L,R}\sum_{j}(t^l_j c^{\dagger}_{0,\sigma,l} d_{j,\sigma} + \text{H.c.}), \tag{4}$$

with leads creation (annihilation) operators $c^{\dagger}_{0,\sigma,l}$ ($c_{0,\sigma,l}$) for an electron with spin direction σ on the first lattice site of the left or right lead (l=L or l=R respectively). Here t^l_j is the tunnel matrix element between a lead l and a quantum dot j.

In this paper we focus on the symmetric coupling case with equal local interaction $U_j = U$ $(j \in \{1,2\})$ in quantum dots. In this case, the inverse of the noninteracting propagator reads as [2]

$$\mathcal{G}_0^{-1}(i\omega,\sigma) = \begin{pmatrix} i\omega - V_g + \sigma\frac{H}{2} + i\frac{\Gamma}{2}\,\text{sgn}(w) & i\frac{\Gamma}{2}\,\text{sgn}(w) \\ i\frac{\Gamma}{2}\,\text{sgn}(w) & i\omega - V_g + \sigma\frac{H}{2} + i\frac{\Gamma}{2}\,\text{sgn}(w) \end{pmatrix}, \tag{5}$$

where we have introduced the frequency independent hybridization function

$$\Gamma = \sum_{j,l}\Gamma^l_j = \pi\sum_{j,l}|t^l_j|^2\rho_{lead}, \tag{6}$$

ρ_{lead} is the local density of states of the leads at the quantum dot site 0.

The full Green function $\mathcal{G}$ is given by the Dyson equation $\mathcal{G}^{-1} = \mathcal{G}_0^{-1} - \Sigma$, where Σ is a self-energy of the system.

3. The method

In this section we will briefly review the recently proposed method of the functional renormalization group [2, 7-9, 11]. It inherits the basic ideas of the Wilsonian RG [12] and allows to express the Wilson mode elimination in terms of an exact functional differential equation for the generating functionals.

In our case, primary quantities of interest are the irreducible vertex functions. More precisely, we are mainly interested in the self-energy (one–particle vertex), since it allows us to obtain directly the Green function of the interacting quantum dots system in the presence of the contacts, which is our main goal. The one–particle irreducible (1PI) scheme [8] is a suitable choice for this purpose. The RG equations in the 1PI scheme are obtained by inserting $\mathcal{G}_0(\Lambda)$ in the generating functional of the one-particle irreducible vertex functions. An infinite hierarchy of differential equations for 1PI vertex functions is obtained by taking the derivative of generating functional with respect to Λ and then expanding it in powers of the external fields.

We write down the explicit equations for the first two vertex functions, namely for the self–energy (one-particle vertex) Σ^Λ:

$$\partial_\Lambda \Sigma^\Lambda(k^{'};k) = -\sum_{q^{'}q} e^{i\omega_2 0^+} \mathcal{S}^\Lambda(q,q^{'})\Gamma_4^\Lambda(k^{'},q^{'};k,q) \tag{7}$$

and the effective two-particle interactions (two-particle vertex) Γ^Λ:

$$\begin{aligned}
\partial_\Lambda \Gamma_4^\Lambda(k_1^{'},k_2^{'};k_1,k_2) = & \sum_{s,s^{'},q,q^{'}} \mathcal{G}^\Lambda(s,s^{'})\mathcal{S}^\Lambda(q,q^{'}) \\
& \times \Big\{\Gamma_4^\Lambda(k_1^{'},k_2^{'};s,q)\Gamma_4^\Lambda(s^{'},q^{'};k_1,k_2) \\
& - \Big[\Gamma_4^\Lambda(k_1^{'},q^{'};k_1,s)\Gamma_4^\Lambda(s^{'},k_2^{'};q,k_2) + (s \leftrightarrows q, s^{'} \leftrightarrows q^{'})\Big] \\
& + \Big[\Gamma_4^\Lambda(k_2^{'},q^{'};k_1,s)\Gamma_4^\Lambda(s^{'},k_1^{'};q,k_2) + (s \leftrightarrows q, s^{'} \leftrightarrows q^{'})\Big]\Big\} \\
& - \sum_{k,k^{'}} \mathcal{S}^\Lambda(k,k^{'})\Gamma_6^\Lambda(k_1^{'},k_2^{'},k^{'};k_1,k_2,k),
\end{aligned} \tag{8}$$

where the indices $k_i(k_i^{'})$ collect all quantum numbers (in our case they are dot indices $\{i,j\dots\}$, frequencies w, and spin indices σ) and $\mathcal{S}^\Lambda$ denotes the so-called single-scale propagator:

$$\mathcal{S}^\Lambda = \mathcal{G}^\Lambda \partial_\Lambda \left(\mathcal{G}_0^\Lambda\right)^{-1} \mathcal{G}^\Lambda. \tag{9}$$

In order to solve this infinite set of coupled differential equations, it must be truncated in some manner. For this, in the following, we completely neglect the flow of the vertex functions Γ_{2m}^Λ for $m \geqslant 3$. Thus, the three-particle contributions Γ_6^Λ vanish and we have a closed system of the RG equations for Σ^Λ and Γ_4^Λ.

Now we need to specify the explicit dependence of the $\mathcal{G}_0^\Lambda$ on parameter Λ, which will be used in our calculations. It is important because due to the truncation of the infinite hierarchy of RG equations different cutoff schemes are not equivalent. For this reason, an incorrect choice of the cutoff can lead to unphysical divergences of the vertex functions. For example, it was noted that the most frequently used cutoff function – sharp multiplicative cutoff may lead to unphysical behavior of the vertex functions.

In order to avoid this problem, in this article we introduce another regularization method for bare propagator, namely

$$\left[\mathcal{G}_0^\Lambda\right]^{-1} = \left[\mathcal{G}_0\right]^{-1} + i\left(\Lambda - |\omega_n|\right)\Theta\left(\Lambda - |\omega_n|\right)\operatorname{sgn}(\omega_n), \tag{10}$$

which is analogous to proposed by Litim [10]. This smooth additive cutoff has the undoubted advantage, compared with sharp cutoff, since it smoothly switched off the propagator at low frequencies.

4. Results and discussion

At first we consider the effect of the intermediate electron–electron interactions on the position of the energy levels $\epsilon_{j\sigma} = V_g - \sigma\frac{H}{2} + \Sigma_j^\sigma$, where the self-energy Σ can be extracted from the functional renormalization group calculations. These quantities are important because their position relative to the Fermi level greatly determines the conductivity of the system. Note that due to the symmetry of the system, $\epsilon_{1\sigma} = \epsilon_{2\sigma}$. In figure 1 (a) the energy levels of quantum dots $\epsilon_{j\sigma}$ are plotted as a function of the gate voltage V_g. There are two major regions of the gate voltage in which energy levels are split differently.

For $|V_g| > |V_g^c|$ the energy difference between spin up and spin down states is of the order of magnitude of the applied field H. This results in the difference between the average occupation numbers of spin up and down electrons which is also proportional to the magnetic field. One

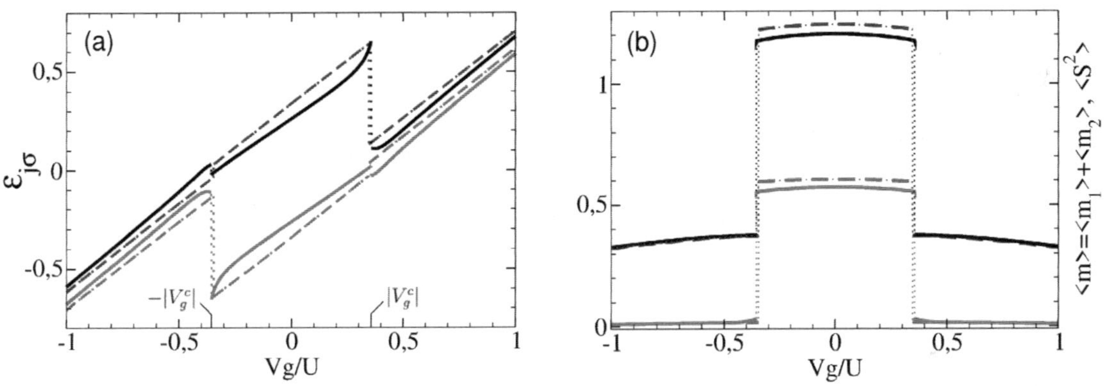

Figure 1. (a) The energy levels $\epsilon_{j\sigma}$ as a function of the gate voltage. Solid up (down) line: $\epsilon_{j\downarrow}$ ($\epsilon_{j\uparrow}$) in the fRG. Dashed up (down) line: $\epsilon_{j\downarrow}$ ($\epsilon_{j\uparrow}$) in the MF. (b) Gate voltage dependence of the average of the square of the total spin $\langle \mathbf{S}^2 \rangle$ and the total magnetization $m = m_1 + m_2$. Upper solid (black) line: $\langle \mathbf{S}^2 \rangle$ in the fRG. Upper dashed (blue) line: $\langle \mathbf{S}^2 \rangle$ in the MF. Lower solid (red) line: m in the fRG. Lower dashed (green) line: m in the MF. The parameters are $U/\Gamma = 1$ and $H/U = 0.08$.

can see that this energy level splitting increases with decreasing the gate voltage up to $|V_g^c|$. Consequently, magnetization $m_j = \frac{1}{2}(\langle n_{j\uparrow}\rangle - \langle n_{j\downarrow}\rangle)$ on each quantum dot begins to grow for $|V_g| \to |V_g^c|$ (figure 1 (b)).

In the range of voltages $|V_g| < |V_g^c|$ the Coulomb interaction leads to the strong splitting of the $\epsilon_{j\uparrow}$ and $\epsilon_{j\downarrow}$ energy levels in an external magnetic field. Now, the energy of the spin up state is much lower than that of the spin down state. Therefore, the occupancy of the spin-up state becomes energetically favoured over the spin-down state. This results in $0 < m_j \sim 1$, $\langle \mathbf{S}^2 \rangle \sim 1$ and a formation of a high spin state on each quantum dot. In this case the RG calculation predicts a very weakly expressed decrease in both the magnetization and the average of the square of the total spin for $|V_g| \to |V_g^c|$.

At the point $|V_g| = |V_g^c|$, when the above two regimes change into one another, the phase transition take place and all the discussed above physical quantities show the jump behavior.

Figure 1 shows the gate voltage dependence of the average of the square of the total spin $\langle \mathbf{S}^2 \rangle$. It shows a sharp increase in $\langle \mathbf{S}^2 \rangle$ as $|V_g|$ falls below $|V_g^c|$. That is a direct indication of the phase transition. Thus, accounting for the intermediate value of the interaction enables us to conclude that in this case we have the same type of phase transition as in the case of the strong Coulomb interaction in quantum dots. Comparison between fRG and mean-field approximation depicted in figure 1 shows qualitative agreement. In the mean-field approximation the phase transition is also accompanied by a jump–like change of the occupation number and the energy of levels. But, one sees that the results of the two methods lead to the different behavior of the curves of the energy levels, as well as the different estimates of the square of the total spin $\langle \mathbf{S}^2 \rangle$ and the total magnetization m for voltage V_g near half-filling.

Knowing the renormalized energy levels and renormalization of the hopping parameter allows us to calculate the quantity of our main interest: the linear-response conductance G as a function of the gate voltage. For the present setup, the linear conductance $G = dI/dV_{sd}$ of an interacting system can be computed as [2]:

$$G = G_0 \frac{\Gamma^2}{8} \sum_{\sigma} \left| \sum_{j,j'} \mathcal{G}^{\sigma}_{j,j'}(0) \right|^2, \tag{11}$$

where $G_0 = \dfrac{2e^2}{h}$ is the conductance quantum.

First we consider the non-interacting case, $U = 0$. After replacing $\mathcal{G} \to \mathcal{G}_0$ in (11) we have:

$$G(V_g) = \frac{e^2}{h} \sum_{\sigma} G_{\sigma}(V_g) = \frac{e^2}{h} \sum_{\sigma} \frac{\Gamma^2}{\left(V_g - \sigma \dfrac{H}{2}\right)^2 + \Gamma^2}, \tag{12}$$

which is a superposition of two Lorentzians with peak positions at $V_g = \sigma \dfrac{H}{2}$.

If we include the interaction between spin up and spin down electrons on both quantum dots, the transport properties of the system changes greatly. Even at the level of mean field

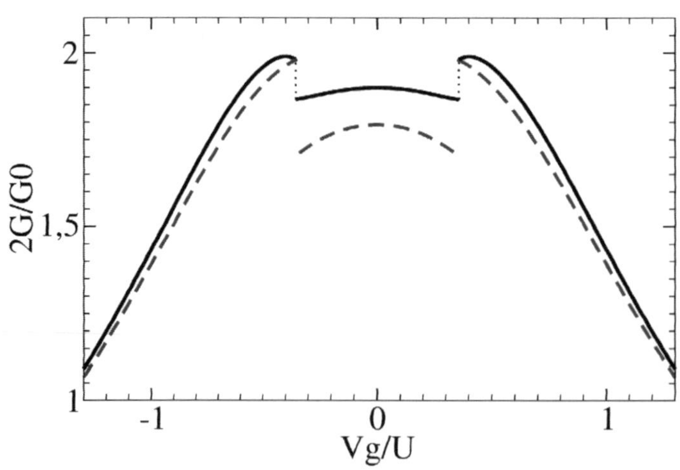

Figure 2. Conductance $2G/G_0$ as a function of V_g/U of the parallel double dot with $U = \Gamma$ and $H = 0.08U$. Solid black line: fRG approach with the Litim cutoff. Dashed blue line: MF approximation.

approximation, the dependence of the conductance on the gate voltage changes its lineshape, the transition between the phases of the different spin symmetries is accompanied by a jump in the conductance (figure 2). We also observe a sharp drop of the conductance in the high spin state. These features of the $G(V_g)$ can be explained via the position of the double quantum dots system energy levels with respect to the Fermi level.

Next, we calculate the gate voltage dependence of the linear conductance using the functional renormalization group scheme. As shown in figure 2, the phase transition is manifested by the appearance of the step discontinuities in the curve of the conductance. Importantly, the drop of the conductance in the high spin state is smaller than in the case of the mean-field approach. Furthermore, even in the low spin (paramagnetic) regime one can see a notable difference in the linear conductance. In this case, the fRG gives higher value of the conductivity at the same gate voltage. We also note that in this case, the conductivity is not completely determined by the position of the $\epsilon_{j\sigma}$ levels. This is due to the fact that the off-diagonal components of the self-energy are also renormalized in a non trivial way. They have a two-fold effect on the conductivity: modify renormalization of $\epsilon_{j\sigma}$ (and vice versa) and explicitly enter into the expression for $G(V_g)$. Thus, the individual contributions of the self-energy to the conductivity are mixed in the obtained results.

5. Conclusion

We found that in the case of an intermediate value of the on-site Coulomb interaction the fRG with the Litim cutoff reproduce correlation effects. The system exhibits a quantum phase transition between the high spin state and a low spin state that is driven by the gate voltage. This phase transition is accompanied by the appearance of the step discontinuities in the conductance, which can be observed in the experiment. In the phase transition region, the magnetization and the energy of levels of the quantum dots computed in the fRG also show a jump–like behavior.

Comparison between fRG and mean-field approximation shows that although they are in qualitative agreement and give the same estimate for the gate voltage of the phase transition, they yield to significantly different behavior of conduction and other considered variables in the spin ordered phase.

Acknowledgments

The work is supported by FASO Russian Federation (theme "Spin" No. 01201463330), grant of the "Dynasty" foundation, and Act 211 Government of the Russian Federation, contract No. 02.A03.21.0006.

References

[1] Andergassen S, Enss T and Meden V 2006 *Phys. Rev.* B **73** 153308
[2] Karrasch C, Enss T and Meden V 2006 *Phys. Rev.* B **73** 235337
[3] Orellana P A, Ladrón de Guevara M L and Claro F 2004 *Phys. Rev.* B **70** 233315
[4] Ladrón de Guevara M L, Claro F and Orellana P A 2003 *Phys. Rev.* B **67** 195335
[5] Zitko R and Bonca J 2007 *Phys. Rev.* B **76** 241305(R)
[6] Bulla R, Costi T and Pruschke T 2008 *Rev. Mod. Phys.* **80** 395
[7] Wetterich C 1993 *Phys. Lett.* B **301** 90
[8] Salmhofer M and Honerkamp C 2001 *Prog. Theor. Phys.* **105** 1
[9] Salmhofer M 1999 *Renormalization* (Berlin: Springer)
[10] Litim D F 2001 *Phys. Rev.* D **64** 105007
[11] Salmhofer M, Honerkamp C, Metzner W, and Lauscher O 2004 *Phys. Rev.* B **112** 943
[12] Wilson K G 1975 *Phys. Rev.* B **47** 773

Tunnel volt-ampere characteristics for colloidal gold quantum dots under dissipative tunneling conditions

V D Krevchik[1], M B Semenov[1], D O Filatov[2], R V Zaytsev[1], P V Krevchik[1], P S Budyansky[1], I A Egorov[1], I M Moyko[1], I K Skorosova[1], M A Sultanov[1] and I M Semenov[1]

[1] Faculty of Engineering, IT and Electronics, Department of Physics, Penza State University, 40 Krasnaya str., Penza 440026, Russia
[2] Faculty of Physics, N I Lobachevsky State University of Nizhny Novgorod, 43 Gagarina Ave., Nizhny Novgorod 603950, Russia

E-mail: physics@pnzgu.ru

Abstract. In this project the tunnel volt-ampere characteristics of colloidal gold quantum dots in a system of combined tunnel and atomic force microscopes were experimentally measured. It is assumed that ionic conductivity contributes for tunnel current the most. A qualitative comparison was made for tunnel volt-ampere characteristics and the theoretical probability curve of 2D-dissipative tunneling under influence of two local wide-band phonon modes. A qualitative agreement was found for theoretical and experimental curves, which indicates the possible contribution of the dissipative tunneling in the current through a quantum dot at the tip of a cantilever, which can be amplified in clusters ranging in size from 1 to 5 nm in thinner films.

1. Introduction

The electronic properties of ultrafine nanoparticles (UFPs) have been attracting researchers' attention throughout several decades. Electronic structure and properties of UFPs differ sharply from those of bulk samples of metals. The properties of UFPs occupy an intermediate position between the individual atoms of metals and bulk samples. In particular, of the fundamental interest is the critical number of atoms in UFPs from which their properties are indistinguishable from those of bulk samples. In the UFP systems a number of fundamental physical effects can be observed, such as changes in the electronic structure of UFP due to size quantization and the associated changes in heat capacity, magnetic susceptibility [1–11]; effects of Coulomb blockade of electron tunneling through UFP, etc. [1–6]. The purpose of this work is to study the tunneling volt-ampere characteristics which were obtained for growing colloidal gold quantum dots in the system of combined atomic-force and scanning tunneling microscopes (AFM/STM), as well as to study the conditions of possible contribution of 2D-dissipative tunneling in the tunneling volt-ampere characteristics. It is necessary to make a qualitative comparison of a tunneling volt-ampere characteristic for a growing colloidal gold quantum dot in the combined AFM/STM system and a theoretical curve for the field dependence of the probability of 2D-dissipative tunneling under the influence of two local phonon modes.

2. Experimental part

The experimental part of the procedure partly corresponds to the method of authors from University of Kobe (Japan), published in [1]. The formation of gold particles in films of Au(III) – SiO_2/TiO_2 is performed using an atomic force microscope. Topographic images of the film surface are recorded before the formation of gold nanoparticles without external electric field. Then a cantilever comes close to a certain surface point. Local ion reduction of Au(III) is performed by applying a negative potential to the indium oxide film with a cantilever as grounding. Ion reduction of Au(III) can be observed by cathode current in the volt-ampere characteristic and the current dependence on distance between the cantilever and the film surface. Formation of gold nanoparticles also occurs during the process of scanning a particular area by a cantilever while applying a negative potential to a film. The scanning speed and cycles are manipulated by changes in density and morphology of generated gold nanoparticles. In the Japanese authors' experimental work [1], semispherical colloidal gold quantum dots grew only at a negative voltage, but in the experiment performed in this work, clusters of size up to 10 nm (figure 1) were formed at a negative voltage, while toroidal colloidal gold nanostructures were formed in a positive one (figure 2). The theoretical work was carried out based on the theory of dissipative tunneling using the instanton method. At the formation of colloidal gold clusters with size of above 10 nm in the combined system of AFM/STM, the main conductivity mechanism is assumed to be ionic. In this case, because the thickness of a gelatinous film is about 100 nm, the golden quantum dot grows up to the 100 nm size, when the dissipative tunneling mechanism is suppressed. Due to the diffusion drift, trivalent gold ions penetrate into the area of high field strength under the tip of a cantilever of AFM\STM, which stimulates the corresponding flow of electrons. Apparently, the mechanism of conductivity occurs if the film thickness is sufficiently large (100 nm) and the size of clusters is large as well (above 10 nm). On the other hand, the dissipative tunneling mechanism can be implemented in systems of sufficiently thin films (less than 10nm thick) and clusters of small sizes (from 1 to 5 nm).

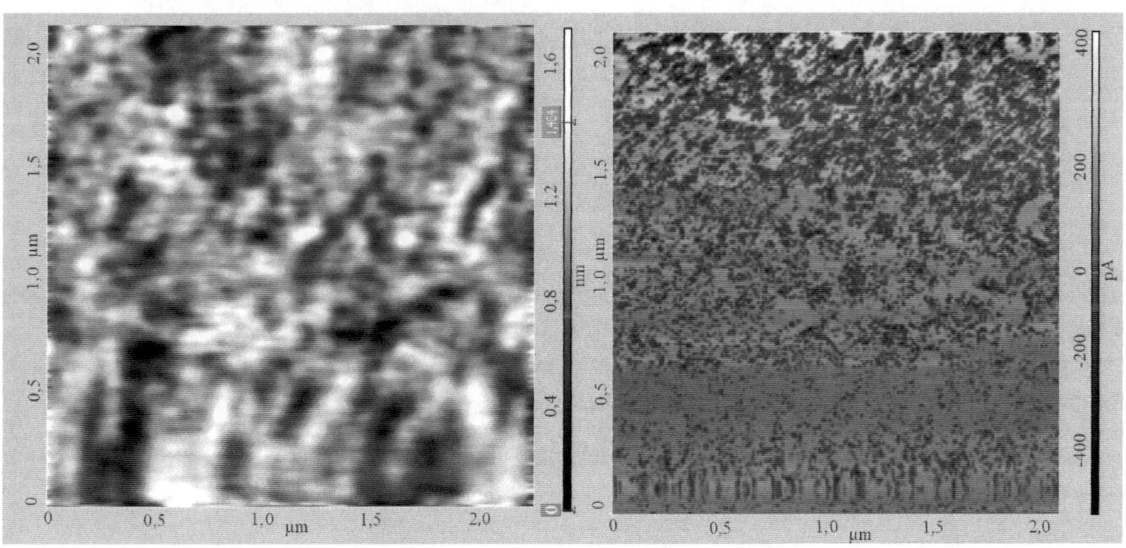

Figure 1. Topography (left) and a current image (right) of the surface area of the film before the modification (before the formation of a colloidal gold cluster). The voltage between the probe and the sample is 0.5 V.

3. Possible conductivity mechanisms and qualitative comparison with the experiment

Indeed, the ion conductivity mechanism can be interpreted as a trivalent gold ion drift, initially equally distributed throughout the volume of a gel film in high external electric field at the needle of the cantilever, followed by the process of hemispherical cluster formation. This ion conductivity mechanism will prevail over the tunnel one, when the electric field of positive ions of gold $|E^*|$ exceeds the value of the external electric field $|E^*|>|E_{ext}|$, or $R_0/d > U/d|E^*|-1$, where R_0 – the cluster radius, L – the gel film thickness, $d = L - R_0$ (the distance from the top of the cluster to the tip of the cantilever), U – the voltage of an external electric field. The reverse inequality correlates with the mechanism of tunneling current, and the previously proposed model of 2D-dissipative tunneling [11] becomes relevant.

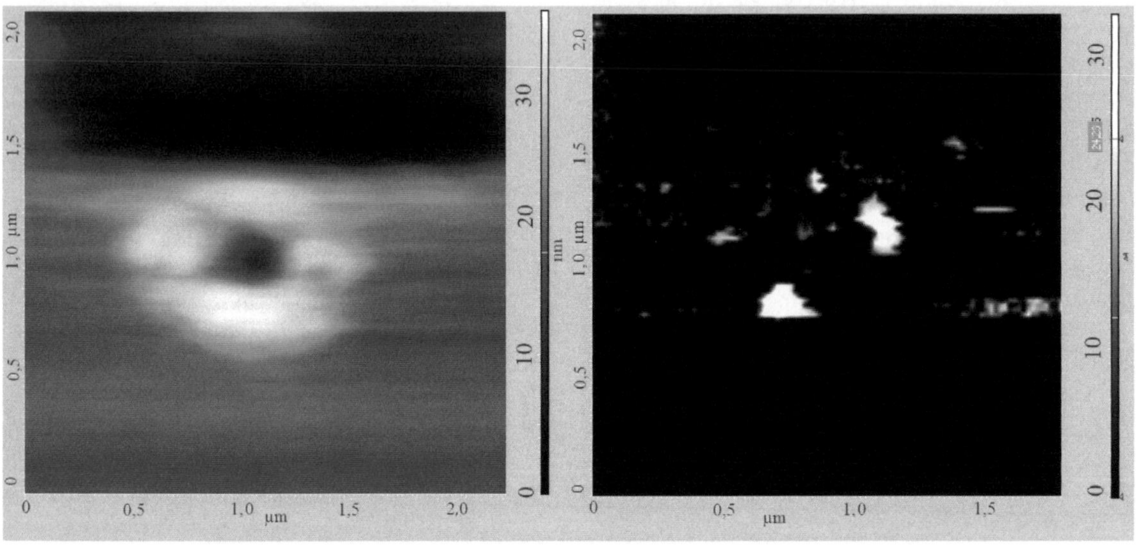

Figure 2. Topography (left) and current image (right) of the surface area of the film after the modification (formation of toroidal nanostructure under the tip of a cantilever) at a positive voltage. The potential difference between the probe and sample is +0.5 V.

For subsequent qualitative comparison with the experimental volt-ampere characteristics, we calculated the probability of 2D-dissipative tunneling using the one – instanton approach in the case of parallel movement of tunneling particles under influence of two local phonon modes in the wide band in the non-oscillatory mode. This probability can be defined as $\Gamma = \exp(-S^*)$, where $S^* = S/(a^2\omega)$. Using the results of our works [5, 7–11], we can derive the following general formula for the action S as a function of parameters τ_1 и τ_2 (instanton centers or barrier breakthrough timings):

$$S = 2a(b+a)(\tau_1+\tau_2)\omega^2 - \frac{\omega^2(a+b)^2(\tau_1+\tau_2)^2}{\beta} - \frac{\omega^4(a+b)^2(\tau_1-\tau_2)^2}{(\omega^2-2\alpha)\beta} - \frac{2\omega^4(a+b)^2}{\beta}\sum_{n=1}^{\infty}\left\{\frac{(\sin\nu_n\tau_1+\sin\nu_n\tau_2)^2}{\nu_n^2(\nu_n^2+\omega^2+\zeta_n)}+\frac{(\sin\nu_n\tau_1-\sin\nu_n\tau_2)^2}{\nu_n^2(\nu_n^2+\omega^2-2\alpha)}\right\}, \quad (1)$$

where ζ_n is determined taking into account the influence of two local wide-band phonon modes $\zeta_n = \nu_n^2 \dfrac{C_2^2}{\omega_2^2\left(\omega_2^2+\nu_n^2\right)} + \nu_n^2 \dfrac{C_3^2}{\omega_3^2\left(\omega_3^2+\nu_n^2\right)}$; a, b – positions of the double-oscillator potential minima along one of the parallel tunneling coordinates in the absence of external electric field; ν_n – the Matsubara frequencies $\nu_n = \dfrac{2\pi n}{\beta}$, $\beta = \dfrac{\hbar}{kT}$ – the return temperature, ω – the frequency of a two-well oscillator along the coordinates of tunneling; ω_2, ω_3 – the frequencies of the local phonon modes, normal to the tunneling coordinate, α – the coefficient of the interaction of the parallel tunneling electrons in the dipole-dipole approximation.

The influence of the electric field on the asymmetric two-well oscillator potential along one of the parallel tunneling coordinates can be represented as:

$$\widetilde{U}(q) = \frac{\omega_0^2}{2}(q-b)^2\theta(q) + \left[\frac{\omega_0^2}{2}(q+a)^2 - \Delta I\right]\theta(-q) - |e|Eq\,, \tag{2}$$

where the parameter $\Delta I = \dfrac{\omega_0^2}{2}(a^2-b^2)$ determines the original asymmetry of potential in absence of the field. In the external electric field, the value of the asymmetry of the two-well oscillator potential, as is well known, changes proportionally to the magnitude of the field.

For the dimensionless quasi-classical action (because of the convenience of numerical comparison of theoretical and experimental curves), we make dimensionless designations: $\varepsilon = \varepsilon^*\omega = (\tau_1 - \tau_2)\omega$; $\tau = 2\tau^*\omega = (\tau_1 + \tau_2)\omega$; $\beta^* = \beta\omega/2$; $\alpha^* = 2\alpha/\omega^2$, $b^* = b/a$, $b \ge a$. Effect of the electric field can be accounted for by renormalization of the parameters: $a = a_0 + |e|\,E/\omega_0^2$, $b = b_0 - |e|\,E/\omega_0^2$. Then, the dimensionless semi-classical action takes the form:

$$S^* = \frac{S}{a^2\omega} = 2\left(b^*+1\right)\tau - \frac{\left(b^*+1\right)^2\tau^2}{2\beta^*} - \frac{\left(b^*+1\right)^2\varepsilon^2}{2(1-\alpha^*)\beta^*} - \frac{\omega^4\left(b^*+1\right)^2}{\beta^*}\cdot\sum_{n=1}^{\infty}A_n,$$

$$where\ \ A_n = \frac{\sin^2\nu_n\tau_1}{\nu_n^2\left(\nu_n^2+\omega^2+\zeta_n\right)} + \frac{\sin^2\nu_n\tau_2}{\nu_n^2\left(\nu_n^2+\omega^2+\zeta_n\right)} + \frac{\sin^2\nu_n\tau_1}{\nu_n^2\left(\nu_n^2+\omega^2-2\alpha\right)} + \tag{3}$$

$$+\frac{\sin^2\nu_n\tau_2}{\nu_n^2\left(\nu_n^2+\omega^2-2\alpha\right)} - \frac{\cos\nu_n\left(\tau_1-\tau_2\right)}{\nu_n^2\left(\nu_n^2+\omega^2-2\alpha\right)} + \frac{\cos\nu_n\left(\tau_1+\tau_2\right)}{\nu_n^2\left(\nu_n^2+\omega^2-2\alpha\right)}$$

The calculations of the presented sums were conducted by a method similar to the one considered by authors [5, 11]. The final form for the action S is very cumbersome, thus, it is not shown in this work. Figure 3 shows a qualitative comparison of the experimental volt-ampere characteristics for the clusters under the cantilever with the field dependence of the probability of 2D-dissipative tunneling.

In figure 3 one can see that the theoretical curve qualitatively corresponds only to one of the two experimental volt-ampere characteristics. Apparently, this is due to the fact that at the early stage of the cluster formation, the tunneling mechanism can still be seen in the diffusion drift. But, after the formation of the cluster greater than 10 nm, the dissipative tunneling mechanism is suppressed (curve 3 in figure 3). When using clusters with sizes greater than 10 nm (the thickness of a gel film

Journal of Physics: Conference Series **690** (2016) 012029 doi:10.1088/1742-6596/690/1/012029

containing Au(III) ions is about 100 nm), the barrier-suppression wavelength becomes greater than the de Broglie one for tunneling particles, and the tunneling probability diminishes exponentially.

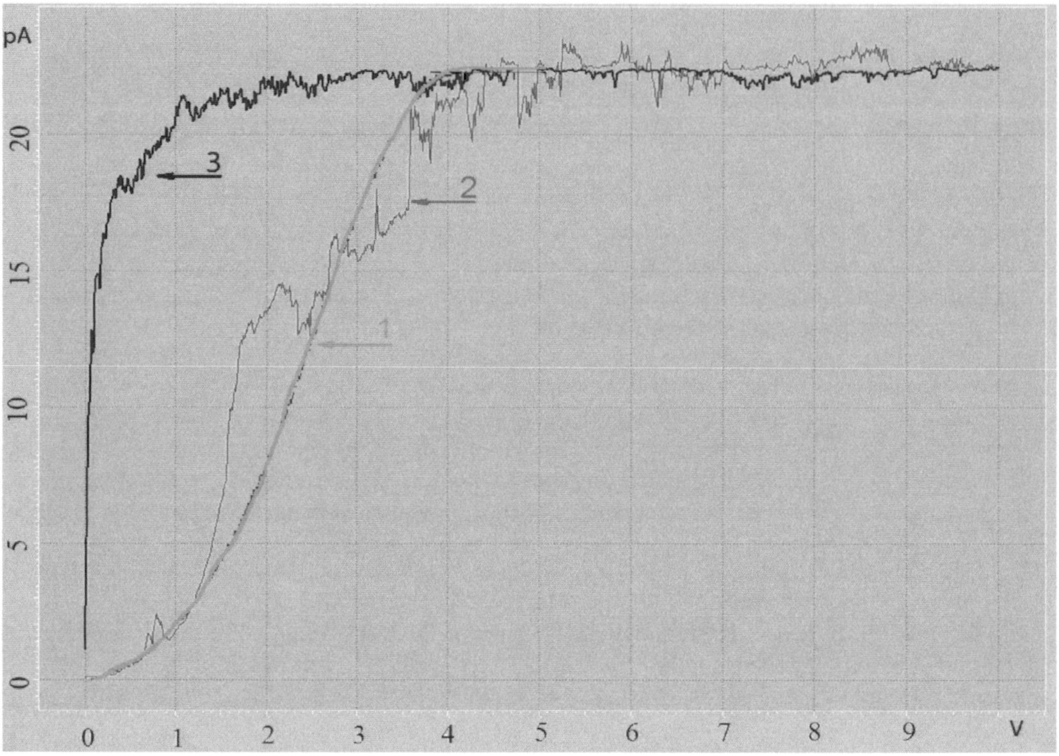

Figure 3. A qualitative comparison of the theoretical curve 1 for the field dependence of the probability of 2D-dissipative tunneling, obtained within the present model, and the experimental volt-ampere characteristic (curve 2) of a colloidal gold quantum dot under the cantilever of the AFM/STM system. Curve 3 describes the reverse tunnel current-voltage characteristic for a golden quantum dot with a size of about 100 nm when the dissipative tunneling mechanism is suppressed, and the ionic conductivity mechanism is dominated.

The "Cluster ripple" associated with the process of diffusion of trivalent gold ions drift, in our opinion, is reflected in the form of a "flicker" – chaotic noise oscillations in the volt-ampere characteristics.

4. Conclusion

Thus, in this experiment with the volt-ampere characteristics for colloidal gold clusters under the cantilever, the primary conductivity mechanism, when the forming cluster's size is greater than 10 nm, is concluded to be ionic. The qualitative comparison of the tunneling volt-ampere characteristics for colloidal gold clusters in the AFM/STM system and the theoretical curve for the dependence of the probability of 2D-dissipative tunneling under influence of two local phonon modes shows the presence of a possible contribution of the dissipative tunneling in the tunneling current through a growing quantum dot at initial stages of growth. It is concluded that the ionic conductivity mechanism will prevail over the tunnel one when the magnitude of the voltage induced by the electric field of positive ions of gold exceeds the value of the external electric field voltage.

This theoretical model of 2D-dissipative tunneling does not allow to reveal flicker-like oscillations of volt-ampere characteristics. For these purposes, further development of the 2D-dissipative tunneling model with the account of the oscillating mode tunneling is needed.

References

[1] Yanagi H and Ohno T Nanofabrication of Gold Particles in Glass Films by AFM-Assisted Local Reduction 1999 *Langmuir* **15** 4773

[2] Antonov D A, Filatov D O, Zenkevich A V, Lebedinskii Yu Yu Investigation of the electronic properties of Au nanoclusters in SiO_2 by combined scanning tunneling/atomic force microscopy 2007 *Bulletin of the Russian Academy of Science. Physics* **71** 56

[3] Lapshina M A, Filatov D O, Antonov D A Current image formation in combined STM/AFM of metal nanoclusters in dielectric films 2008 *Surface: x-ray, synchrotron and neutronic research* **2** 616

[4] Borodin P A, Bukharaev A A, Filatov D O, Isakov M A, Shengurov V G, Chalkov V Yu, Denisov Yu A Investigation of the local density of states in self-assembled GeSi/Si(001) nanoislands by combined scanning tunneling and atomic-force spectroscopy 2011 *Semiconductors* **45** 403

[5] Krevchik V D, Semenov M B 2011 *Controlled dissipative tunneling. Tunnel transport in ultrafine systems, in Russian* (Moscow: Fizmatlit) p 288

[6] Borodin P A, Bukharaev A A, Filatov D O Study of local density of electron states in InGaAs/GaAs quantum rings by combined STM/AFM 2009 *Surface: x-ray, synchrotron and neutronic research* **5** 547

[7] Dakhnovskii Yu I, Ovchinnikov A A, Semenov M B Low-temperature chemical reactions considered as dissipative tunnel systems 1987 *JETP* **65** 541

[8] Zhukovskii V Ch, Dakhnovskii Yu I, Gorshkov O N, Krevchik V D, Semenov M B, Smirnov Yu G, Chuprunov E V, Rudin A V, Skibitskaya N Yu, Krevchik P V, Filatov D O, Antonov D A, Lapshina M A, Yamamoto K, Shenina M E Observed tunnel bifurcation in an external electric field 2009 *Moscow University Physics Bulletin* **64** 475

[9] Zhukovskii V Ch, Dakhnovskii Yu I, Krevchik V D, Semenov M B, Maiorov V G, Kudryashov E I, Shcherbakova E V, Yamamoto K Study of dissipative tunneling controllability in systems of interaction quantum molecules 2007 *Moscow University Physics Bulletin* **62** 73

[10] Zhukovskii V Ch, Dakhnovskii Yu I, Krevchik V D, Semenov M B, Maiorov V G, Kudryashov E I, Yamamoto K Study of tunneling controllability in structures such as "Quantum dot – quantum well" and "Quantum molecule" (in Russian) 2006 *Moscow University Physics Bulletin* **3** 24

[11] Krevchik V D, Ovchinnikov A A, Semenov M B, Aringazin A K, Dahnovsky Yu, Yamamoto K Two-dimensional tunnel correlations with dissipation 2003 *Phys. Rev* B **68** 155426

Carbon nanotubes in the surfactants dispersion: formation of the microenvironment

A O Borovskaya[1], B Z Idiatullin[2] and O S Zueva[1]

[1] Kazan State Power Engineering University, 51 Krasnoselskaya str., Kazan 420066, Russia
[2] Kazan Institute of Biochemistry and Biophysics of Kazan Scientific Center of RAS, 2/31 Lobachevsky str., Kazan 420111, Russia

E-mail: tat355@mail.ru

Abstract. The chemical shifts from the protons belonging to surfactants differing only by the nature of counterions (Li, Na, Cs) are considered in their interaction with the surface of carbon nanotubes. It was shown that the dominant mechanisms of interaction of the surfactant molecules with the nanotubes surface in aqueous solutions depend on the nature of counterions. In particular, sodium dodecyl sulfate molecules interact with the nanotubes surface mainly by their head groups instead of tail groups, and thus it can be assumed that the surface of nanotubes can be coated with a layer of flattened micelles. In the other cases, a structureless random adsorption of surfactant molecules with partially ordered arrangement of head and tail groups, incompact for the molecules of lithium dodecyl sulfate or cling swarm round for the molecules of cesium dodecyl sulfate is more likely.

1. Introduction

In systems of nanoparticles there is a significant percentage of their constituent atoms residing at the surface, whereby the properties of nanomaterials are extremely sensitive to the microenvironment. Any change in the microenvironment, such as surface doping, functionalization of the surface by means of organic functional groups or its modification by coating of molecules layers causes a change in electronic and photophysical properties of nanoparticles. This sensitivity to external influences opens up the possibility not only for creating materials with desired properties, but also for the controlled variation of these properties.

In particular, carbon nanotubes because of the nature of their structure are considered as one of the most versatile nanomaterials. Single-walled carbon nanotubes consisting of atoms constituting the graphene plane rolled into a tube are almost perfect one-dimensional systems in which there is a quantum confinement and the high correlation between the charge carriers. Depending on the arrangement of carbon atoms with respect to their axes (viz. chirality) the nanotubes can be direct bandgap semiconductors, or metals with nearly ballistic conduction. Features of the structure of carbon nanotubes allow using them to create the various devices in photonics and optoelectronics. Of particular interest are the unique optical properties of semiconducting nanotubes – their ability to near-infrared emission of semiconducting species under electrical or optical excitation with certain parameters. Other intensively developed applications are creating nanotransistors, field emission displays, fuel cells, solar cells, photodetectors, switches, gas- and chemosensors based on carbon nanotubes [1–5].

Note that the electronic and optical properties of nanotubes, all of which are surface atoms, are extremely sensitive to the atoms of nearest microenvironment. There is a way to change the electronic properties of nanotubes by purposefully forming atoms of the local environment in a certain way. The appearance of methods that allow to obtain a stable suspension of carbon nanotubes in the surfactants dispersion [6] has led to the possibility of controlling the optical characteristics of nanotubes due to the change of dielectric properties of the local environment (by the surface modification) [7].

Another area of using carbon nanotubes stabilized by the surfactant is creating a system that can limit the nonlinear laser radiation with increasing intensity, which is important in developing practical tools for creating protection for eyes and sensors. Designed suspension limits the laser light in a broad spectral range by scattering of incoming radiation by light-induced inhomogeneities of the medium. Possibilities of parameter control of nonlinear optical limiting by varying the composition and properties of the system components are considered in papers [8, 9].

Nevertheless, despite the fact that carbon nanotubes have attracted extensive research interest, their practical usage in many areas is still hampered by the difficulty of their separation into monodisperse in diameter, length and chirality fraction. In papers [10–12], it was shown as surfactants can be used to achieve the desired separation. However, it is necessary to know the mechanisms of surfactant molecules adsorption on carbon nanotubes to further improve this procedure, as well as for the controlled change of properties of the systems containing the nanotubes.

2. Methods and materials

The method of nuclear magnetic resonance (NMR) which has been successfully applied to solve many problems of colloid chemistry was used as the main method of experimental study of the morphology of multicomponent systems [13–15]. The chemical shifts of NMR signals of proton, carrying information about their immediate environment and the dynamic processes occurring within the system were studied. All experiments were carried out on ^{1}H nuclei using an AVANCE III spectrometer (Bruker, Germany) operating at a proton-resonance frequency of 600.03 MHz. The parameters of spectrum registration were as follows: spectrum width, 7211.5 Hz; number of points, 64K; number of scans, 8; relaxation delay, 5 s; and duration of the 90° pulse, 8.3–10.7 s. Under the experimental conditions due to the short lifetime of the surfactant molecules in various states (~10^{-6} s), the observed signal was a weighted sum of contributions from molecules in different states: a) monomer, b) micellar, c) associated with the nanotube surface. The resonance frequency was determined with an error of 0.1 Hz which corresponds to 0.17 10^{-3} ppm.

For this study we have chosen multiwall carbon nanotubes of carbon nanomaterial "Taunit" dispersed in solutions of three surfactants differing of counterions nature – sodium, lithium and cesium dodecyl sulfates, in deuterated water (Deuteriumoxid 99.9%) at a temperature $T = 30$ °C (Na, Li) and $T = 40$ °C (Cs). For the preparation of suspensions, ~ 5 mg carbon nanotube samples were filled with surfactant solutions in deuterated water of the desired concentration to the volume of 1 ml. The need to use deuterated water was dictated by the possibility of removing the strongest contribution of the signal coming from the protons of ordinary water, which makes it possible to study the signals from protons that are a part of the hydrocarbon chains of surfactant. Solutions with nanotubes were subjected to ultrasonication for 15 minutes by an Elma Sonic S 40H device, then centrifuged at an ELMI centrifuge for 10 minutes at 10,000 g. The upper part of the solution above the dense sediment was selected for measurements.

Surfactant content was varied in a range of 1–100 mM. Sodium dodecyl sulfate, SDS, (Sigma, L4509) and lithium dodecyl sulfate, LiDS, (ACROS Organics) with a main substance content of 99% were used as received. Cesium dodecyl sulfate, CsDS, was obtained from SDS by ion exchange in CsCl (from ECROS) solution. A mixed solution of SDS (0.5 M) and CsCl (1.0 M) was prepared at 50 °C. Then the solution was allowed to stand at 50 °C for 2 h and at room temperature for next 12 h. The sediment was separated from the solution using a Buchner funnel and again dissolved in CsCl solution. After three consecutive recrystallizations, the sediment was washed with acetone and dried over sulphuric acid till constant mass was obtained. The product yield was about 85%. The degree of ion exchange for Na^+ and Cs^+ was determined by flame photometry (air–acetylene mixture) at 2,300°C and it was not lower than 96%.

3. Experiment

In this paper, an experimental study of the processes occurring on the surface of carbon nanotubes when modifying them by the surfactants was carried out. As a result of the measurement, the data array on the individual chemical shifts of the protons resonant lines from investigated surfactant in the concentration range of 1 mM to 100 mM was formed in the absence and in the presence of carbon nanotubes. Analysis of the concentration dependencies of proton chemical shifts of surfactants and response of the chemical shifts to the presence of the carbon nanotubes gave the opportunity to explore the adsorption of surfactant molecules on the surface of nanotubes and their self-assembly into micellar formation appearing in the dispersions of the surfactants, and possibly on the surface of carbon nanotubes.

The structural formula of dodecyl sulfate macroion can be written as $^-SO_4$-CH_2-CH_2-$(CH_2)_9$-CH_3. The observed chemical shifts were caused by the existence of four nonequivalent positions of protons in the molecule of surfactant, namely: a) the protons of α-CH_2 group which are closest to the head group SO_4 of the hydrocarbon chain, b) the protons of β-CH_2 group, c) the protons of the middle molecule methylene units which are almost equivalent, d) the protons of the tail group CH_3 of the molecule.

In the case of SDS (NaDS) dispersions, the maximum deviations of the chemical shifts due to the presence of carbon nanotubes were observed for the protons of α-CH_2 group of alkyl chain which are closest to the head group SO_4. The concentration dependence of difference of the proton chemical shifts for α-CH_2 groups of SDS in the presence and absence of carbon nanotubes is shown in figure 1 (left). Some, but much weaker, changes were observed for the group of β-CH_2. At the same time, no changes in proton chemical shifts for other groups were observed.

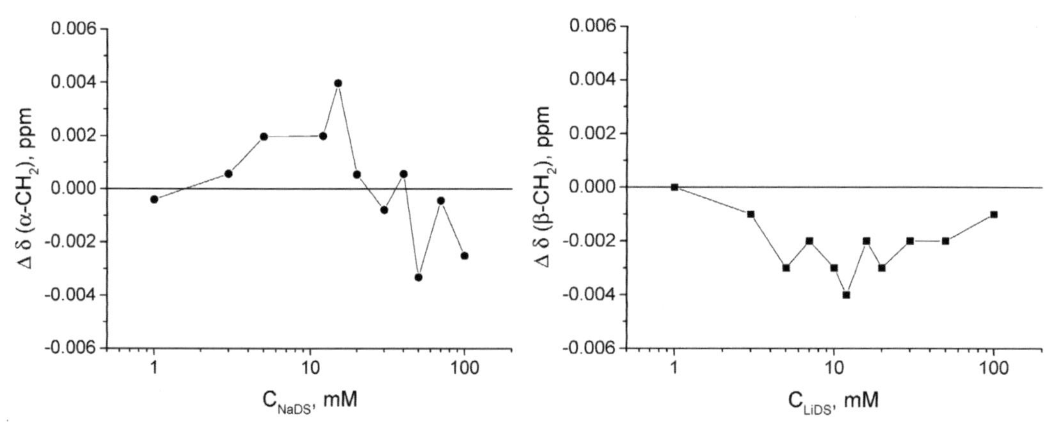

Figure 1.The difference of the proton chemical shifts for α-CH_2 group of SDS (left) and of the proton chemical shifts for β-CH_2 group of LiDS (right) between systems with and without carbon nanotubes in the solution at the same concentration of surfactants.

In the case of LiDS dispersions, the maximum deviations of the chemical shifts due to the presence of carbon nanotubes were observed for the protons of β-CH_2 group rather than for the protons of α-CH_2 group. The concentration dependence of the maximum response of proton chemical shifts for the proton β-CH_2 groups in the presence of carbon nanotubes is shown in figure 1 (right).

The proton resonance lines shifts corresponding to all four non-equivalent proton positions are clearly seen in cesium dodecyl sulfate (see figure 2). It should be noted that the maximum response of the proton chemical shifts due to the presence of the carbon nanotubes in this case is much larger, and hence, the interaction of molecules with the surface of nanotubes CsDS is much stronger.

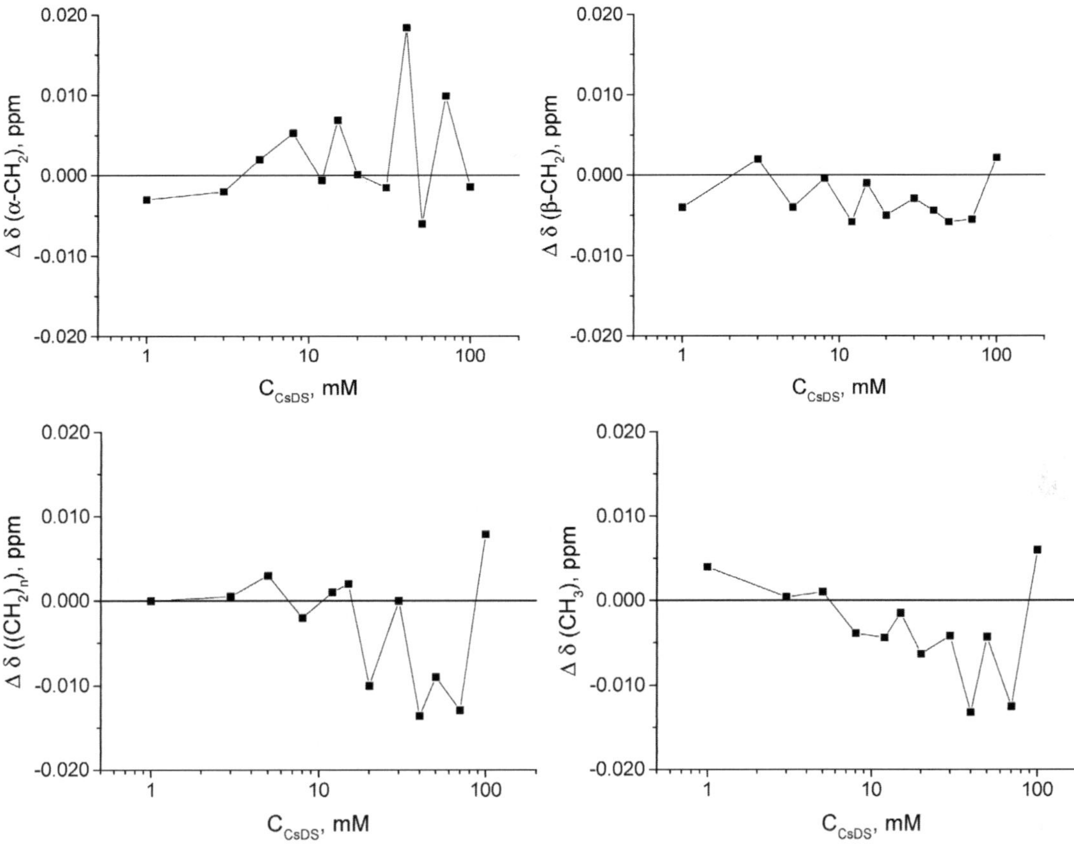

Figure 2. The difference of the proton chemical shifts corresponding to all non-equivalent proton positions for CsDS between systems with and without nanotubes in the solution at the same concentration of the surfactant

4. Results and discussion

The fact that in the case of sodium dodecyl sulfate the maximum deviations of chemical shifts were observed for protons of α-CH_2 group of alkyl chain which are closest to the head group SO_4 suggests that the adsorption of SDS occurs not as it was originally proposed in [6] and is still considered valid by many authors, for example, [16]. Previously, it was thought for reasons of chemical affinity that the hydrophobic molecules of SDS interact by their hydrocarbon tails with the hydrophobic surface of carbon nanotubes, shielding them from contact with water, and the hydrophilic head groups are turned to the aqueous surfactant phase, facilitating solubilization of the nanotubes. Some authors suggest that structureless random adsorption with no preferential arrangement of the head and tail groups is responsible for the stabilization of the dispersions [17, 18].

On the one hand, since the nanotubes are dispersed well enough with SDS [19], we may conclude that the head groups of the surfactant molecules are really rotated to the aqueous medium. On the other hand, our data suggest that the surfactant molecules interact with the surface of nanotubes by means of their head groups. Tail groups are not involved into the interaction. This means that the dominant adsorption mechanism is another one. It can be assumed that at concentrations above the critical micelle concentration, the surface of nanotubes is coated with a layer of flattened micelles, like during adsorption of molecules SDS on the aluminum oxide surface [20]. The images of similar formations on the nanotube surface which is functionalized by another amphiphilic compound are presented in [21]. Incidentally, the dense arrangement of flattened micelles is to some extent similar to a bilayer. In addition, we can assume the presence of adsorption on carbon nanotubes not only of surfactant macroions, but also of a certain amount of sodium counterions with possible recovery of integrity of some dissociated macromolecules. A significant relative change in the conductivity of SDS dispersions observed by us (unpublished data) at low concentrations of surfactants in the presence of nanotubes can only be explained by a significant decrease in the number of carriers of both signs in the solution. It should be noted that the adsorption of positive sodium counterions is not surprising because the surface of nanotubes in the aqueous solutions charged negatively in accordance with existing literature data. Sodium counterion adsorbed due to Coulomb interaction attracts surfactant macroion recovering for some time (since all the processes possess dynamic nature) integrity of the molecule. That is why the micelles may have a flattened shape being essentially transitional option from gemimicelles to micelles.

In the case of lithium dodecyl sulfate the following picture can be proposed. Largely hydrated lithium counterions can not be adsorbed on the surface of nanotubes. However, the presence of lithium counterions near the surface causes the head groups to come off from the surface to interact with them. That is why the maximum response of chemical shift was observed not for α-CH_2 groups but for β-CH_2 groups. For the rest of surfactant ions, the most probable is the structureless random adsorption with partially ordered arrangement of head and tail groups, incompact for the molecules of lithium dodecyl sulfate because the head and tail parts are not strongly associated with the nanotubes.

A slightly different picture was observed for cesium dodecyl sulfate. The addition of cesium ions to SDS can isolate the carbon nanotubes from direct contact with water molecules more effectively, as noticed in [7]. According to our data, the dominant mechanism of adsorption of CsDS is characterized by a strong interaction of all parts of the molecule with the nanotubes surface. This means that some counterions can be adsorbed on the surface of nanotubes, thus connecting with the surfactant ion sufficiently to form a stable and electrically neutral molecule. Then, the surfactant molecules are located almost parallel to the plane of the carbon sheet due to weaker interactions. As a result, the surfactant ions densely lie on the surface characterizing the cling swarm round structureless adsorption. Therefore, the use of DSC leads to surface modification of carbon nanotubes at much lower concentrations.

5. Conclusion

The study of chemical shifts from the protons belonging to the three surfactants differing only by the nature of counterions in their interaction with the surface of carbon nanotubes was carried out. It was shown that the dominant mechanisms of interaction of the surfactant molecules with the nanotubes surface depend on the nature of counterions. In particular, sodium dodecyl sulfate molecules interact with the nanotubes surface mainly by their head groups instead of tail groups, and thus it can be assumed that the surface of nanotubes can be coated with a layer of flattened micelles. In the other cases, the most likely is a structureless random adsorption of surfactant molecules with partially ordered arrangement of head and tail groups, incompact for the molecules of lithium dodecyl sulfate or cling swarm round for the molecules of cesium dodecyl sulfate. It was suggested that a modification of nanotube surfaces is related to the processes of a dynamic nature: adsorption of some of the counterions and restoring the integrity of a certain number of dissociated surfactant molecules.

Acknowledgments
This work was supported by grants of the Russian Foundation for Basic Research (RFBR 13-02-97055a).

References

[1] Avouris P, Freitag M and Perebeinos V 2008 *Nat. Photonics* **2** 341
[2] Brandao-Silva A C, Lima R M A, Fantini C, Jesus-Silva A, Alencar M A R C, Hickmann J M, Jain R M, Strano M S and Fonseca E J S 2014 *Carbon* **77** 939
[3] Valadão D R B, Pires D G, Alencar M A R C, Hickmann J M, Fantini C, Pimenta M A and Fonseca E J S 2012 *Opt. Mater. Express* **2** 749
[4] Zeng Q, Wang S, Yang L, Wang Z, Pei T, Zhang Z, Peng L-M, Zhou W, Liu J, Zhou W and Xie S 2012 *Opt. Mater. Express* **2** 839
[5] Fam D W H, Palaniappan Al, Tok A I Y, Liedberg B and Moochhala S M 2011 *Sensors Actuat B-Chem.* **157** 1
[6] O'Connell M J, et al. 2002 *Science* **297** 593
[7] Duque J G, Densmore C G and Doorn S K 2010 *J. Am. Chem. Soc.* **132** 16165
[8] Venediktova A V , Vlasov A Yu , Obraztsova E D , Videnichev D A , Kislyakov I M and Sokolova E P 2012 *Appl. Phys. Lett.* **100** 251903
[9] Vlasov A Yu, Venediktova A V, Videnichev D A, Kislyakov I M, Obraztsova E D and Sokolova E P 2012 *Phys. Status Solidi B* **249** 2341
[10] Arnold M S, Green A A, Hulvat J F, Stupp S I and Hersam M C 2006 *Nat. Nanotechnol.* **1**, 60
[11] Hersam M C 2008 *Nat. Nanotechnol.* **3**, 387
[12] Niyogi S; Densmore C G and Doorn S K 2009 *J. Am. Chem. Soc.* **131** 1144
[13] Zuev Yu F, Gnezdilov O I, Zueva O S and Us'yarov O G 2011 *Colloid Journal* **73** 59
[14] Gnezdilov O I, Zuev Yu F, Zueva O S, Potarikina K C and Us'yarov O G 2011 *Appl. Magn. Reson.* **40** 91
[15] Idiyatullin B Z, Potarikina K S, Zuev Yu F, Zueva O S and Us'yarov O G 2013 *Colloid Journal* **75** 532
[16] Duan W H, Wang Q and Collins F 2011 *Chem. Sci.* **2 (7)** 1407
[17] Vaisman L, Wagner H D and Marom G 2006 *Adv. Colloid Interface Sci.* **128-130** 37
[18] Tummala N R and Striolo A 2009 *ACS Nano* **3** 595
[19] Zueva O S, Osin Y N, Salnikov V V and Zuev Yu F 2014 *Fundamental research, Russian* **11** 1021
[20] Li N, Thomas R K and Rennie A R 2012 *J. Colloid Interface Sci.* **369** 287
[21] Mackiewicz N, Surendran G, Remita H, Keita B, Zhang G, Nadjo L, Hagège A, Doris E and Mioskowski C 2008 *J. Am. Chem. Soc.* **130** 8110

Study of porous sol-gel nanocomposites based on silicon dioxide and tin dioxide modified by fullerenol $C_{60}(OH)_n$ (n = 22–24)

E V Maraeva[1], M S Istomina[1], V A Moshnikov[1,2], S S Nalimova[1], A I Maximov[1], N I Alexeev[1] and K N Semenov[3]

[1] Department of Micro-and Nanoelectronics, Saint Petersburg Electrotechnical University "LETI", 5 Prof. Popova str., St. Petersburg 197376, Russia
[2] Department of Integral Electronics, St.Petersburg Polytechnic University, 29 Polytechnicheskaya str., St.Petersburg 195251, Russia
[3] Department of Chemical Thermodynamics and Kinetics, Saint-Petersburg State University, 7-9 Universitetskaya emb., St. Petersburg 199034, Russia

E-mail: *jenvmar@mail.ru*

Abstract. In this study silicon dioxide – tin dioxide nanomaterials modified by fullerenol were obtained through the sol-gel technology. The results of nitrogen thermal desorption, atomic force microscopy and X-ray diffraction measurements are discussed. It was shown that introducing fullerenol $C_{60}(OH)_n$ (n = 22–24) at the stage of preparing a sol greatly affects the microstructure of composites and results in the formation of an additional system of pores with an average radius of 16 nm.

1. Introduction

Fullerenols are among the most important and promising fullerene derivatives which can be easily synthesized with the properties readjusted by varying the number of hydroxyl groups. The analysis of the literature shows a wide range of possible applications of fullerenols in such areas as mechanics, construction, medicine, pharmacology, material science [1-3]. In particular, high adhesion to metals, alloys and semiconductor (A^3B^5, A^2B^6, A^4B^4) surfaces, high transparency in the visible and infrared ranges of the spectrum, high chemical and thermal stability allow their active use in conjunction with other nanomaterials in micro- and optoelectronics [1]. The aim of this study was to investigate the possibility of modifying the sol-gel system based on silicon dioxide and tin dioxide by introducing water-soluble forms of fullerene – fullerenols $C_{60}(OH)_n$ (n = 22–24).

2. Experiment

In this study a $C_{60}(OH)_n$ (n = 22–24) inorganic salt $SnCl_2 \cdot 2H_2O$ as a precursor of tin dioxide and tetraethoxysilane as a precursor of silicon dioxide were selected. Buthanol C_4H_9OH was used as a solvent. The modifying agent (fullerenol $C_{60}(OH)_n$ (n = 22–24) was introduced at the stage of preparing the solution, i.e. a sol, and the concentration of fullerenols in the alcohol was 1 mg/l.

To obtain powders, 10 % aqua ammonia (NH_4OH) was added to the solutions-sols. Adding ammonia is accompanied by changes in pH of the solution ranged from 2 to 8, and an immediate condensation of the hydrolysis products takes place resulting in a much faster gelation process. After the gelation, evaporation of the solvent at indoor temperature was performed, and then the powders received were heat-treated for 30 minutes at the temperature of 600 °C.

To control the properties of the synthesized nanomaterial surface, the nitrogen thermal desorption and atomic force microscopy were used. To control the phase composition the X-ray diffraction method was used.

AFM experiments were performed using an NTEGRA-Therma nanolaboratory (NT-MDT, Zelenograd, Russia). Commercial etched silicon tips NSG 01 with typical resonance frequency of 150 kHz were used as AFM probes. X-ray diffraction experiments were carried out using a «DRN Farad» (*Cr-Kα*) diffractometer.

Specific surface area measurements were made using a Sorbi № 4.1 analyzer (CJSC «META», Novosibirsk, Russia) that realizes physical adsorption of noble gas by the sample to be studied [4, 5]. Before the measurements, all the samples were pretreated in a SorbiPrep sample preparation station. The principle of operation of the station is based on the degassing dispersed and porous materials by heating in a stream of an inert gas (helium). The station allows you to set the temperature and time of heating in the ranges of 50–400° C and 0–99 minutes, respectively.

3. Results and discussion

During the experiments, a series of samples of the "silicon dioxide – tin dioxide" system was obtained using the sol-gel method which was both modified and not modified by fullerenols, with different ratios of components (mol. %): 90% SiO_2 – 10% SnO_2 ; 50% SiO_2 – 50% SnO_2.

The samples received were examined by the X-ray phase analysis. As an example, figure 1 shows the results of phase composition investigation for the powders containing 90% SiO_2 – 10% SnO_2 not modified (1) and modified by fullerenols (2). The value 1 in the vertical axis corresponds to the intensity of x-ray reflexes and depends on measurement process conditions.

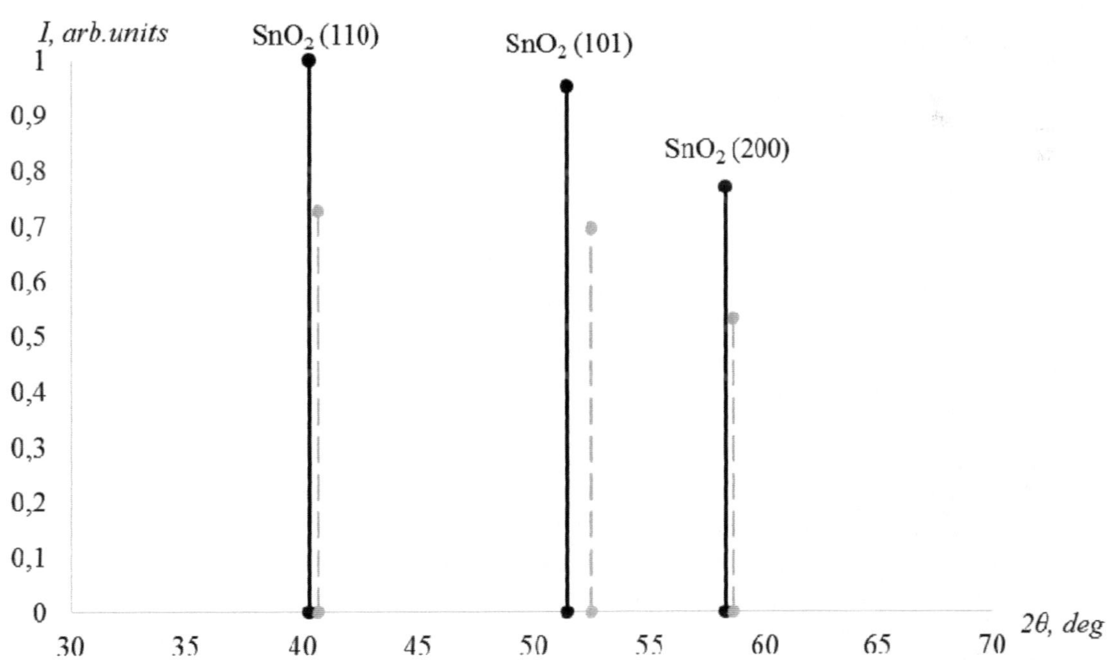

Figure 1. Line diagrams of powder pattern of the samples containing 90% SiO_2 – 10% SnO_2 not modified (1) and modified by fullerenols (2)

As can be seen from figure 1, the X-Ray diffraction pattern shows peaks corresponding to the crystalline structure SnO_2. A set of diffraction reflections (110), (101), (200) seen on the X-Ray diffraction pattern gives evidence of a tetragonal (space group $P4_2$ / nmm) crystalline structure of the rutile type. At the same time, other crystalline structures were not observed. As can be seen from the figure, a reflex corresponding to the (101) planes, containing fullerenol $C_{60}(OH)_n$ as a sample, was shifted toward larger angles, which corresponds to a decrease in the interplanar space in the direction of the axis 4 of the order in the crystalline structure of the rutile type. It should be noted that for the samples containing 50% SiO_2 – 50% SnO_2, the color change from white to grey was observed after the oxygen-free heat treatment in a SorbiPreb sample preparation station. This is presumably due to the formation of tin oxide (SnO) phase in an amorphous state.

The microstructure of the synthesized samples was investigated by the atomic force microscopy method in the tapping mode in an NTEGRA nanolaboratory. Figure 2 illustrates the morphology of the sample surface containing 10 mol. % SnO_2 and 90 mol. % SiO_2. Figure 3 shows an AFM image of a sample of the same composition synthesized with the addition of the $C_{60}(OH)_n$ (n = 22–24) fullerenol. It has been found that the addition of the fullerenol greatly affects the morphology of the surface of the oxide composite: the density of grain distribution and their average size increases. The atomic force microscopy data indicate that on the surface of the sample prepared with the addition of the fullerenol the aggregates of ~2 microns, observed in figure 2(a), are not formed. The results are confirmed by cross-section profiles of the AFM images on enhanced lines shown in figure 2(b) and 3(b).

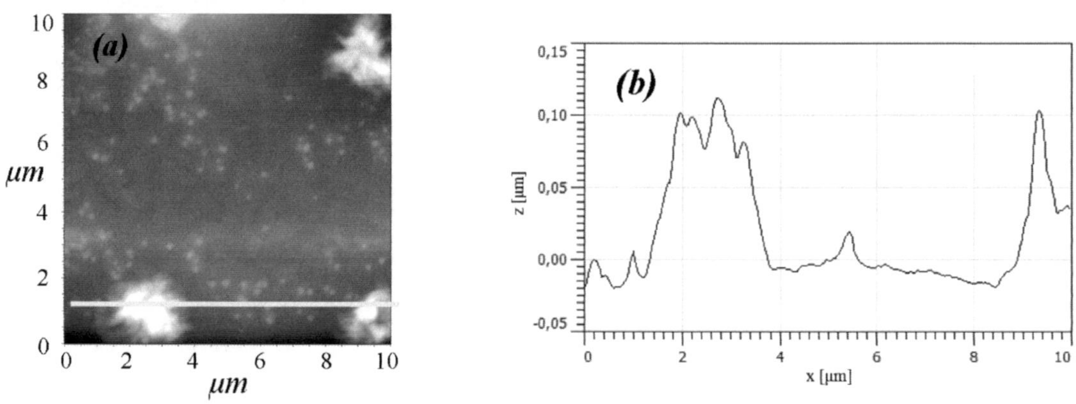

Figure 2. AFM images of the sample containing 90 SiO_2 – 10 SnO_2 (mol. %) synthesized without adding the fullerenol: (a) – the surface relief (scan space size - 10 × 10 μm^2), (b) – the cross-section profile.

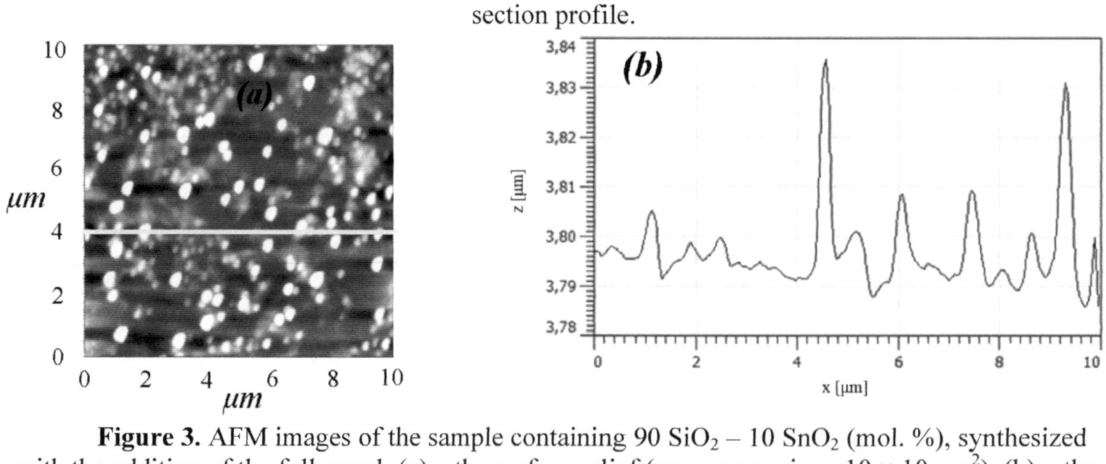

Figure 3. AFM images of the sample containing 90 SiO_2 – 10 SnO_2 (mol. %), synthesized with the addition of the fullerenol: (a) – the surface relief (scan space size - 10 × 10 μm^2), (b) – the cross-section profile.

To investigate the influence of fullerenol insertion on the porosity of silicon dioxide – tin dioxide nanocomposites, the powders were examined with the use of the capillary condensation method. The features of the method are discussed in [4]. The results are shown in tables 1-3.

Table 1. Sizes of pores of the 90% SiO_2 – 10% SnO_2 powder taken by a full adsorption isotherm

R_i, nm	V_i/V_{total}, %
1.8	13
2.2	8
3	25
4.2	21
8	34

Table 2. Sizes of pores of the 50% SiO_2 – 50% SnO_2 powder taken by a full adsorption isotherm

R_i, nm	V_i/V_{total}, %
2.2	20
2.9	27
4	23
7	27
12	3

Table 3. Sizes of pores of the 50% SiO_2 – 50% SnO_2 powder modified by fullerenol taken by a full adsorption isotherm

R_i, nm	V_i/V_{total}, %
2.2	16
2.9	27
4	23
7	23
12	8
16	3

It was determined that all the patterns have a multilevel pore system. This fact correlates with the previous investigation of the authors [6-9]. As it can be seen from the table 1, a sample of 90% SiO_2 – 10% SnO_2 has the smallest pores. Estimation of pore size distribution showed that in a sample of 50% SiO_2 – 50% SnO_2 the volume fraction of mesopores with an average radius of 7 nm with respect to the total pore volume is the largest (27 %). Modification of a sample of 50% SiO_2 – 50% SnO_2 results in the formation of an additional system of pores with an average radius of 16 nm that is apparently caused by burning out of fullerenols. Such systems can be promising for injection of lead chalcogenide and cadmium chalcogenide quantum dots.

4. Conclusions

Porous nanocomposites based on silicon dioxide – tin dioxide both modified and not modified by fullerenols were obtained using the sol-gel method. It was found that introducing fullerenol $C_{60}(OH)_n$ (n = 22–24) in silicon dioxide – tin dioxide nanocomposites results in the formation of an additional system of pores with an average radius of 16 nm. The fullerenols also greatly affect the microstructure and crystalline structure of composites. The obtained results are promising for optimizing the production technology of porous matrices with required parameters.

Acknowledgments
The reported study was supported by Russian Ministry of Education state task № 16.2112.2014 / K.

References

[1] Semenov K N, Charykov N A and Keskinov V N 2011 *Journal of Chemical & Engineering Data* **56** 230
[2] Semenov K N, Charykov N A, Murin I V and Pukharenko Y V 2015 *Journal of Molecular Liquids* **202** 1
[3] Sidorov L N, Yurovskaya M A, Borschevskii A Y, Trushkov I V and Ioffe I N 2005 *Fullerenes, in Russian* (Moscow: Ekzamen)
[4] Belorus A O, Maraeva E V, Spivak Yu M and Moshnikov V A 2015 *Journal of Physics: Conference Series* **586** 012017
[5] Lenshin A S and Maraeva E V 2011 *Izvestiya Sankt-Petersburgskogo Electrotechnicheskogo Universiteta, in Russian* **6** 9
[6] Maximov A I, Moshnikov V A, Tairov Yu M and Shilova O A 2008 *Basics of sol-gel technology of nanocomposites, in Russian* (St. Petersburg: Elmor)
[7] Gracheva I E, Moshnikov V A, Karpova S S and Maraeva E V 2011 *Journal of Physics: Conference Series* **291** 012017
[8] Gracheva E I, Moshnikov V A, Maraeva E V, Karpova S S, Alexandrova O A, Alekseyev N I, Kuznetsov V V, Olchowik G, Semenov K N, Startseva A V, Sitnikov A V and Olchowik J M 2012 *Journal of Non-Crystalline Solids* **358** 433
[9] Moshnikov V A, Gracheva I E, Kuznezov V V, Maximov A I, Karpova S S and Ponomareva A A 2010 *Journal of Non-Crystalline Solids* **356** 2020

Metamorphic InGaAs photo-converters on GaAs substrates

D V Rybalchenko, S A Mintairov[1], R A Salii, N K Timoshina, M Z Shvarts and N A Kalyuzhnyy

Ioffe Institute, 26 Polytekhnicheskaya str., 194021 St. Petersburg, Russia

[1]E-mail: mintairov@scell.ioffe.ru

Abstract. Metamorphic heterostructure photoconverters (PC) based on InGaAs have been grown on GaAs substrates by the MOCVD technique. It has been shown that using the multilayer metamorphic buffer with layer thickness of 120 nm and In composition step of 2.5 % provides a good quality of the bulk layers grown on the buffer with up to 24 % of In. A PC with photosensitivity up to 1350 nm and the quantum efficiency of 80 % in the range of 1050–1100 nm have been made. The 34.5 % efficiency of laser radiation conversion at 1064 nm has been demonstrated.

1. Introduction

To date, solar cells (SCs) based on III-V heterostructures are the most effective devices in the field of converting renewable energy into electricity. The efficiency of the modern monolithic metamorphic triple-junction GaInP/GaAs/GaInAs SCs exceeds 44 % [1] and that of mechanically stacked 4-junction GaInP/GaAs-GaInAsP/GaInAs SCs is 46 % [2].

Application of the metamorphic growth of the wide band gap GaInP/GaInAs tandem [3-5] allows raising the SC efficiency compared to a pseudomorphic structure [6]. In this case, the lattice mismatch between subcells and a germanium substrate is compensated by a composition gradient set of metamorphic buffer layers. After relaxation of the buffer layers, a transition to a new lattice constant takes place and dislocations turn up on heterointerfaces. Owing to this, only the buffer layer remains strongly defective and penetration of dislocations into photoactive layers appears to be very little.

Another area of application of metamorphic PCs is wireless fiber optic systems for transmitting laser radiation power or information data. For this, one can use conventional 800–1300 nm lasers having different applications in medicine, industry and telecommunication systems. Laser photoconverters (LPC) for this wavelength range is the most in demand as an optical pair to a conventional powerful solid-state laser on Nd:YAG irradiating at 1064 nm. For this purposes a Si based LPC with about 40 % conversion efficiency for 1064 nm has been demonstrated [7]. However, the efficiency was low without using very complex special pyramid-like profiling of facial surface and maximal conversion efficiency was demonstrated at low radiation power. So, the development of efficient III-V LPC also promising for SHF applications is in great demand for modern optoelectronics.

In the present work, metamorphic InGaAs heterostructures with a wide composition range have been fabricated on GaAs substrates by the MOCVD technique, and PCs based on them have been created.

2. Experimental setup

Metamorphic PC structures were grown in a low pressure MOCVD reactor under 100 mbar on *n*-GaAs (100) substrates misoriented by 6° towards the [111] direction. Metal alkyls were used as sources of the III-group atoms: threemethylgallium, threemethylaluminum, threemethylindium. Arsine was used as a source of As atoms. Monosilane and diethylzinc were used to dope layers with *n*-type and *p*-type impurity, respectively.

The PC structure represents a *p*-*n* diode with a back surface barrier and a wide-band gap window (figure 1). Variation of the In portion in the layers of $In_xGa_{1-x}As$ metamorphic buffer (MB) was carried out stepwise with a step $x \sim 2.5$ % at a thickness of one sublayer of ~ 120 nm. A set of $In_xGa_{1-x}As$ structures with In concentration x of the order of 10, 14, 19, 24 and 29 % has been created with a number of sublayers in the MB region of 4, 6, 8, 10 and 12, respectively. The PCs based on these structures have been fabricated (figure 1).

Spectral dependencies of the quantum efficiency (QE) and reflection were recorded on a specialized installation in a widen wavelength range up to 1400 nm [8]. Study of structural parameters of PC structures was performed with using a raster SEM CamScan Series 4 DV100.

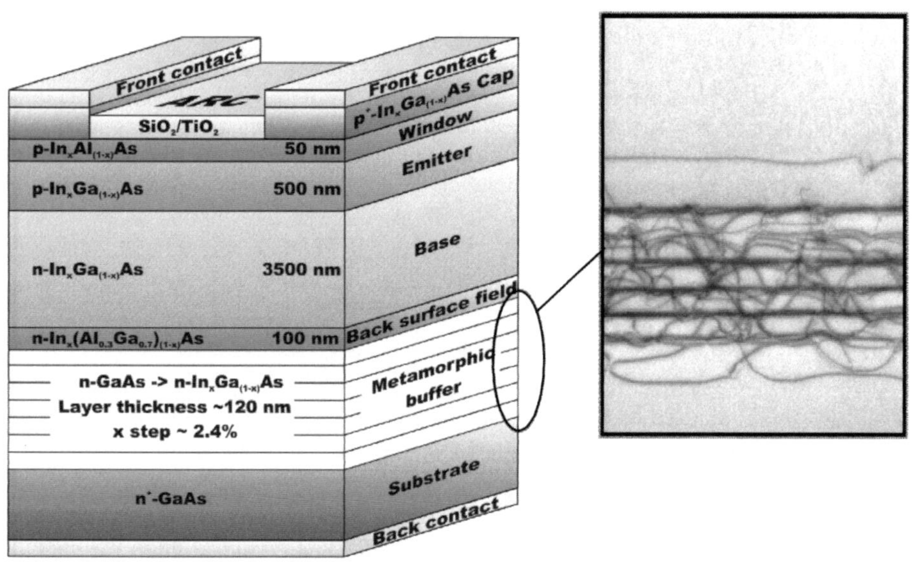

Figure 1. Schematic of the metamorphic $In_xGa_{1-x}As$ LPC cross-section (left) and of a metamorphic buffer between GaAs and $In_xGa_{1-x}As$, with In concentration $x \sim 14$ % obtained by the transmission electron microscopy (right).

3. Results and discussion

Growth of MB layers of thickness greater than the critical one and relaxation of elastic stresses were resulting in forming the large amount of misfit dislocations, which were turning up on the buffer interfaces (figure 1, right). In the picture, a dislocation grid lying predominantly in the interfaces between layers with a different In content is clearly seen.

Application of MB has allowed obtaining a satisfactory $In_xGa_{1-x}As$ LPC surface morphology with the In content up to 24 %, which is confirmed by the surface SEM pictures (figure 2a, b and c). Raising the In concentration x up to 29 % resulted in a sharp deterioration of the surface morphology (figure 2d). This indicates insufficient relaxation of the elastic stresses in the MB layers and one can expect an increase of the defect concentration in the LPC photoactive layers.

It has been proved by investigations of the QE of the PCs created on the base of the grown structures (figure 3). Increase of In concentration resulted in broadening of the spectral response band because of decreasing the band gap of PCs. At the same time, the changing of QE in the range of 400–600 nm due to increase of absorption in a window layer has been observed.

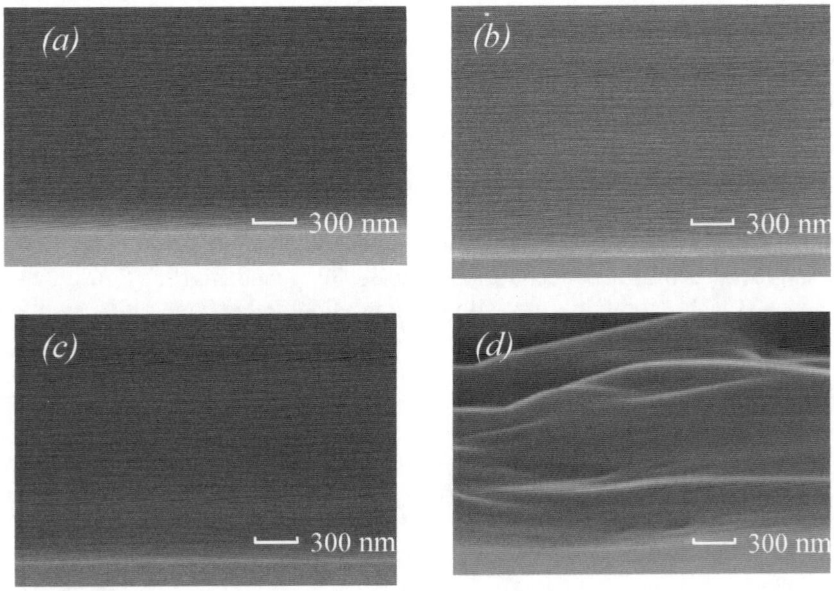

Figure 2. Pictures of metamorphic $In_xGa_{1-x}As$ LPCs surface with the In concentration x: (a) – 10 %, (b) – 19 %, (c) – 24 %, (d) – 29 %, obtained by the SEM.

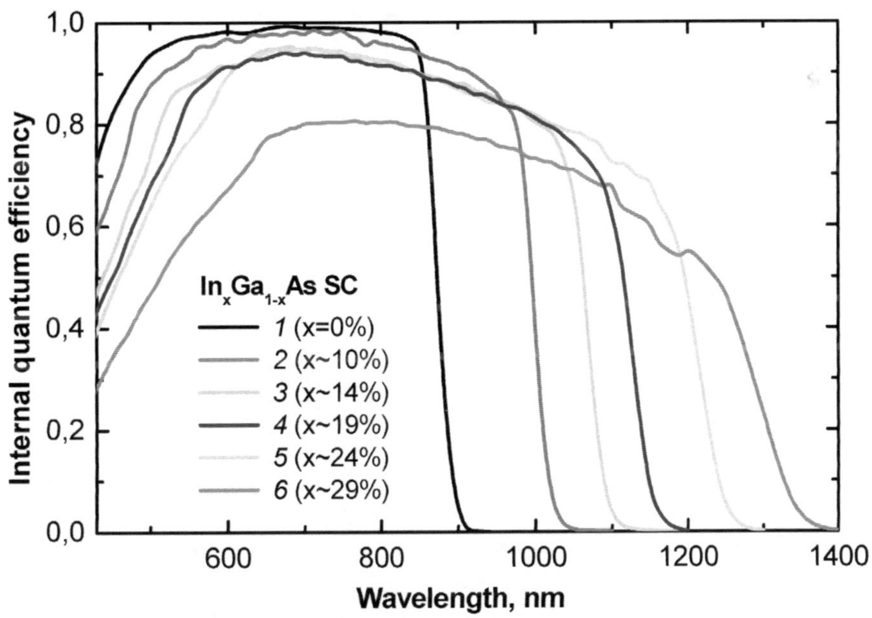

Figure3. Spectra of the internal QE for $In_xGa_{1-x}As$ PC with the In concentration x: 1 – 0 % (pure GaAs), 2 – 10 %, 3 – 14 %, 4 – 19 %, 5 – 24 %, 6 – 29 %.

The maximum level of the QE spectra depended weakly on the In concentration at $x < 25$ % and was more than 90 % (figure 3, curves 2-5). Further increase of the In concentration x up to 29 % resulted in a drastic drop of the maximum level of QE spectrum down to 81 % (figure 3, curve 6). It was caused by an increase of defects concentration in the photoactive layers due to insufficiently effective relaxation of stresses.

The QE in the range of 800–1000 nm depends weakly on the In concentration (figure 4, curves 1,2) while for longer wavelengths, the dependency becomes much stronger. However, the developed structures give a possibility for creating effective LPC for wavelengths of up to 1100 nm. For 1064 nm the LPCs with the In concentration $x \sim 24$ % will have the highest efficiency since such LPCs ensure the maximum QE of about 80 % (figure 4, curve 3). The weak dependence of the QE on In composition for wavelengths of 810 and 980 nm (figure 4, curves 1 and 2) will also allow creating wide band LPCs based on a structure with the In content $x \sim 24$ % for effective conversion of laser beams in the range of 800–1100 nm.

It should be noted that for effective conversion of radiation at wavelengths in the range greater than 1250 nm it is necessary to raise the In concentration x up to 30 % and higher (figure 4, curve 4). However, the developed MB technology does not allow suppressing defect penetration in photoactive LPC layers. Thus, it is necessary to optimize the design of MB and conditions for its growth to create effective LPCs for the given wavelengths.

PCs fabricated of a structure with the In content $x \sim 24$ % have demonstrated a high value of the efficiency for the 1064 nm wavelength (that of the Nd:YAG laser) which was 34.5 % at the radiation density of 3–4.5 W/cm^2 (figure 5). Further decrease of efficiency at higher radiation density occurs due to increase of series resistivity losses at higher photocurrent.

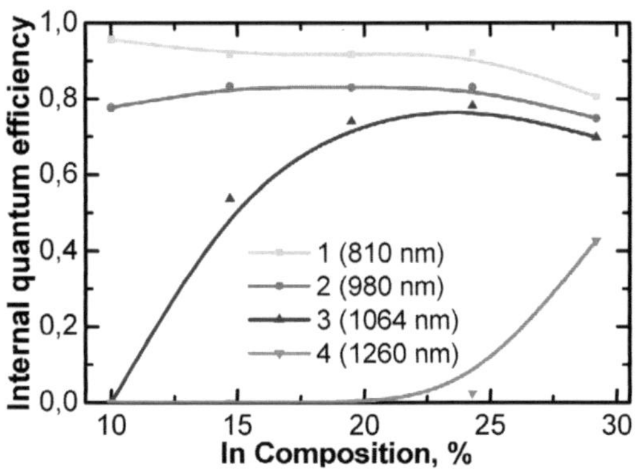

Figure 4. Dependencies of internal QE on $In_xGa_{1-x}As$ LPC composition at wavelengths of conventional powerful lasers.

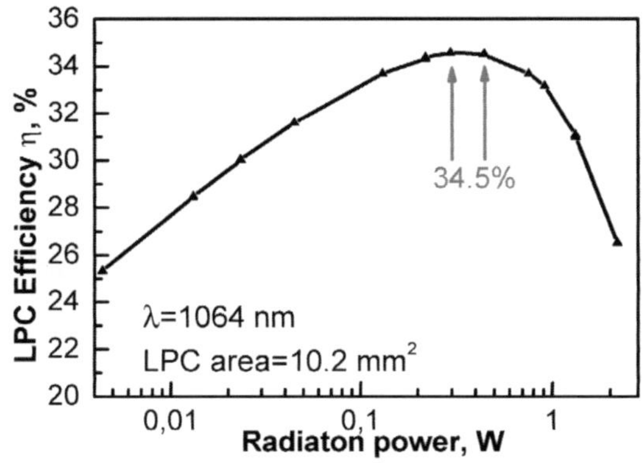

Figure 5. Dependence of the efficiency of converting 1064 nm laser beam on its power.

4. Conclusion

Investigations of metamorphic LPCs based on $In_xGa_{1-x}As$ fabricated by the MOCVD on GaAs substrates have been carried out. It has been shown that the use of a multilayer metamorphic buffer region with a step by layer thickness of 120 nm and by In content of 2.5 % ensures good relaxation of elastic strain and an effective turn up of dislocations on heterointerfaces up to the In concentrations $x \sim 24$ %. With increasing the indium concentration up to 29 %, such buffer does not allow growing bulk layers with a satisfactory surface morphology. LPCs based on $In_xGa_{1-x}As$ layers with the In concentration $x \sim 24$ % demonstrated the maximum internal QE of about 95 % with its further decrease down to 80 % in the wavelength range of 1050–1100 nm. The 1064 nm laser beam conversion efficiency was 34.5 % at the radiation density of 3–4.5 W/cm^2.

Acknowledgments

This work has been supported by the Presidium of the Russian Academy of Sciences (Resolution of the Presidium of RAS № 176 of December 23, 2014).

References

[1] Sasaki K, Agui T, Nakaido K, Takahashi N, Onitsuka R and Takamoto T. 2013 *AIP Conf. Proc.* **1556** 22

[2] Dimroth F, Grave M, Beutel P, Fiedeler U, Karcher C, Tibbits T, Oliva E, Siefer G, Schachtner M, Wekkeli A, Bett A, Krause R, Piccin M, Blanc N, Drazek C, Guiot E, Ghyselen B, Salvetat T, Tauzin A, Signamarcheix T, Dobrich A, Hannappel T and Schwarzburg K 2014 *Prog. Photovoltaics: Res. Appl.* **22** 277

[3] King R R, Law D C, Edmondson K M, Fetzer C M, Kinsey G S, Yoon H, Sherif R A and Karam N H 2007 *Appl. Phys. Lett*. **90** 183516

[4] Guter W, Schone J, Philipps S, Steiner M, Siefer G, Wekkeli A, Welser E, Oliva E, Bett A and Dimroth F 2009 *Appl. Phys. Lett*. **94** 223504

[5] Law D C, Liu X Q, Boisvert J C, Redher E M, Fetzer C M, Mesropian S, King R R, Edmondson K M, Jun B, Woo R L, Krut D D, Chiu P T, Bhusari D M, Sharma S K and Karam N H 2012 *Proc. 38th IEEE Photovoltaic Specialists Conf.* 003146

[6] King R R, Boca A, Hong W, Liu X Q, Bhusari D, Larrabee D, Edmondson K M, Law D C, Fetzer C M, Mesropian S and Karam N H 2009 *Proc. 24th European Photovoltaic Solar Energy Conf. and Exhibition* 55

[7] Green M, Zhao J, Wang A, Wenham S 1992 *IEEE Electron Device Lett.* **13** 317

[8] Shvarts M Z, Chalov A E, Ionova E A, Larionov V R, Malevskiy D A, Rumyantsev V D, Titkov S S 2005 *Proc. 20th European Photovoltaic Solar Energy Conf.* 278.

Distribution of pairing functions in superconducting spin-valve switching modes

R R Gaifullin[1], R G Deminov[1], L R Tagirov[1], T Yu Karminskaya[2], M Yu Kupriyanov[1,2,3] and A A Golubov[3,4]

[1] Institute of Physics, Kazan Federal University, 18 Kremlevskaya str., Kazan 420008, Russia
[2] Skobeltsyn Institute of Nuclear Physics, Moscow State University, 1-2 Leninskie Gory, Moscow 119992, Russia
[3] Moscow Institute of Physics and Technology, 9 Institutskiy per., Dolgoprudny 141700, Russia
[4] Faculty of Science and Technology and MESA+ Institute of Nanotechnology, University of Twente, P.O. Box 217, 7500 AE Enschede, Netherlands

E-mail: gaifullin.rashid@gmail.com

Abstract. We investigated the critical temperature T_c of SF1F2 trilayers (S is a singlet superconductor, F1 and F2 are ferromagnetic metals), where the long-range triplet superconducting pairing is generated at canted magnetizations of the F layers. We examined the spin-singlet and spin-triplet pairing distributions and their amplitudes as a function of the layers thicknesses under different values of the angle α in the SF1F2 structure to clarify which one of the pairing distributions and how may affect the superconducting T_c.

1. Introduction

We investigated the critical temperature T_c of SF1F2 trilayers (S is a singlet superconductor, F1 and F2 are ferromagnetic metals), where the long-range triplet superconducting pairing is generated at noncollinear magnetizations of the F layers [1]. An asymptotically exact numerical method [2] is employed to calculate T_c as a function of the trilayer parameters, such as mutual orientation of magnetizations, interfaces transparencies, and the layers thicknesses. Earlier we demonstrated that T_c in the semi-infinite SF1F2 structures can be a non-monotonic function of the angle α between magnetizations of the two F layers [3], contrary to the monotonic $T_c(\alpha)$ behavior calculated for the F1SF2 superconducting spin-valve design [4]. The existence of the anomalous dependence of the spin-triplet correlations on the angle α in FFS structures in limit of thin F films was shown recently [5]. We examined the spin-singlet and spin-triplet pairing distributions and amplitudes as a function of the layers thicknesses at different values of the angle α in the SF1F2 structure to clarify which one of the pairing distributions and how may impact on the superconducting T_c.

2. The model and numerical method

At first we found a nonmonotonic dependence of T_c in a SF1F2 trilayer as a function of the angle α between the exchange fields of the two F layers (figure 1).

The S layer is of the thickness d_S ($-d_S < x < 0$), the middle F1 layer is of the thickness d_{F1} ($0 < x < d_{F1}$), the outer F2 layer is of the thickness d_{F2} ($d_{F1} < x < d_{F1}+d_{F2}$), the x axis is normal to the plane of the layers. The exchange field in the middle F1 layer is along the z direction, $\mathbf{h} = (0, 0, h)$, while the exchange field in the outer F2 layer is in the yz plane: $\mathbf{h} = (0, h \sin\alpha, h \cos\alpha)$. The angle α varies between 0 (parallel configuration, P) and π (antiparallel configuration, AP).

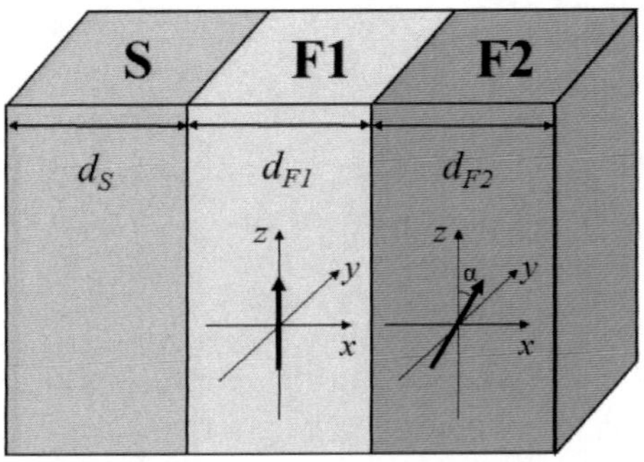

Figure 1. SF1F2 trilayer. The S/F1 interface corresponds to $x = 0$. The thick arrows in the F layers denote direction of the exchange fields **h** lying in the (y, z) plane. The angle between the in-plane exchange fields is α.

We consider SF1F2 structure in the dirty limit, which is described by the Usadel equations. Near T_c, the Usadel equations are linearized and contain only the anomalous Green function $\hat{f}$ [1]:

$$\frac{D}{2}\nabla^2 \hat{f} - |\omega| \hat{f} - \frac{i \operatorname{sgn} \omega}{2}\left\{\hat{\tau}_0 (\mathbf{h}\hat{\boldsymbol{\sigma}}), \hat{f}\right\} + \Delta \hat{\tau}_1 \hat{\sigma}_0 = 0. \tag{1}$$

Here, D is the diffusion constant, $\hat{f}$ is 4×4 matrix, $\omega = \pi T_c(2n + 1)$ where the integer n is the Matsubara frequency, $\hat{\tau}_i$ and $\hat{\sigma}_i$ are the Pauli matrices in the Nambu-Gor'kov and spin spaces, respectively. $\hat{\tau}_k \hat{\sigma}_l$ is direct product of these matrices. The order parameter Δ is real-valued in the superconducting layer, while in the ferromagnetic layers it is zero. In general, the diffusion constant D acquires a proper subscript, S or F, when equation (1) is applied to the superconducting or ferromagnetic layers, respectively.
The Green function has the following components:

$$\hat{f} = \hat{\tau}_1 \left(f_0 \hat{\sigma}_0 + f_3 \hat{\sigma}_3 + f_2 \hat{\sigma}_2 \right). \tag{2}$$

Here, f_0 is the singlet component, f_3 is the triplet with zero projection on the z axis, and f_2 is the triplet with ±1 projections on z (the latter is present only at $\alpha \neq 0, \pi$).
There are the following symmetries:

$$\begin{aligned} f_0(-\omega) &= f_0(\omega), && f_0 \text{ is purely real} \\ f_3(-\omega) &= -f_3(\omega), && f_3 \text{ is purely imaginary} \\ f_2(-\omega) &= -f_2(\omega), && f_2 \text{ is purely imaginary,} \end{aligned} \tag{3}$$

which makes it sufficient to consider only positive Matsubara frequencies, $\omega > 0$.
Problem of calculating T_c can be reduced to an effective set of equations for the singlet component in the S layer: the set includes the self-consistency equation and the Usadel equation,

$$\Delta \ln \frac{T_{cS}}{T_c} = 2\pi T_c \sum_{\omega > 0} \left(\frac{\Delta}{\omega} - f_0 \right), \tag{4}$$

$$\frac{D}{2}\frac{d^2 f_0}{dx^2} - \omega f_0 + \Delta = 0, \tag{5}$$

with the boundary conditions:

$$\left.\frac{df_0}{dx} = 0\right|_{x=-d_S}, \qquad \left. -\xi_S \frac{df_0}{dx} = W f_0 \right|_{x=0}. \tag{6}$$

Here, T_{cS} and $x_S = \sqrt{D_S / 2pT_{cS}}$ are the superconducting transition temperature and coherence length for an isolated S layer, respectively. This is exactly the problem for which the multimode method (as well as the method of fundamental solution) was developed in [2] and then applied to F1SF2 [4] and semi-infinite SF1F2 [3] spin valves. We only need to determine the explicit expression for W in equation (6), solving the boundary problem for the SF1F2 structure.
The Usadel equation (1) generates the following characteristic wave vectors:

$$k_\omega = \sqrt{\frac{2\omega}{D}},\quad k_h = \sqrt{\frac{h}{D}},\quad \tilde{k}_h = \sqrt{k_\omega^2 + 2ik_h^2}. \tag{7}$$

In the S layer the solution of equation (1) is (A and B are purely imaginary):

$$\begin{pmatrix} f_0 \\ f_3 \\ f_2 \end{pmatrix} = \begin{pmatrix} f_0 \\ 0 \\ 0 \end{pmatrix} + \begin{pmatrix} 0 \\ A \\ B \end{pmatrix} \cdot \frac{\cosh\left(k_\omega (x + d_S)\right)}{\cosh\left(k_\omega d_S\right)}. \tag{8}$$

Note that here we keep $f_0(x)$ as is, because we cannot solve the equation for this component since it contains $\Delta(x)$. In the middle F1 layer the solution of equation (1) has the following form (C_1 and S_1 are purely imaginary, $C_3 = -C_2^*$, $S_3 = -S_2^*$):

$$\begin{pmatrix} f_0 \\ f_3 \\ f_2 \end{pmatrix} = C_1 \begin{pmatrix} 0 \\ 0 \\ 1 \end{pmatrix} \cosh\left(k_\omega x\right) + C_2 \begin{pmatrix} 1 \\ 1 \\ 0 \end{pmatrix} \cosh\left(\tilde{k}_h x\right) + C_3 \begin{pmatrix} -1 \\ 1 \\ 0 \end{pmatrix} \cosh\left(\tilde{k}_h^* x\right)$$

$$+ S_1 \begin{pmatrix} 0 \\ 0 \\ 1 \end{pmatrix} \sinh\left(k_\omega x\right) + S_2 \begin{pmatrix} 1 \\ 1 \\ 0 \end{pmatrix} \sinh\left(\tilde{k}_h x\right) + S_3 \begin{pmatrix} -1 \\ 1 \\ 0 \end{pmatrix} \sinh\left(\tilde{k}_h^* x\right). \tag{9}$$

In the outer F2 layer (E_1 is purely imaginary, $E_3 = -E_2^*$):

$$\begin{pmatrix} f_0 \\ f_3 \\ f_2 \end{pmatrix} = E_1 \begin{pmatrix} 0 \\ -\sin\alpha \\ \cos\alpha \end{pmatrix} \cosh\left(k_\omega x\right) + E_2 \begin{pmatrix} 1 \\ \cos\alpha \\ \sin\alpha \end{pmatrix} \cosh\left(\tilde{k}_h x\right) + E_3 \begin{pmatrix} -1 \\ \cos\alpha \\ \sin\alpha \end{pmatrix} \cosh\left(\tilde{k}_h^* x\right)$$

$$+ H_1 \begin{pmatrix} 0 \\ -\sin\alpha \\ \cos\alpha \end{pmatrix} \sinh\left(k_\omega x\right) + H_2 \begin{pmatrix} 1 \\ \cos\alpha \\ \sin\alpha \end{pmatrix} \sinh\left(\tilde{k}_h x\right) + H_3 \begin{pmatrix} -1 \\ \cos\alpha \\ \sin\alpha \end{pmatrix} \sinh\left(\tilde{k}_h^* x\right). \tag{10}$$

The boundary conditions at the free interface F2 of the structure take the form:

$$\left.\frac{df_i}{dx} = 0\right|_{x=d_{F1}+d_{F2}}. \tag{11}$$

The boundary conditions at the SF1 and F1F2 interfaces have the form [6]:

$$\left.\left(f_i + \gamma_B \xi \frac{df_i}{dx}\right)\right|_{\text{left}} = f_i\big|_{\text{right}},\quad \left.\left(\gamma\xi \frac{df_i}{dx}\right)\right|_{\text{left}} = \left.\left(\xi \frac{df_i}{dx}\right)\right|_{\text{right}}, \tag{12}$$

where g_B and g are the spin-independent suppression parameters:

$$\gamma_{BSF1} = R_{BSF1} A_B / \rho_S \xi_S,\quad \gamma_{SF1} = \rho_{F1}\xi_{F1} / \rho_S \xi_S,$$
$$\gamma_{BF1F2} = R_{BF1F2} A_B / \rho_{F1} \xi_{F1},\quad \gamma_{F1F2} = \rho_{F2}\xi_{F2} / \rho_{F1}\xi_{F1}. \tag{13}$$

R_{BF1S}, R_{BF2F1} and A_B are the resistance and the area of the SF1 and F1F2 interfaces, r_S, r_{F1} and r_{F2} are the resistivities of the S, F1 and F2 layers, respectively.

We choose the simplest formulation: all the interfaces are transparent ($g_B = 0$), the diffusion constants and the conductivities are the same ($g = 1$), the absolute values of the exchange fields in the two F layers coincide, f_0 is a constant within the S layer. Our strategy now is to obtain the effective boundary conditions (6) for $f_0(x)$ by eliminating all other components in the three layers.

The boundary conditions (11) and (12) give 14 equations. We are mainly interested in one of them, determining the derivative of the singlet component on the S side of the SF1 interface ($x = 0$):

$$\left.\frac{df_0}{dx}\right|_{x=0} = \tilde{k}_h S_2 - \tilde{k}_h^* S_3. \tag{14}$$

The remaining 13 boundary conditions form a system of 13 linear equations for 13 coefficients in equations (8)-(10). The solution of this system is nonzero due to $f_0(x)$ in equation (8). Then, we substitute the S_2 coefficient into equation (14) and thus explicitly find W in equation (6). All the information about the two F-layers is contained in the single real-valued function $W(\alpha)$.

3. Results and discussions

The results of numerical calculations of T_c as a function of the angle α, and pairing distributions under different values of the layers thicknesses in SF1F2 trilayer are given in figures 2 - 4.

Figure 2 (a) demonstrates the *direct* spin-valve effect ($T_c^{AP}(\alpha = 180°) > T_c^{P}(\alpha = 0°)$). The basic physical reason of the difference $\Delta T_c = T_c^{AP} - T_c^{P}$ is partial compensation of the pair-breaking ferromagnetic exchange field, when the magnetizations of the F1 and F2 layers are aligned antiparallel. As far as the F-layers are thin compared with the coherent lengths, the compensation is pretty good providing large ΔT_c. The both triplet pairing components f_2 and f_3 tend to have a maximum mostly at the outer surface of the F2 layer.

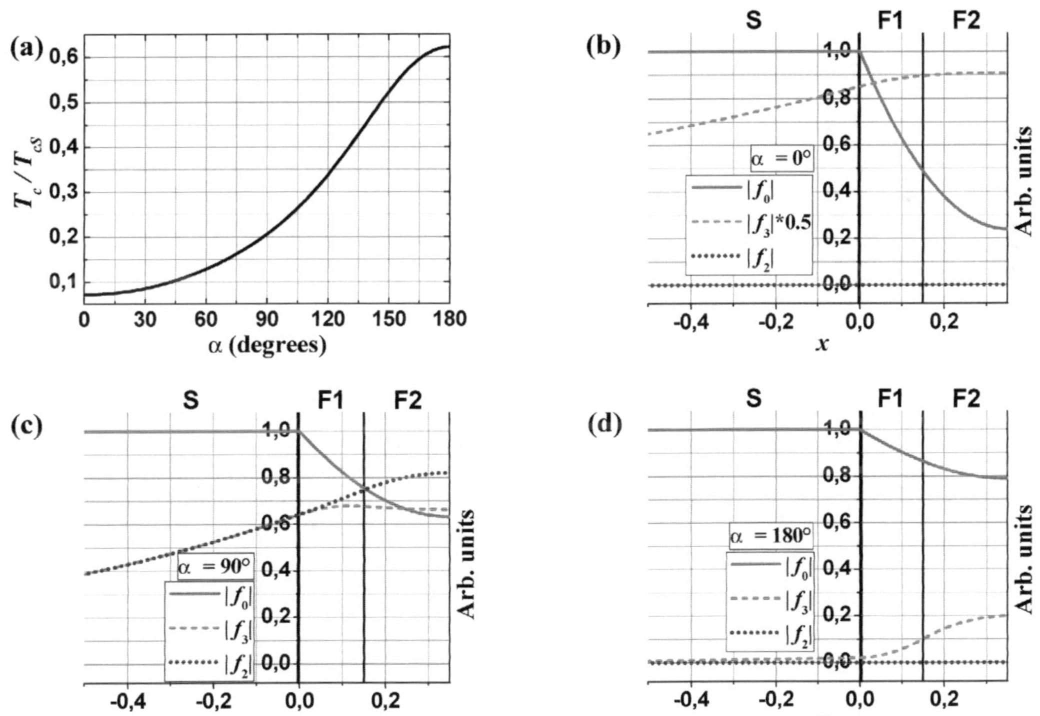

Figure 2. Critical temperature T_c vs. misalignment angle α (a). Spin-singlet and spin-triplet pairing distributions at α = 0° (b), 90° (c) and 180° (d) for n = 2; $d_{F1}/\xi_{F1} = 0.15$, $d_{F2}/\xi_{F2} = 0.2$, and table 1; the functions moduli have been drawn here and hereafter.

Figure 3(a) demonstrates the *triplet* spin-valve effect (T_c(non-collinear) < T_c(α = 180°), T_c(α = 0°)). In this mode, the oscillating behavior of the singlet superconducting pairing in the F layers is observed. The minimal critical temperature T_c corresponds to the long-range triplet pairing f_2 changing sign in the outer F2 layer, and double crossing zero by the singlet pairing component f_0 in the F layers (figure 3(c)), which does not present at the maximum temperatures (figures 3(b) and 3(d)). The zero spin-projection triplet component f_3 is located close to the S/F1 interface.

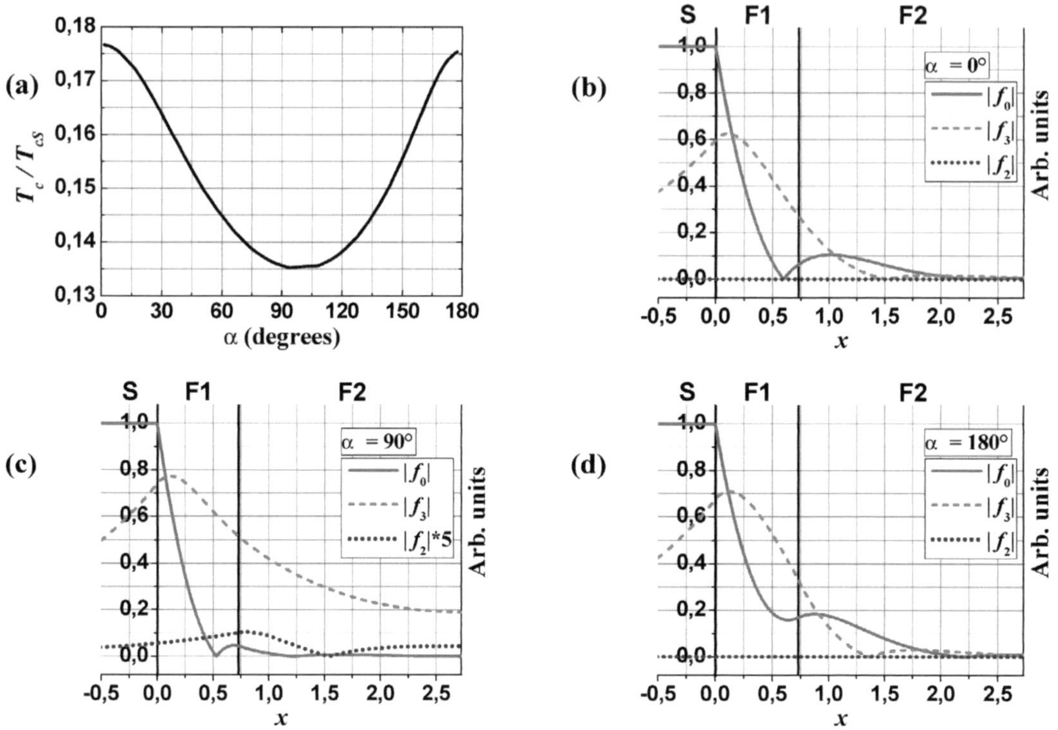

Figure 3. Critical temperature T_c vs. misalignment angle α (a). Spin-singlet and spin-triplet pairing distributions at α = 0° (b), 90° (c) and 180° (d) for n = 2; $d_{F1}/\xi_{F1} = 0.73$, $d_{F2}/\xi_{F2} = 2$, and table 1.

Figure 4(a) demonstrates the *inverse* spin-valve effect (T_c(α = 0°) > T_c(α = 180°)). In this mode the long-range triplet pairing component f_2 has a maximum at the outer surface of the F2 layers, while the zero spin-projection triplet component f_3 is always located close to the S/F1 interface. The minimum critical temperature T_c corresponds to the peculiar behavior – double crossing of zero by the singlet superconducting component f_0 in the F layers (figure 4(d)) which is not the case at α = 0° (figure 4(b)).

Table 1. Common parameters of the SF1F2 trilayer.

Parameter	value
d_S/ξ_S	2.75
γ	1
γ_B	0
ξ_F/ξ_S	1

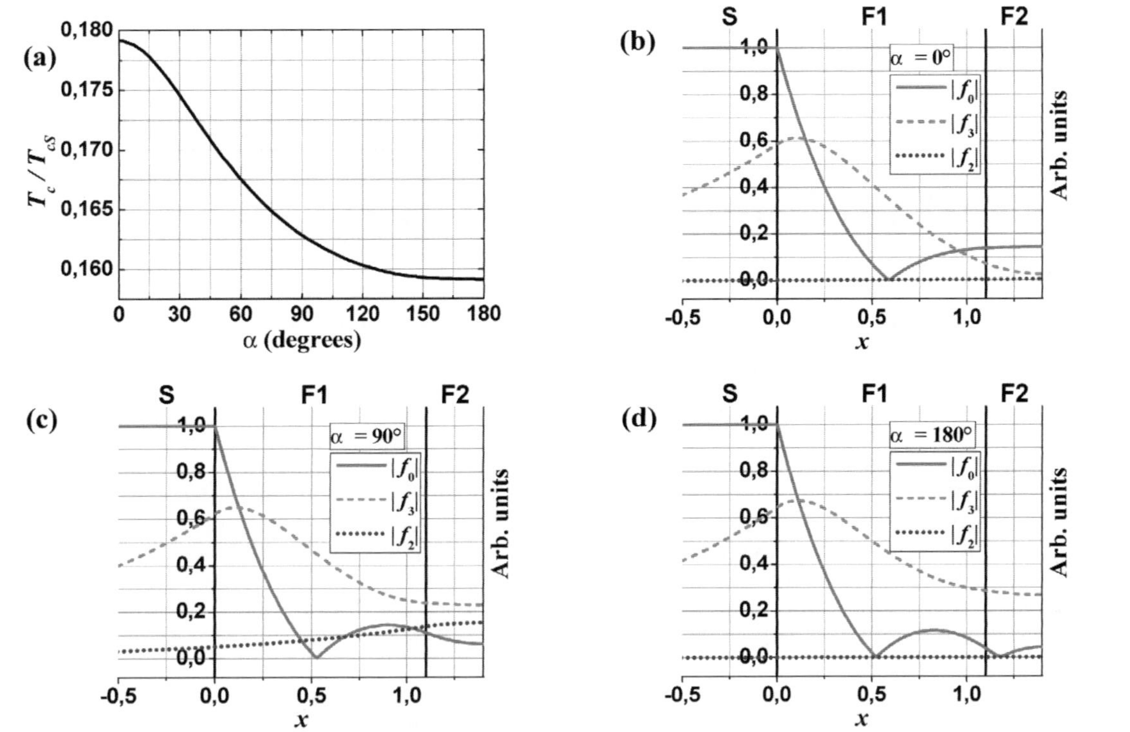

Figure 4. Critical temperature T_c vs. misalignment angle α (a). Spin-singlet and spin-triplet pairing distributions at α = 0° (b), 90° (c) and 180° (d) for n = 2; $d_{F1}/\xi_{F1} = 1.1$, $d_{F2}/\xi_{F2} = 0.3$, and table 1.

4. Conclusion

We have considered a finite-thickness SF1F2 spin valve. We visualized distributions of spin-singlet and spin-triplet superconducting pairing components and attributed their peculiarities to different spin-valve switching modes of the SF1F2 trilayer heterostructure. This may be important to plan experimental detecting of the triplet pairings and the inverse proximity effect in SF hybrids by, for example, polarized neutron reflection [7, 8].

Acknowledgments
The support by RFBR (grants No. 16-02-01171-a, 14-02-31002-mol_a, 15-32-20362-bel_a_ved, 15-52-10045_KO_a), DFG and by the Program of Competitive Growth of Kazan Federal University, funded by the Russian Government, is gratefully acknowledged.

References

[1] Bergeret F S, Volkov A F and Efetov K B 2005 *Rev. Mod. Phys.* **77** 1321
[2] Fominov Ya V, Chtchelkatchev N M and Golubov A A 2002 *Phys. Rev.* B **66** 014507
[3] Fominov Ya V, Golubov A A, Karminskaya T Yu, Kupriyanov M Yu, Deminov R G and Tagirov L R 2010 *JETP Lett.* **91** 308
[4] Fominov Ya V, Golubov A A and Kupriyanov M Yu 2003 *JETP Lett.* **77** 510
[5] Karminskaya T Yu, Golubov A A and Kupriyanov M Yu 2011 *Phys. Rev.* B **84** 064531
[6] Aarts J, Geers J M E, Brück E, Golubov A A and Coehoorn R 1997 *Phys.Rev.* B **56** 2779
[7] Khaydukov Yu N, Aksenov V L, Nikitenko Yu V, Zhernenkov K N, Nagy B, Teichert A, Steitz R, Rühm A and Bottyán L 2011 *J. Supercond. Nov. Magn.* **24** 961
[8] Khaydukov Yu N *et al.* 2013 *JETP Lett.* **98** 107

Journal of Physics: Conference Series **690** (2016) 012034 doi:10.1088/1742-6596/690/1/012034

The role of Purcell effect for third harmonic generation

T A Voytova, A V Yulin, A E Krasnok, K V Baryshnikova and P A Belov

ITMO University, 49 Kronverksky Ave., St. Petersburg, 197101, Russia

E-mail: t.voytova@phoi.ifmo.ru

Abstract. Third-harmonic generation from a nonlinear layer in a dielectric resonator is considered. It is shown theoretically that the Purcell effect can significantly affect the efficiency of harmonic generation. To describe THG by complex cavities, a homogenization approach is developed. It is shown that this approach works well for modelling the emission from a nanostructured surface consisting of subwavelength-size nonlinear nanoparticles.

1. Introduction

Nonlinear optics has been attracting the interest of researchers for several decades. Numerous nonlinear optical phenomena have been discovered, the most important of them are frequency mixing, higher-harmonic generation, supercontinuum generation, optical solitons, etc., see [1–4]. The theory of new frequencies generation at the interface between linear and nonlinear media has been developed since the pioneering works by Bloembergen et al. [5]. In particular, the expressions for reflected and transmitted second-harmonic light from a thin nonlinear dielectric slab were obtained. It was demonstrated that the refraction on the boundary between media with different dielectric susceptibility ϵ will affect the intensity of the emitting radiation at higher harmonics. The possibility of enhancement of the nonlinear effects by using optical resonators, in other words by exploiting the Purcell effect, was considered in [6–10]. It is well known that the Purcell effect contributes to the rates of spontaneous emission. In the case of higher harmonics generation by nonlinear processes, the efficiency also depends on the local density of photonic states at the location of the nonlinear polarization source [11]. It is shown [12] that the Purcell effect can increase the photoluminescence in Si ring resonators by at least 50 times. There are several papers devoted to the consideration of new harmonic generation from a planar nonlinear dielectric film placed inside a dielectric resonator [13, 14]. The research on the new frequencies generation by the arrays of nonlinear nanoparticles has been recently reported in a number of papers [15–19]. The optical properties of nanoscale structures differ significantly from the properties of bulk materials because of the strong confinement of the electromagnetic field inside nanoparticles resulting in geometrical resonances. In the recent work [20], the emission of the third harmonic from dielectric (silicon) nanodisks placed on a glass layer was reported. A pronounced reshaping of the third-harmonic spectra and the increase of THG intensity by two orders of magnitude with respect to bulk silicon was explained in the terms of the interplay between the electric and magnetic dipole resonances.

2. THG from a thin nonlinear film

The main goal of the paper is to consider how the interplay of two effects, the enhancement of the pump field and the Purcell effect, influences the process of third harmonic generation. For this purpose we consider a very simple physical system consisting of a dielectric Fabry-Perrot resonator with a thin nonlinear layer. The resonator is pumped by external radiation at the fundamental frequency $\omega_1 = \Omega$. In our model a nonlinear layer has a cubic (Kerr) nonlinearity and thus the incident radiation excites the current at the frequency triple of that of the pumping wave ($\omega_3 = 3\Omega$). This current causes the emission of the third harmonic. We assume that the nonlinearity is weak and we can disregard the influence of the third harmonic on the propagation of the fundamental mode. Because of the same reason, we disregard generation of higher harmonics.

In this case the problem can be split into two parts. The first part is finding a solution for the distribution of the field at the fundamental frequency. For this we need to solve the Helmholtz equation with the appropriate boundary conditions at the infinity (no radiation coming from minus infinity and a fixed incident wave coming from plus infinity).

$$\hat{H}(\omega_1)E_1 = 0. \tag{1}$$

The second part of the problem is the solution of the Helmholtz equation for the third harmonic E_3 with no incident waves.

$$\hat{H}(\omega_3)E_3 = \frac{4\pi}{c^2}\omega_3^2 P_3, \tag{2}$$

where the nonlinear polarization vector is determined as $P_3 = \hat{\chi}^{(3)} | E_1 |^2 E_1$. It is important to note that equation (2) is the usual linear wave equation with driving force associated with the oscillations of the polarization at the third harmonic of the fundamental frequency $\omega_3 = 3\Omega$ in the right-hand side of the equation. Let us note that the operator $\hat{H}$ depends on the frequency if dielectric permittivity ε is a function of frequency.

The amplitude of the third harmonic field E_3 is the solution of equation (2) and can be expressed through Green functions:

$$E_3 = \int \hat{G}(r, r')f(r')dr', \tag{3}$$

where $f = \frac{4\pi}{c^2}\omega_3^2 \hat{\chi}^{(3)} | E_3 |^2 E_3$. The tensor Green function for the equation (3) $\hat{G}(r, r')$ is defined as a solution of the equation:

$$\hat{H}(\omega_3)\hat{G}(r, r') = \delta(r - r')\hat{I}, \tag{4}$$

where $\hat{I}$ is the unity tensor. The important point is that the radiation at the third harmonic depends on both the distribution of the fundamental field (it gives the emitting source) and the spatial structure of the Green function (which defines the coupling of the source and the emitting wave). The latter we refer to as the Purcell effect.

To demonstrate how the Purcell effect can influence third harmonic generation, let us consider a one-dimensional model that can be easily studied analytically. The system is a Fabry-Perot resonator formed by air, a glass layer of thickness d and a silicon substrate. We assume that the resonator is pumped by the radiation at normal incidence (along the z-axis) and its polarization is perpendicular to the plane of incidence. The schematic view of the resonator under consideration is shown in figure 1. In this system at the distance h from the air-glass interface we place a very narrow layer (δ-layer) of a nonlinear material with cubic nonlinearity.

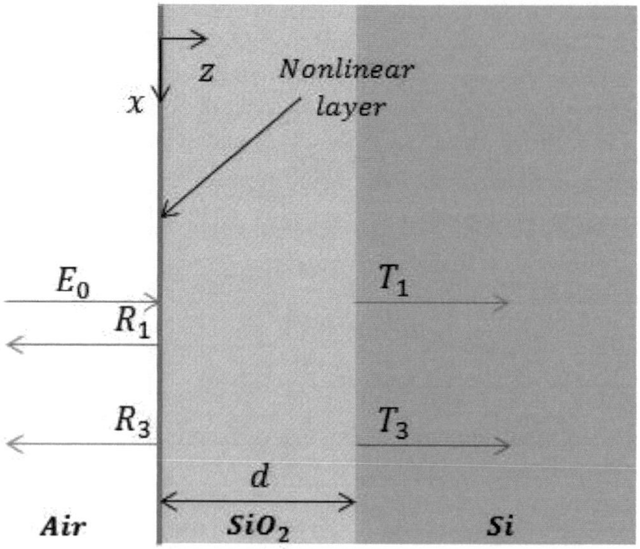

Figure 1. Geometry of our problem: THG by a nonlinear layer in the Fabry-Perot.

According to our plan, we start with propagation of the wave at the fundamental frequency. Using the well-known transfer matrix method it is easy to obtain a solution of the equation (1) for the pumping field amplitude along the propagation axis *z*:

$$E_1(z) = \begin{cases} 1/q\, E_0 (r_{12} e^{ik_2 d} + r_{23} e^{-ik_2 d}) e^{-ik_1 z} \equiv R_1(z) & \text{if } z \le 0; \\ 1/q\, E_0 t_{12} (e^{ik_2 (d-z)} + r_{23} e^{-ik_2 (d-z)}) & \text{if } 0 < z \le d; \\ 1/q\, E_0 t_{12} t_{23} e^{-ik_3 z} \equiv T_1(z) & \text{if } z > d, \end{cases} \tag{5}$$

where k_i is the wavevector in the medium with index *i*, $r_{ij} = (n_i - n_j)/(n_i + n_j)$, $t_{ij} = 2n_i/(n_i + n_j)$, index "1" corresponds to air, "2" - to glass, "3" - to silicon; and we used the notation $q = (e^{ik_2 d} + r_{12} r_{23} e^{-ik_2 d})$.

The reflection and transmission coefficients for the pumping wave for a Fabry-Perot resonator with glass thickness $d = 2$ μm are shown in figure 2. Here, R_1 and T_1 are the reflected and transmitted parts of pump energy; the curves are plotted as a function of the fundamental frequency. It is obvious that the reflection and transmission coefficients oscillate depending on the fundamental wavelength.

It is straightforward to calculate the Green function for the third harmonic excitation. In our model, it is suggested that the nonlinear source is the δ-layer located in the vicinity of the air/glass boundary at a distance *h*. Thus, it is easy to find an analytical expression for the Green function:

$$E_1(z) = \frac{1}{ik_2} \begin{cases} (1 - r_{12}^2)/t_{12}\; u/w\, e^{-ik_1 z} & \text{if } z \le 0; \\ u/w\, (e^{ik_2 z} - r_{12} e^{-ik_2 z}) & \text{if } 0 < z \le z'; \\ v/w\, (e^{ik_2 (d-z)} - r_{12} e^{-ik_2 (d-z)}) & \text{if } z' < z \le d; \\ t_{23}\; v/w\, e^{-ik_3 z} & \text{if } z > d, \end{cases} \tag{6}$$

where we used the notations: $u = (e^{ik_2(d-z')} + r_{23}e^{-ik_2(d-z')})$, $v = (e^{ik_2z'} - r_{12}e^{-ik_2z'})$, $w = (r_{23}e^{-ik_2(d-z')} - e^{ik_2(d-z')})(e^{ik_2z'} - r_{12}e^{-ik_2z'}) - (e^{ik_2z'} + r_{12}e^{-ik_2z'})(r_{23}e^{-ik_2(d-z')} + e^{-ik_2(d-z')})$.

An important fact is that the Green function oscillates, the oscillation period is approximately three times shorter comparing to the period of the fundamental harmonic. It is also worth mentioning that, because of the dispersion, the positions of the maxima of the imaginary part of the Green function and the nonlinear polarization do not coincide. The dependence of the fundamental wave intensity at the location of the δ - layer and the Green function $\hat{G}(z, z' = h)$ on the propagation coordinate z are shown in the inset in figure 2. The calculations are done for the parameters $d = 3$ μm, $\Omega = 1.25$ μm.

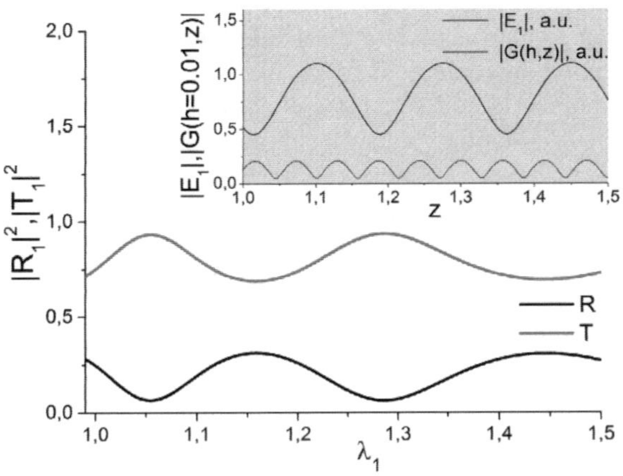

Figure 2. Reflection (black line) and transmission (red line) coefficients as functions of wavelength for the fundamental harmonic. The inset: Dependence of the amplitude of the fundamental wave (blue) and the Green function (green) $\hat{G}(z, z' = h)$ for the fixed on the propagation coordinate; the fundamental frequency $\Omega = 1$ μm.

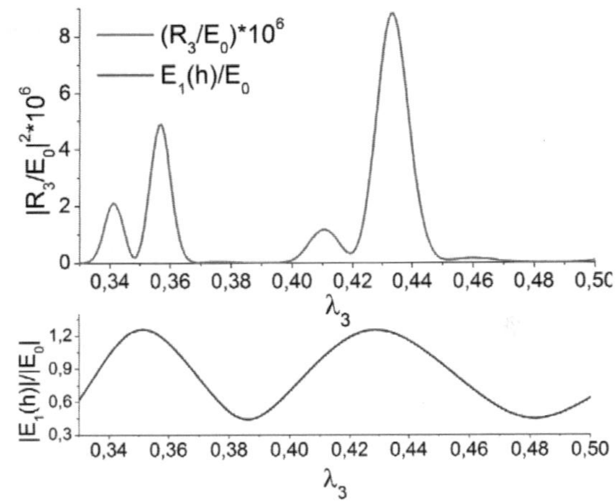

Figure 3. The dependence of the third harmonic radiation intensity on wavelength (green curve) for the case when the nonlinear layer is situated at the air-glass interface. The blue line shows the intensity of the pump wave at the position of the layer $E_1(z = h)$. $d = 3$ μm, $h = 0.01$ μm.

In the considered case the expression in the right-hand side of the inhomogeneous wave equations for the third harmonic can be written as $f(z) = 4\pi / c^2\ \omega_3^2 \chi^{(3)} (E_1)^3 \delta(z-h)$. Therefore, it is easy to obtain the amplitude of the third harmonic:

$$E_3(z) = \hat{G}(z,h) \frac{4\pi}{c^2} \omega_3^2 \chi^{(3)} (E_1)^3 . \tag{7}$$

As seen from equation (7), the radiation intensity for the third harmonics is given by the product of the nonlinear polarization and the Green function. In other words, the radiation intensity is proportional to the intensity of nonlinear current and this component has its maximum when the fundamental harmonic is in resonance. In the same time, the radiation intensity is also proportional to the coupling of the current to the radiating mode. This dependence is given by the Green function. This phenomenon known as the Purcell effect makes the radiation intensity frequency-dependent with pronounced maxima at certain frequencies.

The bottom panel in figure 3 shows the dependence of the pump field intensity at the layer point ($z = h$) on wavelength. The induced nonlinear current at the third harmonic is proportional to the cube of this value. On the top panel in figure 3, the normalized radiation intensity of the third harmonics (for fixed $z<0$) is shown as a function of wavelength. It is seen that the maxima of the radiation intensity are close to the maximum of the nonlinear current. However, the radiation intensity has three maxima. In a way it can be interpreted as an analogy of Fano resonance when a wider resonance (Fabry-Perot resonance for the fundamental frequency) is modulated by sharper resonances (the Green function for the third harmonic). The additional peaks in the curve are not symmetrical due to the dispersion.

3. THG from nanostructured interface

The third-harmonic generation by nanostructured nonlinear surfaces can be described using the model discussed above. By applying the homogenization scheme, we have demonstrated the Purcell effect in THG by the discrete array of nonlinear nanodisks. We examined the silicon discs with thickness d_3 and diameter d_2 arranged at a distance d_1 from each other. Nanodisks are placed on a glass plate with a thickness d_4, laying on a silicon substrate. In our calculations, we considered thin disks ($d_3 \ll \lambda_3/n_3$) located at a close distance (this condition means that the period of arrangement $d_1 + d_2$ does not exceed the third harmonic wavelength λ_3).

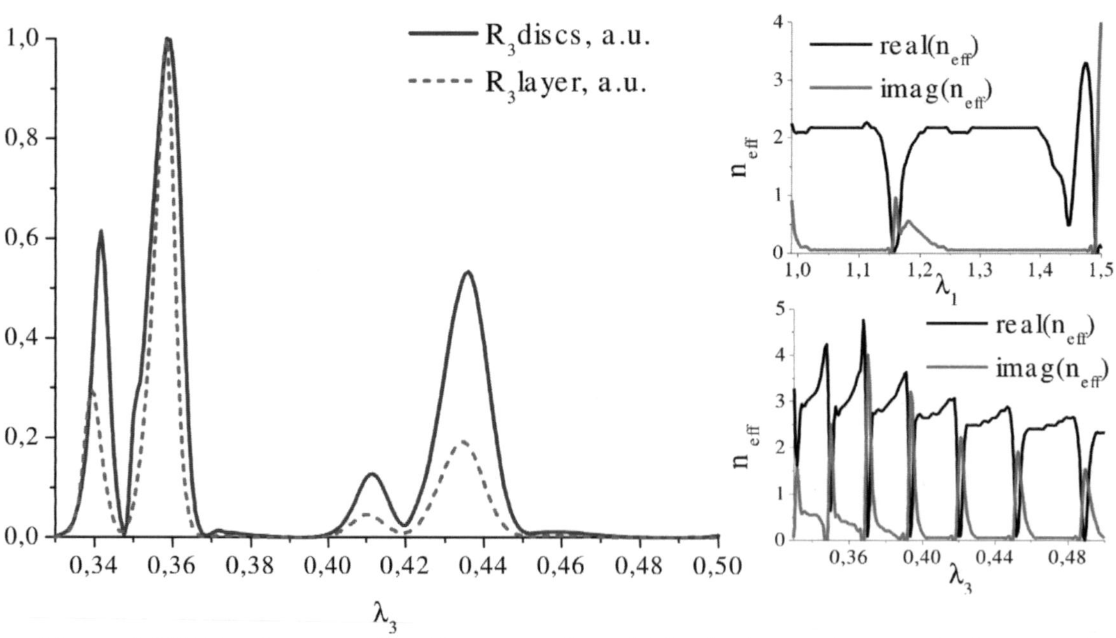

Figure 4. THG by the array of nonlinear nanodiscs obtained using the homogenization procedure. Parameters of the samples are listed in the inset: the distance between discs $d_1 = 200$ nm; the diameter $d_2 = 100$ nm; the thickness $d_3 = 20$ nm; the thickness of the glass layer $d_4 = 2$ μm. The normalized third harmonic radiation intensity by the effective homogeneous layer (solid blue curve) and by the thin silicon film with thickness $b = 1$ nm (dashed green curve) vs wavelength. In the right panels the effective parameters of the homogeneous layer are shown.

We modeled third-harmonic generation by nanodiscs using a program CST Microwave Studio. Comparing the results of numerical calculations with analytical results for multilayered structures we can obtain the effective parameters of a homogenized layer, an example is shown in the right panels of figure 4. Then, the dependence of THG by the effective nonlinear layer can be calculated straightforwardly. It is clearly seen that the dependence of the normalized harmonic intensity for the nonlinear array of nanodisks (solid blue curve) and for homogeneous silicon layer (dashed green curve) are in good agreement. This means that theory developed in this paper is applicable to the description of the harmonic generation by nanoparticle arrays of subwavelength size.

4. Conclusion

Now we briefly summarize the main results of the paper. It has been demonstrated that the Purcell effect is of great importance for third-harmonic generation. The Fabry-Perot resonance for the fundamental frequency increases the nonlinear current when exciting the third harmonic. However, the intensity of third harmonic generation depends on the product of the current and the coupling of the current to the emitting mode. In the general case, the maximum of the driving current and the maximum of the coupling are achieved at different wavelengths. This can lead to the appearance of additional maxima on the dependence of the third harmonic power on wavelength. This can be an alternative explanation of the multi-hump intensity-wavelength dependencies observed in experiments [20] and usually attributed to the closely situated electric and magnetic dipole resonances.

Acknowledgments

This work was financially supported by the Russian Science Foundation (15-19-30023) and Dynasty Foundation.

References

[1] Boyd R W 1992 *Nonlinear Optics* (New York: Academic)
[2] Bass M, Franken P A, Hill A E, Peters C W and Weinreich G 1962 *Phys. Rev. Lett.* **8** 18
[3] Franken P A, Hill A E, Peters C W and Weinreich G 1961 *Phys. Rev. Lett.* **6** 118
[4] Drazin P and Johnson R 1989 *Solitons: An Introduction* (Cambridge: Cambridge University Press)
[5] Bloembergen N and Pershan P 1962 *Phys. Rev.* **128** 606
[6] Bravo-Abad J, Rodriguez A, Bermel P, Johnson S G, Joannopoulos J D, Soljacic M 2007 *Optics Express* **15** 16161
[7] Soljaycic M and Joannopoulos J 2004 *Nature Materials* **3** 211
[8] D'Aguanno G, Centini M, Sibilia C, Bertolotti M, Scalora M, Bloemer M J and Bowden C M 2001 *Phys. Rev. E.* **64** 016609
[9] Hamam R, Ibanescu M, Reed E J, Bermel P, Johnson S G, Ippen E, Joannopoulos J D and Soljaycic M 2008 *Optics Express* **16** 12523
[10] Bermel P 2007 *Phys. Rev. Lett.* **99** 053601
[11] Purcell E 1946 *Phys. Rev.* **69** 681
[12] Lin S, Kobayashi Y, Ishikawa I and Wada K 2008 *Appl. Phys. Lett.* **92** 021113
[13] Saeta P and Miller M 2001 *Appl. Phys. Lett.* **79** 2704
[14] Kolthammer W, Barnard D, Carlson N, Edens A D, Miller N A and Saeta P N 2005 *Phys. Rev. B* **72** 045446
[15] Monteiro-Filho J and Gomez-Malagon L 2012 *J. Opt. Soc. Am. B* **29** 1613
[16] Hache F, Ricard D and Flytzanis C 1986 *J. Opt. Soc. Am. B* **3** 1647
[17] Rativa D, de Araujo R and Gomes A 2008 *Opt. Express* **16**, 19244
[18] Podlipensky A, Seifert J L and Graener H 2003 *Opt. Lett.* **28** 716
[19] Zhang J, Liu H, Wang Z and Ming N 2007 *Appl. Phys. Lett.* **90** 163122

[20] Shcherbakov M R, Neshev D N, Hopkins B, Shorokhov A S, Staude I, Melik-Gaykazyan V, Decker M, Ezhov A A, Miroshnichenko A E, Brener I, Fedyanin A A and Kivshar Yu S 2014 *Nano Lett.* **14** 6488

Bulk wave excitation from the surface of a $LiNbO_3$ crystal by a system of planar electrodes

A V Muromets, V B Voloshinov and I A Kanonin

Lomonosov Moscow State University, Faculty of Physics, Leninskie Gory, Moscow 119991, Russia

E-mail: anastasiagirl@mail.ru

Abstract. Ultrasound generation using planar electrodes deposited on the surface of a lithium niobate crystal is investigated. The analysis is carried out in an acousto-optic device based on the XY cut of a lithium niobate crystal and in an acousto-optic cell designed as a prototype of a modulator of non-polarized radiation. The experimental investigation is conducted at different frequencies of the driving electrical signal. The generated acoustic field is visualized by the acousto-optic method at different polarizations of the incident optical beam in case of slow shear and longitudinal acoustic modes. A relation between diffraction efficiencies for ordinary and extraordinary polarized optical radiation is analyzed. The obtained experimental results on the diffraction efficiencies are compared to the theoretically predicted values. It is demonstrated that the polarization characteristics of the devices differ from the conventional characteristics by the dependence on the structure of the acoustic beam.

1. Introduction

Acousto-optic (AO) devices are widely used in signal processing, laser beam control, frequency shifting, spectral filtering, etc. [1]. Improvement of optoelectronic technology stimulated a more detailed analysis of the AO device characteristics [2]. One of the desired requirements for these devices is related to simplicity of manufacturing. The most difficult part in fabrication of AO devices is transducer bonding: low-loss attachment of a piezoelectric transducer plate to a crystal. In order to make the AO devices cheaper and easier for fabrication, we use the method of bulk wave excitation without application of the piezoelectric transducer plate. The acoustic wave is generated from the surface of a piezoelectric crystal due to its inherent piezoelectric effect [3].

It is known that the efficiency of the AO interaction is primarily determined by the velocity of sound and photo-elastic constants [1–2]. A paratellurite crystal exhibits good AO characteristics [1–2] and can be used in AO devices with surface generation of bulk acoustic waves [9–12]. The efficiency of surface excitation of ultrasound is determined by piezoelectric properties of the crystal [1–3], therefore, in our further investigations we use a lithium niobate crystal because of its strong piezoelectric effect [1–3,5]. There has been a considerable interest in the excitation of bulk acoustic waves in the lithium niobate cells using interdigital transducers [5–7]. The study was predominantly aimed at the conversion of electric energy into acoustic energy [6]. The AO interaction was employed to study acoustic fields. A significant difference in the diffraction efficiencies for optical beams with different polarizations was demonstrated in [7]. In this paper, we define an incident optical beam with the polarization direction collinear with the polarization direction of an "ordinary" intrinsic mode of a crystal as an ordinary polarized optical beam. Similarly, if the polarization direction of the incident optical beam is collinear with the polarization direction of the "extraordinary" intrinsic mode of a crystal, we call this optical beam extraordinary polarized. The polarization sensitivity is one of the key problems of optoelectronic devices [8–16]. The aim of the research is to investigate polarization characteristics of the AO interaction in the devices with surface generation of bulk acoustic waves in a lithium niobate crystal and their dependence on the acoustic field structure.

2. Theoretical model

The intensity of the diffracted light depends on acoustic driving power and the AO figure of merit (AOFM) M_2 [1–2]. In a crystalline media, AOFM is determined by the expression [1,11]:

$$M_2 = \frac{1}{4n_0 n_1 \rho V^3}\left[d_i^{(1)} \varepsilon_{ij} \varepsilon_{kl} d_l^{(0)} p_{jkmn}\left(r_m m_n + r_n m_m\right)\right]^2 \tag{1}$$

where n_0 and n_1 – refractive indices of the incident and diffracted plane optical waves, respectively, ε_{ij} – components of dielectric permittivity tensor, $d_l^{(0)}$ and $d_i^{(1)}$ – corresponding components of a unit electric displacement field vector, p_{jkmn} –components of photoelastic tensor, r_m and m_n – components of acoustic polarization and wave normal vectors, ρ – medium density and V – phase velocity of ultrasound.

A geometry of the AO interaction and a choice of a crystal cut have a great impact on the diffraction efficiency of an optical beam [1–2]. In order to decrease the polarization sensitivity of AO devices, we found a crystal cut of lithium niobate where the diffraction efficiencies for the optical beams with ordinary and extraordinary polarizations were close to each other [11]. In this optimal geometry of the AO interaction, an acoustic wave propagates at an angle of 13° relative to the Z axis in the YZ plane of lithium niobate while the optical beam propagates in the YZ plane orthogonally to the wave vector of ultrasound. AOFM for ordinary and extraordinary polarized optical beams are $M_e = 5.73 \cdot 10^{-18}$ s^3/g and $M_o = 2.99 \cdot 10^{-18}$ s^3/g, respectively, i.e. the ratio of intensities for the diffracted ordinary and extraordinary polarized optical beams is $I_e/I_o = M_2^{(o)}/M_2^{(e)} = 1.92$.

3. Experiment

3.1. Investigation of an AO cell based on the XY cut of a lithium niobate crystal

First, we study a prototype of the AO device based on the simple XY cut of a lithium niobate crystal. The facets of the crystal are parallel to the principal crystallographic planes. Three metal electrodes that are parallel to the optical axis Z are deposited on the upper facet of the sample that is orthogonal to the X axis as shown in figure 1. The driving electric signal is applied to the central electrode while the two side electrodes are grounded.

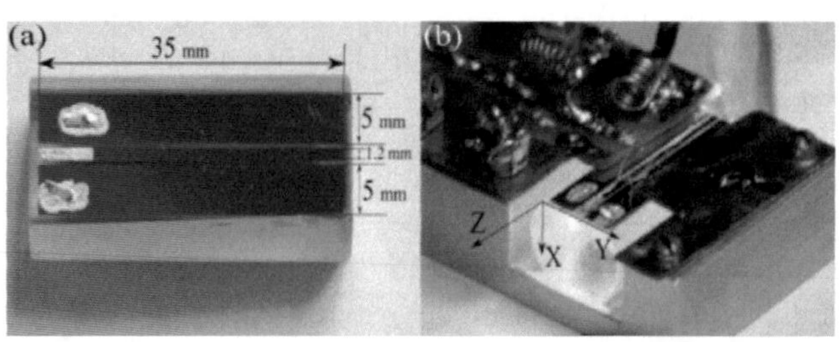

Figure 1. (a) - view of the upper facet of the crystal with electrodes. The gap between the central and side electrodes is 0.3 mm.;
(b) - the AO cell with orientation of crystalline axes. The width of the crystal in X direction is 10 mm.

The scheme of the experiment is shown in figure 2. The output signal of a pulsed oscillator (1) controls a high-frequency (HF) oscillator (2), whose output signal is fed to the central electrode of the AO cell (3) via a network providing matching of electric parameters. The side electrodes are connected with the ground. The waveform of the HF signal can be monitored using an oscilloscope (4). An oscilloscope (5) is used for observation of the time dependence of the diffracted intensity. It is seen that the output linear polarized beam of a laser (6) with the wavelength $\lambda = 532$ nm passes through the cell (3). A semiconductor detector (7) interfaced with an oscilloscope (5) is used to detect the diffracted optical beam intensity. The circuit allows the operation of the cell in a frequency band of 7.0 MHz in the vicinity of the central frequency of 200 MHz. The fraction of the electric power that is reflected from the transducer is about 5%.

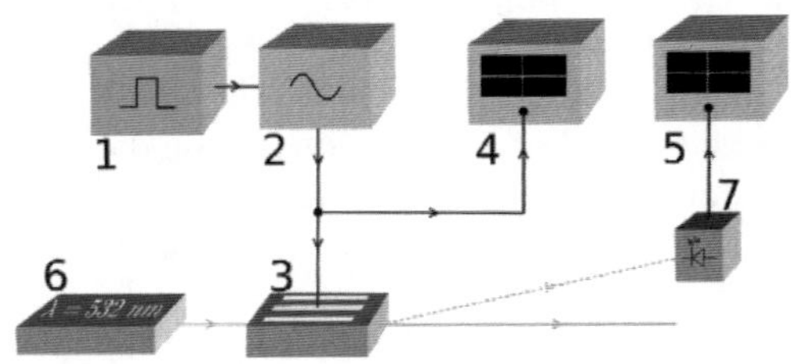

Figure 2. Block scheme of the experimental setup to study the surface excitation of the bulk acoustic waves: (1) – pulsed oscillator, (2) – HF oscillator, (3) – AO cell, (4) – oscilloscope for monitoring of the control voltage, (5) – oscilloscope for monitoring the diffracted intensity, (6) – laser, and (7) – photodetector.

Three acoustic modes (longitudinal, fast quasi-shear and slow quasi-shear) are simultaneously generated in the interelectrode gap and propagate in the bulk of lithium niobate along the X crystal axis. The optical radiation is diffracted by the acoustic field that is excited from the crystal surface. In this paper, we are investigating the isotropic AO interaction when the polarization of the incident optical beam coincides with the polarization of the diffracted optical beam [1]. Figure 3 shows the measured diffracted intensity for several positions of the laser beam probing. The optical wave is directed orthogonally to the XY plane of the crystal. Equation (1) shows that the efficiency of the AO interaction and AOFM M_2 depend on the velocity of sound V as $M_2 \sim V^{-3}$. The measured characteristics prove this dependence.

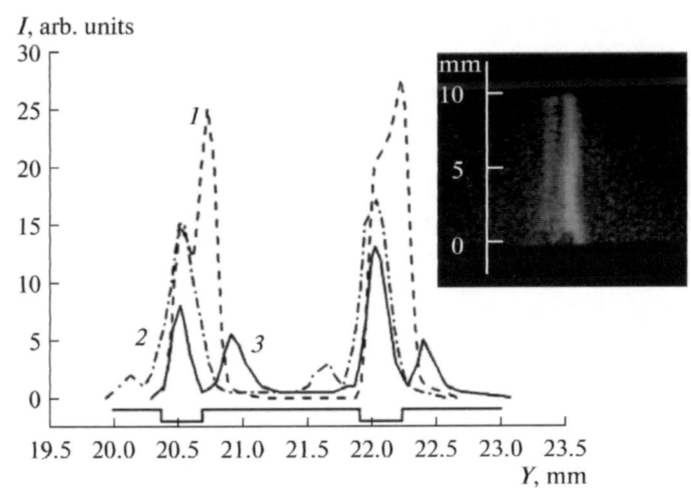

Figure 3. Intensity distribution of the radiation that is diffracted by the acoustic field excited from the surface of the XY cut of a lithium niobate crystal: (1) – slow quasi-shear acoustic mode, (2) – fast quasi-shear mode, and (3) – longitudinal mode. The dashed line in the vicinity of the abscissa axis shows the electrodes, and the inset shows the result of the AO visualization of the acoustic beam in the XY plane of the crystal.

3.2. Investigation of a AO cell based on the (Y-13°) cut of a lithium niobate crystal
In the second stage, we study an AO cell based on a lithium niobate crystal having a cut in accordance with the theoretical analysis [11,17,18]. The direction of the sound phase velocity must be oriented in the *YZ* plane at an angle of 13° relative to the *Z* optical axis. The analysis is carried out for two directions of optical radiation: along the *X* ([100]) crystal axis and at the angle of 13° with respect to the *Y* crystal axis in the *YZ* plane (almost along the [010] axis). The scheme of an AO cell based on the lithium niobate crystal of (*Y*–13°) cut is shown in figure 4. The size of each electrode is 1.9×7mm. The bottom facet of the crystal is tilted to the upper facet preventing appearance of a standing wave. It was shown that the polarization characteristics of the AO interaction are dependent on the acoustic field structure [19]. In the mentioned work, the ultrasound is generated from the surface of a lithium niobate crystal by using an interdigital transducer while the geometry of the AO interaction is the same as mentioned above. Therefore, the diffraction efficiency of the optical beam is dependent on whether the optical beam propagates along the electrodes or in the orthogonal direction. In order to investigate the AO interaction in both cases, two similar systems of planar golden electrodes were deposited on top of the lithium niobate crystal as shown in figure 1.

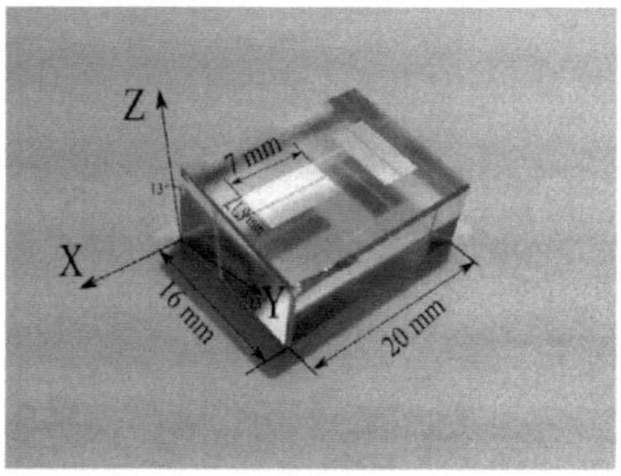

Figure 4. Orientation of crystalline axes in the AO cell based on a lithium niobate crystal having a (*Y*–13°) cut.

We investigated the relation between diffraction efficiencies for the ordinary and the extraordinary polarized optical radiation at different directions of the incident light propagation [19, 20]. In this paper, we measure these diffraction efficiencies in case of different acoustic beam apertures matching the AO cell circuit at different frequencies of ultrasound.

The scheme of the experiment is the same as shown in figure 2 with addition of a polarizator in front of the AO cell and an analizator behind the cell. The intensity of the diffracted light is measured by a detector (7) and the magnitude of the intensity is displayed on an oscilloscope (5). Every measured intensity value is normalized to the intensity of incident optical beam.

Experimental investigation of polarization characteristics of the AO interaction is carried out in case of the slow quasi-shear mode with the velocity $V_s = 3590$ *m/s* at different driving electric frequencies. At the acoustic frequency $f = 54$ MHz, the ratio between intensities of diffracted extraordinary and the incident ordinary polarized radiation propagating along the *X* crystal axis equals to $I_{e[100]}/I_{o[100]} = 16$. For the radiation propagating in the *YZ* crystal plane at 13^0 to the *Y* axis (i.e. approximately along the *Y* crystalline axis [010]), the relation is equal to $I_{e[010]}/I_{o[010]} = 0$. At the driving electric frequency $f = 83$ MHz, the ratios for the different directions of the incident radiation propagation are equal to $I_{e[100]}/I_{o[100]} = 5.42$ and $I_{e[010]}/I_{o[010]} = 0.45$, respectively. We also carried out the calculations and a comparison of the experimental data with the theory. The theoretical values do not depend on the structure of the acoustic beam and are equal to the following values: $I_{e[100]}/I_{o[100]} = 0.37$ and $I_{e[010]}/I_{o[010]} = 1.7$. The registered discrepancy between the obtained experimental and theoretical results shows that the acoustic field structure generated in the crystal affects the AO polarization characteristics.

In order to investigate the acoustic field structure generated by the system of planar electrodes, we carried out a series of experiments where the incident laser beam is sent normally on an input facet of the crystal [21]. This optical radiation diffracts on the generated acoustic beam along its propagation direction. The efficiency of the diffracted light is registered during this laser probing of the acoustic field. Data in figure 5 show that the slow quasi-shear mode is generated between the electrodes while the fast longitudinal acoustic mode is concentrated under the electrodes. The optical radiation having elliptical and linear states of polarization was diffracted by different parts of the generated acoustic beams.

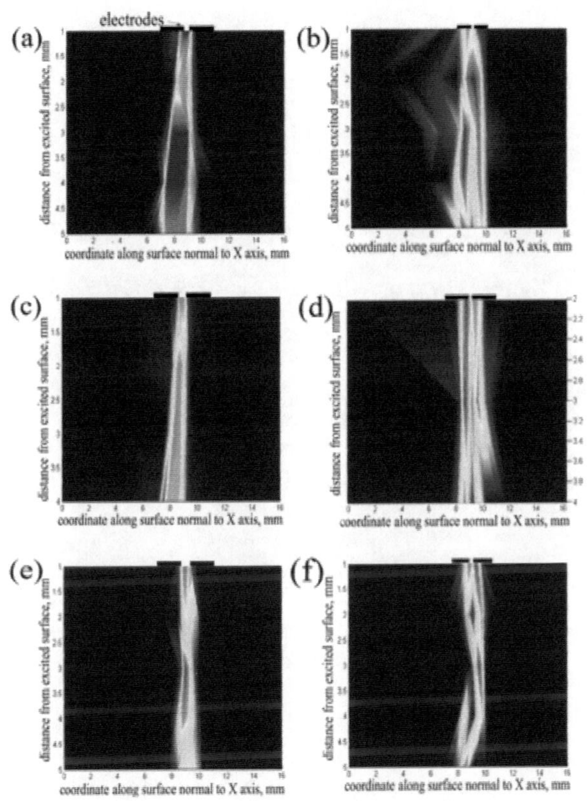

Figure 5. AO visualization of acoustic field structure generated by a system of planar electrodes: 2(a), 2(c), 2(e) – components of the acoustic field for the slow quasi-shear mode interacting with the elliptically polarized, extraordinary and ordinary polarized incident light, respectively; 2(b), 2(d), 2(f) – components of the acoustic field for the fast longintudinal mode ($V_l = 7300$ *m/s*) interacting with the elliptically polarized, extraordinary and ordinary incident radiation, respectively.

4. Conclusion

The surface generation of bulk acoustic waves using a system of planar electrodes deposited on the top facet of a lithium niobate crystal is analyzed in case of the *XY* and (*Y*–13°) crystal cuts. The polarization characteristics of the AO interaction are obtained for the elliptically, ordinary and extraordinary polarized radiation diffracted by the slow quasi-shear and the fast longitudinal acoustic waves. The experiment is conducted at different frequencies of ultrasound. It is shown that the structure of the generated acoustic field has a noticeable impact on the ratio of the diffraction efficiencies for the ordinary and the extraordinary polarized radiation.

Acknowledgements
The research is supported by a grant of Russian Science Foundation (grant N 14-22-00042).

References

[1] Balakshy V I, Parygin V N, Chirkov L E 1985 *Physical Fundamentals of Acousto-optics* (Moscow: Radio i Svyaz)
[2] Xu J, Stroud R 1992 *Acousto-Optic Devices* (New York: Wiley)
[3] Baransky K N 1957 *Doklady Akademii Nauk* **114**(3) 517-9
[4] Schmidt R V 1972 *J. Appl. Phys.* **43** 2498-501
[5] Shaskol'skaya M P 1982 *Acoustic Crystals. Manual* (Moscow: Nauka)
[6] Milsom R F, Reilly N H, Redwood M 1977 *IEEE Trans. Sonics Ultrason.* **24** 147
[7] Yashiro K, Goto N 1978 *IEEE Trans. Sonics Ultrason.* **25** 146
[8] Goruk W S, Normandin R, Vella P J, Stegman J I 1976 *J. Phys. D: Appl. Phys.* **9** 999
[9] Palmieri L, Socino G, Verona E 1985 *Appl. Phys. Lett.* **47** 463
[10] Zadorin A S, Serebrennikov L Ya, Sharangovich S N 1991 *Soviet Avtometriya* **1** 13-8
[11] Balakshy V I, Voloshinov V B, Solodovnikov V V, 2009 *Proc. Sci. Pract. Conf.* (Moscow: Phys. Dep. MSU) 42
[12] Voloshinov V B, Nikitin P A, Trushin A S, Magdich L N 2011 *Tech. Phys. Lett.* **37** 754
[13] Rimeika R, Ciplys D, Kazdailis P, Shur M S 2007 *Appl. Phys. Lett.* **90** 181935
[14] Rimeika R, Ciplys D, Kazdailis P, Balakauskas S 2011 *Opt. Lett.* **36** 2581
[15] Shibaev S S, Novikov V M, Rozdobud'ko V V 2010 *Fiz. Voln. Protsessov Radiotekh. Sist.* **13**(1) 55
[16] Dobrolenskii Yu S, Voloshinov V B, Zyuryukin Yu A 2008 *Kvantovaya Elektron.* (Moscow) **38**(1) 46
[17] Balakshy V I, Mikheev L I 2010 *Proc. of 12-th Russian School-seminar "Wave phenomena in inhomogeneous medium"* (Zvenigorod) **5** 18–21
[18] Balakshy V I, Mikheev L I 2013 *Uchen. Zap.of Kazan State Univ.* (Kazan) **155** 113-8
[19] Trushin A S, Nikitin P A, Muromets A V 2012 *AIP Conf. Proc.* **1433** 102-5
[20] Muromets A V, Trushin A S 2015 *Physics Procedia.* (Elsevier) **70** 721-4
[21] Korpel A 1966 *Appl. Phys. Lett.* **9** 425-7

Investigation of InGaP/Ga(In)As/Ge solar cells characteristics in the temperature range of 300 – 80 K

E V Kontrosh, A V Malevskaya, N M Lebedeva,V S Kalinovskiy and V M Andreev

Ioffe Institute, 26 Polytekhnicheskaya str., St. Petersburg 194021, Russia

E-mail: kontrosh@mail.ru

Abstract. Forward dark and load *I – V* characteristics of triple-junction GaInP/Ga(In)As/Ge solar cells (SCs) in the temperature range 300 – 80 K have been studied. At temperatures below 200 K, jumps of current and voltage in, respectively, dark and load characteristics were observed experimentally and attributed to the existence of a counter potential barrier formed by isotype heterolayers between the tunnel diode and the Ge *p-n* junction in the InGaP/Ga(In)As/Ge SC. An analysis of the forward dark characteristics of GaInP/Ga(In)As/Ge SCs, recorded at 80 K, enabled evaluation of the potential and real conversion efficiencies of incident sunlight. The influence exerted by the shape of the side mesa surface of GaInP/Ga(In)As/Ge SCs on the dominant current flow mechanisms was analysed. A method for single-step separate etching was suggested and studied. This method allows one to reduce surface leakage currents and raise the yield of suitable SCs with an efficiency greater than 35 % at low sunlight concentrations (C = 10 - 100, T= 300 K). The suggested post-growth technology reduces the number of fabrication operations and the SC production cost and improves the reliability of the SC operation in a wide temperature range.

1. Introduction

Multijunction solar cells (MJ SCs) based on binary semiconductor compounds presently have record-breaking conversion efficiencies of incident sunlight [1]. Possible ways to raise the MJ SC efficiency are whether by lowering the operating temperature of the photoactive $p – n$ junction or by diminishing the parasitic effect of the tunnel-trap (overabundant) and recombination current flow mechanisms [2, 3] in MJ SCs. In this context, characteristics of MJ SCs in the temperature range 300 – 80 K were examined and the effect of the side mesa surface of these cells on leakage currents was analysed in the present study.

2. Experimental

To carry out experiments, specimens of triple-junction InGaP/Ga(In)As/Ge SCs grown by the MOCVD technique were used. The SCs to be tested were fabricated by two post-growth techniques of separate etching: multi-step (traditional) and single-step. Forward dark and load *I -V* characteristics of a triple-junction InGaP/Ga(In)As/Ge SC were recorded in the temperature range 300 – 80 K. The forward dark *I-V* characteristics were measured at currents in the range from 10^{-14} to 1 A and the voltages of 0 – 3 V. The load *I-V* characteristics were recorded with a pulsed sunlight simulator (AM0 spectral range) at incident power density of 1367 W/m^2. Figure 1 presents two dark $J-V$ characteristics obtained at 300 and 80 K. The experimental $J-V$ characteristics can be expressed as a sum of three typical exponential portions related to the following current flow mechanisms: the first of these (curves 3 and 8) corresponds to the tunnel-trap (Esaki) mechanism; the second (curves 4 and 9), to the overall recombination (Sah-Noyce-Shockley) and diffusion (Shockley) mechanisms, the third (curves 5 and 10), to the diffusion (Shockley) mechanism. The values of the pre-exponential factors J_{0tr}, J_{0rd}, J_{0d} and the diode coefficients A_{tr}, A_{rd}, A_d, found for these typical portions are presented in Table 1 [4].

Figure 2 shows a set of load $I-V$ characteristics recorded in the temperature range 300 – 80 K. The maximum efficiency achieved for the InGaP/Ga(In)As/Ge SCs being tested is 35,0 % at 204 K (AM0, $C=1$), curve 1.

Current and voltage jumps are observed in the experimental dark and load characteristics at temperatures below than 200 K. According to the results of the estimating calculations performed in the study, the sharp rise in current at voltages of about 2 V (in figure 1, curve 6) is associated with a counter potential barrier arising in isotype heterolayers between the tunnel diode and the Ge *p-n* junction in the InGaP/Ga(In)As/Ge SC.

The dependences of the potential and real conversion efficiency of incident sunlight on the generation current density were calculated by using the procedure described in [2-5] and parameters obtained from the forward dark $J-V$ characteristics of InGaP/Ga(In)As/Ge SCs.

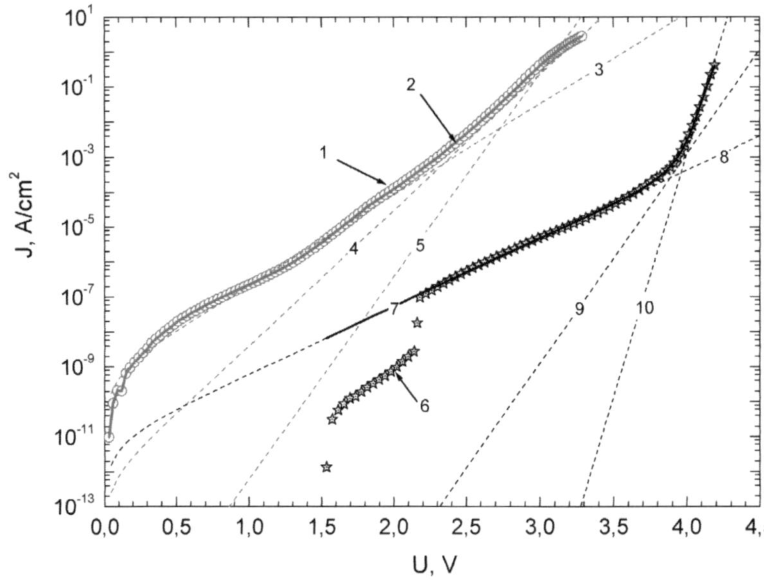

Figure 1. Forward dark $J-V$ characteristics of triple-junction InGaP/Ga(In)As/Ge SCs: (1, 6) – Experimental dark *J-V* characteristics at 300 and 80 K, respectively; (2, 7) – calculated dark characteristics at 300 and 80 K, respectively. Portions corresponding to the following mechanisms: (3, 8) – the tunnel-trap (J_{0tr}, A_{tr}); (4,9) – mixed recombination and diffusion (J_{0rd}, A_{rd}); (5,10) – diffusion (J_{0d}, A_d) (values of J_{0i} and A_i are shown in Table 1)

Figure 3 presents dependences of the potential (curve 1) and real (curve 2 with consideration for the resistance of the structure) efficiencies on the generation current density calculated with the SC parameters (in Table 1) at 80 K. It can be seen that the calculated curve fits well the experimental efficiency (in figure 3, curve 3) found from the load *I-V* characteristic (in figure 2).

According to the dependences presented in figure 1 and figure 3, the efficiency of InGaP/Ga(In)As/Ge SCs fabricated by the conventional (multi-step) method of separate etching is mainly limited at cell operating temperature of 80 K by the following current flow mechanisms:

tunnel-trap (overabundant) mechanism at generation current densities of up to 2×10^{-3} A/cm^2 and recombination mechanism at generation current densities of (2×10^{-3} - 1×10^{-1}) A/cm^2.

Table 1: Pre-exponential factors and diode coefficients

Diode coefficient, A_i	Pre-exponential factors: J_{0tr}, J_{0rd}, J_{0d}, (A/cm^2)	
	$T = 300$ K	$T = 80$ K
$A_{tr} > 6$	1×10^{-9}	8×10^{-12}
$A_{rd} = 4$	3×10^{-13}	1×10^{-18}
$A_d = 3$	1×10^{-27}	6×10^{-78}

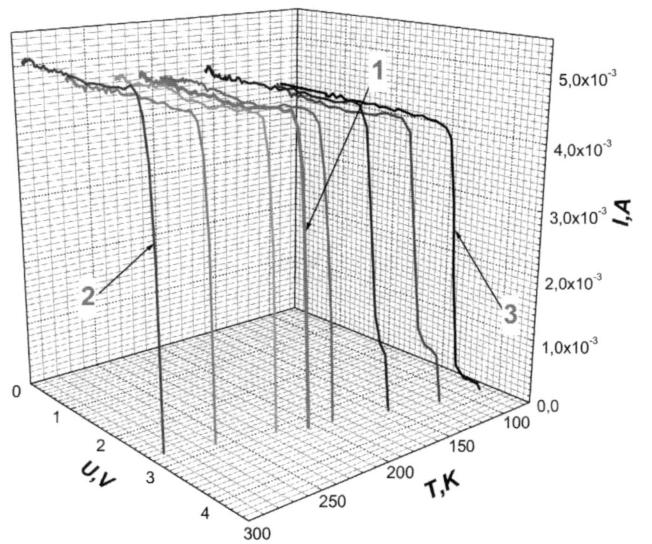

Figure 2. Load *I-V* characteristics of triple-junction InGaP/Ga(In)As/Ge SCs in the temperature range 300 – 80 K. (1) – Efficiency 35 %, ($C = 1$, AM0, 1367 W/m^2) $T = 204$ K; (2) – efficiency 30 %, ($C = 1$, AM0, 1367 W/m^2) $T = 300$ K; (3) – efficiency 32%, ($C = 1$, AM0, 1367 W/m^2) $T = 80$ K.

The results obtained in the present study and in [6] suggest that the increase in the influence of the tunnel-trap (overabundant) and recombination current flow mechanisms at variations of operating temperature of InGaP/Ga(In)As/Ge SCs fabricated by the multi-step method of separate etching is due to a complex (branched) profile of the side surface of the structure (in figure 4 (a)). This profile gives no way of reliably depositing a high-quality passivating coating and results in a substantial rise in the surface leakage currents. To diminish the effect of the tunnel-trap (overabundant) and recombination current flow mechanisms, a method of single-step separate etching has been elaborated, which allows levelling of the InGaP/Ga(In)As/Ge SC side surface (in figure 4 (b)), with the quality of the passivating coating thereby improved and the effect of the surface leakage currents diminished.

For the InGaP/Ga(In)As/Ge SCs fabricated by the multi-step method and by the single-step method of separate etching developed in [7], thermal-cycling tests were made in the temperature range from 300 to 80 K. Figure 5 presents the dependences of the pre-exponential factors for the tunnel-

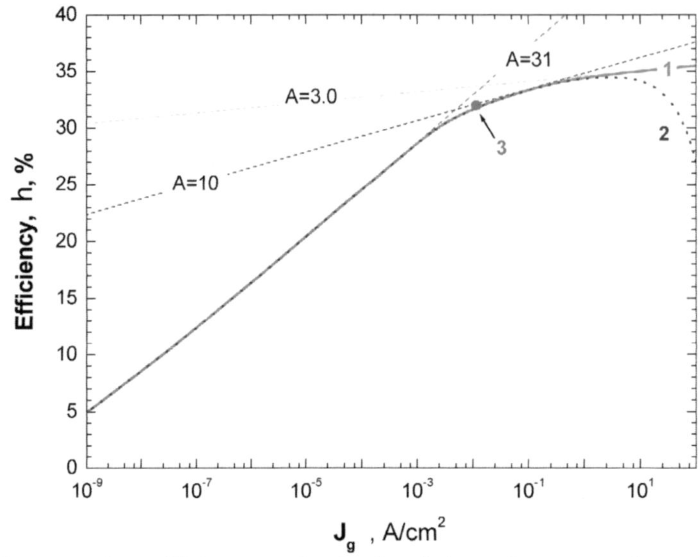

Figure 3. Experimental conversion efficiency and calculated dependences on the generation current density at 80K. (1) – R_s = 0 Ohm/cm^2, (2) – R_s = 0.01 Ohm/cm^2, (3) – experimental efficiency 32 %, (C = 1, AM0, 1367 W/m^2, T = 80 K). Portions corresponding to the following mechanisms: (4) – the tunnel-trap (J_{0tr}, A_{tr}); (5) – mixed recombination and diffusion (J_{0rd}, A_{rd}); (6) – diffusion (J_{0d}, A_d) (values of J_{0i} and A are shown in Table 1)

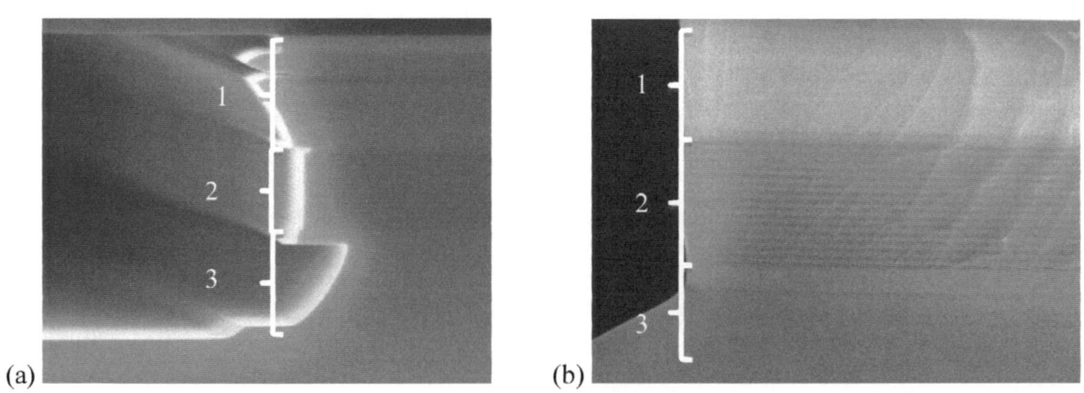

Figure 4. SEM pictures of the side surface profile of the InGaP/Ga(In)As/Ge structure: (1)– InGaP/Ga(In)As heterostructure, (2) – Bragg reflector, (3) – Ge *p-n* junction and Ge substrate; (a) – multistep etching, (b) – single-step etching.

trap (overabundant) – (J_{0t}), and mixed recombination-diffusion (J_{0rd}) current flow mechanisms on the number of thermal cycles for InGaP/Ga(In)As/Ge SCs fabricated by the single-step and multistep separate etching methods.

Figure 6 presents the calculated (dotted line) and experimental dependences of the efficiency of InGaP/Ga(In)As/Ge SCs on the concentration of incident sunlight. It is clear from the curves that the efficiency of triple-junction InGaP/Ga(In)As/Ge SCs fabricated by the single-step method of separate etching substantially exceeds, at the sunlight concentrations of 1 – 100, that of the

SCs fabricated by the multi-step technique and reaches values of 35 % and higher for the AM0 sunlight spectrum.

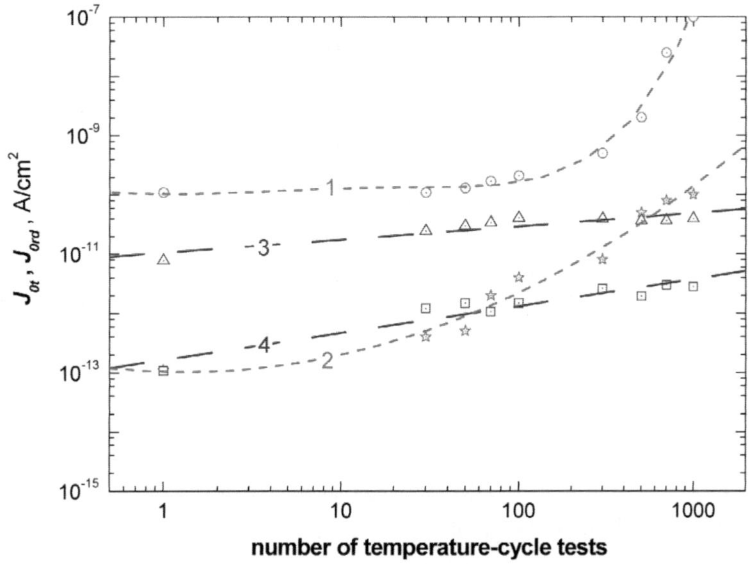

Figure 5. Degradation of the pre-exponential factors for (1, 3) tunnel-trap (overabundant) (J_{0t}) and (2, 4) mixed recombination (Sah-Noyce-Shockley) diffusion (Shockley) (J_{0rd}) current flow mechanisms calculated from experimental MJ SC $J-V$ characteristics recorded directly on the SC epitaxial wafer fabricated by (1, 2) multistep etching and (3, 4) single-step wet process in thermal cycling .

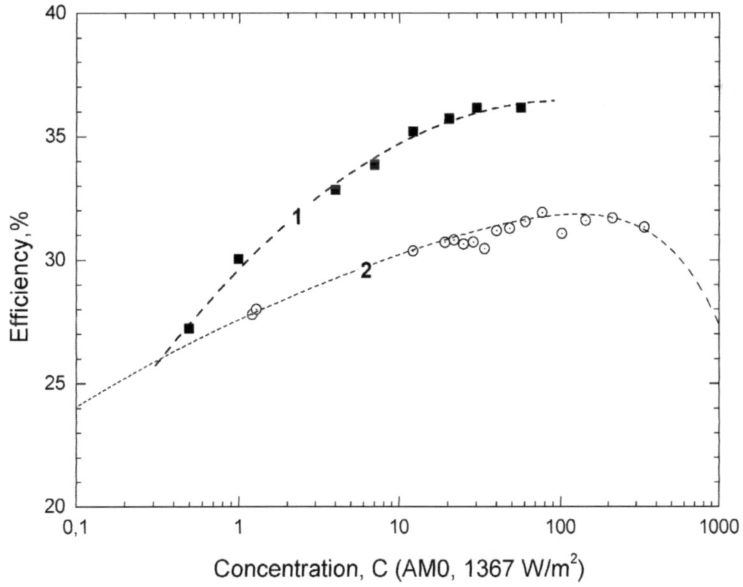

Figure 6. Experimental and calculated (dotted lines) dependences of the efficiency of InGaP/Ga(In)As/Ge SCs fabricated by (1) – single-step method and (2) multistep post-growth techniques of separate etching on the concentration of AM0 sunlight.

3. Conclusion

For InGaP/Ga(In)As/Ge SCs fabricated by the MOCVD technique, dark *J – V* characteristics were recorded and analysed in the temperature range 300 – 80 K. The potential and real SC efficiencies at 80 K were calculated on the basis of an analysis of the dark characteristics in relation to the generation current density.

Load *I – V* characteristics of the SCs were recorded and analysed in the temperature range 300 – 80 K. For a characteristic recorded at 204 K, the maximum efficiency was 35 % ($C = 1$, AM0, 1367 W/m^2).

Current and voltage jumps were observed experimentally in the dark and load *I - V* characteristics of InGaP/Ga(In)As/Ge SCs in the temperature range from 200 to 80 K and attributed to the effect of counter potential barriers at interfaces of isotype heterolayers.

The influence exerted by the shape of the side surface on the dominant current flow mechanisms was examined at various illuminance level and temperatures. It was found that the single-step technique suggested for forming the InGaP/Ga(In)As/Ge SC side mesa surface makes it possible to reduce the effect of the tunnel-trap (overabundant) and recombination current flow mechanisms on the efficiency of InGaP/Ga(In)As/Ge SCs by inhibiting the rise in the surface leakage currents upon thermal cycling (300 – 80 K), improving the quality of the side surface of the chip mesa via passivation, and rising the yield of suitable SCs with efficiency exceeding 35 % under exposure to AM0 sunlight with 10 –100 -fold concentration.

References

[1] Dimroth F, Grave M,, Beutel P, Fiedeler U, Karcher C, Tibbits T N D, Oliva E, Siefer G, Schachtner M, Wekkeli A, Bett A W, Krause R, Piccin M, Blanc N, Drazek C, Guiot E, Ghyselen B, Salvetat T, Tauzin A, Signamarcheix T, Dobrich A, Hannappel T, and Schwarzburg K 2014 *Progress in Photovoltaics: Research and Applications* **22** 277–282

[2] Andreev V M, Evstropov V V, Kalinovskiy V S, Lantratov V M and Khvostikov V P 2009 *Semiconductors* **43** 644

[3] Kalinovskiy V S Evstropov V V and et al 2009 *Proc. 24th European Photovoltaic Solar Energy Conference* (Humburg) 733

[4] Andreev V M Kalinovskiy V S and et al 2010 *Proc. 25th European Photovoltaic Solar Energy Conference and 5th World Conference on Photovoltaic Energy Conversion* (Valencia) 979

[5] Kalinovskiy V S, Kontrosh E V, Dmitriev P A, Pokrovsky P V, Chekalin A V and Andreev V M 2014 *AIP Conference Proceedings* **8** 1616

[6] Andreev V M, Grebenshchikova E A, Dmitriev P A, Il'inskaya N D, Kalinovskiy V S, Kontrosh E V, Malevskaya A V and Usikova A A 2014 *Semiconductors* **48** 1271

[7] Andreev V M, Grebenshchikova E A, Zadiranov U M, Il'inskaya N D, Kalinovskiy V S, Malevskaya A V and Usikova A A 20.06.2013 A method of manufacturing heterostructures chips and etchant *Patent 2485628* (Russian Federation) H01L 31/18

Influence of the composition of the GaAsP solid solution on the Ag-GaAsP barrier height

I A Lamkin, S A Tarasov, A V Solomonov and I I Mikhailov

Saint-Petersburg Electrotechnical University "LETI", 5 Prof. Popova str., St. Petersburg 197376, Russia

E-mail: ialamkin@mail.ru

Abstract. The height of the potential barrier for the Ag-GaAsP has been measured experimentally. The dependence of the Schottky barrier height of the Ag-GaAsP structure on the composition of the solid solution is correlated with the dependence of the bandgap GaAsP on its composition. We calculate the equation describing the dependence of the height of the Ag-GaAsP Schottky barrier on the composition of GaAsP for indirect and direct-gap solid solutions.

1. Introduction

GaAsP solid solutions are widely used in modern optoelectronics when creating devices for the visible, ultraviolet and infrared ranges of the spectrum [1, 2]. One of the most important applications of such materials is the realization of photodetectors with improved noise characteristics. Changing the composition of the solid solution GaAsP allows to create photodetectors with a given position of the long-wave boundary of photosensitivity. Using Schottky barriers on the basis of silver in these detectors allows for achieving the sensitivity and selectivity and improving their other characteristics while lowering manufacturing costs. The potential barrier between the metal and the semiconductor has the most important influence on the properties of such structures, therefore the study of factors that affect the height of the barrier Ag-GaAsP is topical.

2. Samples and experimental technique

The structures on the basis of gallium arsenide phosphide solid solutions have been investigated. Sulfur doped *n*-GaAsP epitaxial layers with a free carrier concentration of $(0.5\ldots5)\times10^{16}$ cm^{-3} were grown by vapour phase epitaxy. Schottky barriers were formed by vacuum thermal evaporation of silver on the semiconductor surface preliminarily cleaned in tetrachloromethane environment. The ohmic contacts were formed by Au deposition in accordance with the standard technique [3, 4]. To study the effect of composition of solid solution on the height of Schottky barrier formed by the contact Ag-GaAsP, a series of samples with different composition of the solid solution had been created.

3. Experimental results and discussions

The height of the barrier metal-semiconductor can be calculated by the formula $q\varphi_B = q(\varphi_m - \chi)$, where $q\varphi_B$ is the height of the potential barrier metal-semiconductor, χ – the electron affinity of the semiconductor, φ_m - the electron work function from metal. However, it is known that the real height of the Schottky barrier depends on many other factors, primarily on the surface state of the semiconductor. In addition, the solid solutions of GaAsP may be both a direct-gaps and indirect-gaps, which leads to different influences of the over-barrier transferring of charge carriers in structures of metal-solid solution GaAsP.

The investigated samples were subjected to various chemical pretreatment of the semiconductor surface before spraying of metallic Ag contact. To determine the height of the potential barrier of the metal-semiconductor we used two methods: the photoelectric method and the current-voltage characteristics. Spectra of the photosensitivity of samples are presented in figure 1.

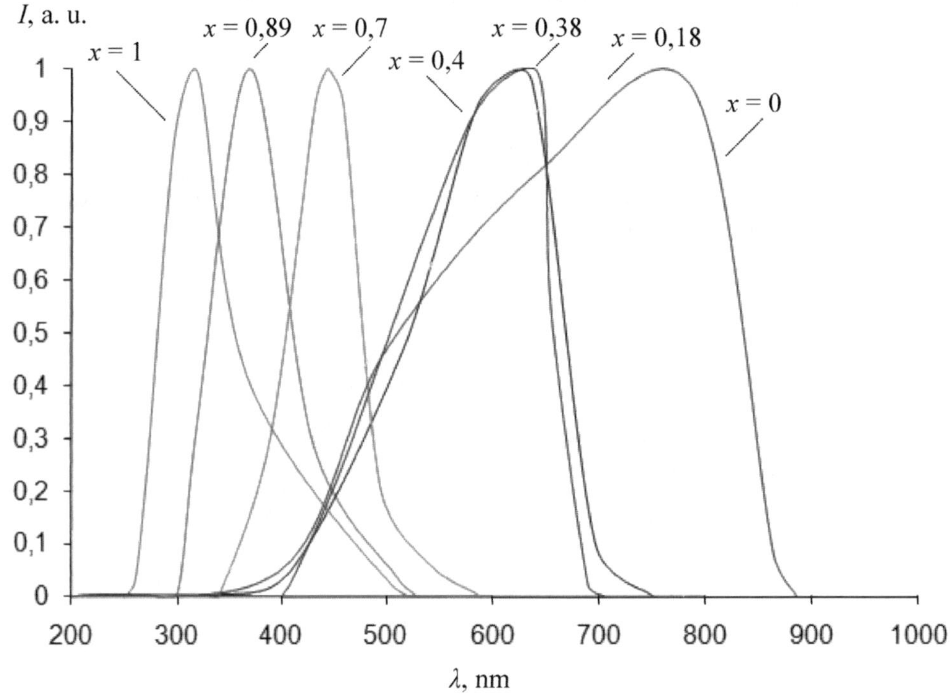

Figure 1. The dependence of the photosensitivity on the wavelength for samples with different composition of GaAsP.

As shown in figure 1, photodiodes on the basis of GaAsP with a composition approaching pure gallium arsenide have a more smooth decline in photosensitivity in the area of long wavelengths, which is caused by indirect-gap of semiconductor. On the basis of obtained spectral characteristics, the dependencies of the root of photocurrent on the energy are built, by which the height of the potential barrier metal-semiconductor was determined by the photoelectric method. To determine the height of the barrier by the current-voltage characteristics, the forward part of current-voltage characteristics has been approximated in accordance with the formula $J = J_S \exp [qU/(nkT)]$, and the coefficient of non-ideality n and saturation current J_S have been determined. Then, according to the formula $\varphi_B = (kT / q) \ln[A^{**}T^2 / J_S]$, the barrier height has been calculated, where A^{**} is the effective Richardson constant. The coefficient of non-ideality n shows the deviation of the slope angle of the graph from the ideal exponential function, and it is an indicator of sample quality. For good structures it did not exceed 1.1. Based on the analysis of obtained values of the potential barrier height, we plotted the dependence of the height of the Schottky potential barrier on the composition of solid solution GaAsP (figure 2).

It can be seen that two dependencies had appeared: the first characterizes the samples with a clean surface (1 in figure 2), this dependence of the Schottky barrier's height in the structure of the Ag-

GaAsP on solid solution's composition correlates with the dependence of the band gap GaAsP on its composition (figure 3).

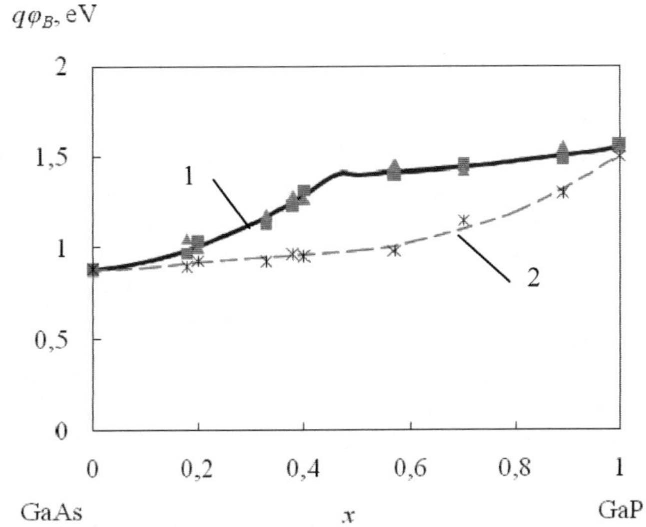

Figure 2. The dependence of the barrier height on the GaAsP solid solution's composition: 1 – clean surface, 2 – for large number of surface states.

The second kind of dependence is primarily determined by a large number of surface states, samples of this type were processed in various chemical reagents (2 in figure 2).

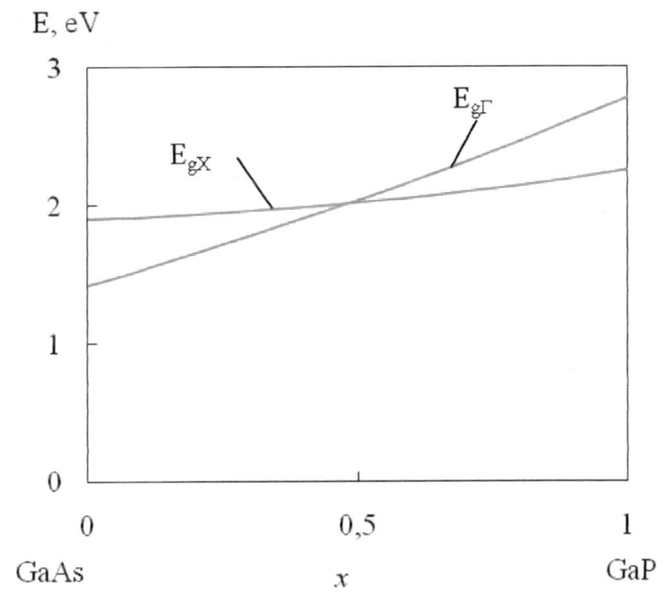

Figure 3. The dependence of the band gap of GaAsP on the solid solution's composition: E_{gX} – bandgap at X point; $E_{g\Gamma}$ – bandgap at Γ point.

As a result, the dependencies of the Ag-GaAsP potential barrier's height on the composition of GaAsP have been approximated for direct-gap and indirect-gap GaAsP. The height of the potential Schottky barrier in the structure of Ag-GaAsP obeys the following dependencies on the solid solution's composition: $q\varphi_B$ [eV] $= 1{,}96x^2 + 0{,}24x + 0{,}88$, at $x \leq 0{,}46$; $q\varphi_B$ [eV] $= 0{,}30x^2 - 0{,}17x + 1{,}42$, at $x \geq 0{,}46$, with an error of 0.1, where x is a mole fraction of GaP in the solid solution GaAsP.

4. Conclusion

In the process of investigation of Ag-GaAsP samples with different rates of GaP, the measurements of their spectra of photosensitivity were carried out. The changing of the position of the long-wave border while increasing the content GaP in the solid solution GaAsP was demonstrated. On the basis of obtained spectral characteristics, the height of the potential Ag-GaAsP barriers for different compositions of the solid solution had been determined. In addition, the height of the metal - semiconductor barrier has been calculated by the current - voltage characteristics. Using the obtained data, the dependence of the Ag-GaAsP potential barrier's height on the composition of the solid solution was plotted.

The height of the potential Schottky barrier in the structure of Ag-GaAsP obeys the following dependencies on the solid solution's composition:

$q\varphi_B$ [eV] $= 1{,}96x^2 + 0{,}24x + 0{,}88$, at $x \leq 0{,}46$

$q\varphi_B$ [eV] $= 0{,}30x^2 - 0{,}17x + 1{,}42$, at $x \geq 0{,}46$,

with an error of 0.1, where x is a mole fraction of GaP in the solid solution GaAsP.

Acknowledgements

The work was supported by the Ministry of Education of the Russian Federation in the framework of the project of the state task in the field of scientific activity, project number 16.1307.2014K.

References

[1] Vinokurov D A, Nikolaev D N, Pikhtin N A, Stankevich A L, Shamakhov V V, Bondarev A D, Rudova N A and Tarasov I S 2011 Diode lasers emitting at 1190 nm with a highly strained GaInAs quantum well and GaAsP compensating layers MOCVD-grown on a GaAs substrate *Semiconductors* **45** 1227

[2] Kurin S Yu, Antipov A A, Roenkov A D, Barash I S, Helava H I , Menkovich E A, Tarasov S A, Lamkin I A, Shmidt N M and Makarov Yu N 2013 UV LEDs for high-current operation *Journal of Physics: Conference Series* **461** 012028

[3] Pikhtin A N, Tarasov S A and Kloth B 2003 Ag-GaP Schottky photodiodes for UV sensors *IEEE Transactions on Electron Devices* **50** 215

[4] Pikhtin A N, Tarasov S A and Kloth B 2002 New values of the Ag-n-GaP potential barrier *Technical Physics Letters* **28** 872

Effect of a wideband heteroepitaxial emitter on dynamics of turn-off switching of high-voltage power GaAs p-i-n diodes

N M Lebedeva[1], **V A Kozlov**[2], **F Yu Soldatenkov**[1,2] **and A A Usikova**[1]

[1]Ioffe Physical-Technical Institute, 26 Politekhnicheskaya str., St. Petersburg, 194021 Russia
[2]Power Semiconductors Ltd., 27 Gzhatskaya str., St. Petersburg, 195220 Russia

E-mail: natali_lebedeva@mail.ioffe.ru

Abstract. The possibility of improving the dynamic characteristics of turn-off switching of high-voltage power GaAs p-i-n diode by applying a heteroepitaxial AlGaAs emitter is investigated in this work. Using a wideband AlGaAs n^+-emitter in manufacturing of power GaAs p-i-n diodes allows to vary the coefficient K value which characterizes the recovery softness of diode blocking properties when switching off in the range of 0.1 to 2 or more. The diodes with blocking voltage of up to 700 V, the reverse recovery time of about 40 ns and $K > 1$ were manufactured.

1. Introduction

Application of heterostructures in manufacturing of high voltage power devices on the base of GaAs allows improving the set of their main static and dynamic characteristics [1-4]. Using wideband heterostructure emitters can significantly extend the dynamic range of current densities within which the emitter injection efficiency remains close to unity. This has implications for improving the conductivity modulation of the diode base layers at high current densities and reducing the static energy losses in the device in a conducting state. Beside static characteristics, using of heterostructure emitters also allows changing the diode switching-off process dynamics, which is especially important for fabricating diodes with a soft and sharp recovery character.

One of the most important dynamic characteristics of high-power pulse bipolar diodes is the reverse recovery time t_{rr} (the time in which the blocking properties of a p-n junction are restored after the diode was switched from the conducting to blocking state). During this time the charge accumulated in the base layers dissipates resulting in reverse diode current. When a diode is switched off the two phases of reverse resistance recovery can be distinguished: (i) the stage of a high reverse conductivity t_s in which the p-n junction remains forward-biased despite the change of the polarity of the applied external voltage (at this stage the diode reverse current grows from zero to its maximum value I_{rmax}) and (ii) the stage of recovery of the high reverse resistance t_f, in which a space-charge region is formed near the p-n junction and, while extending, this region blocks the voltage applied to the diode (t_f is the time during which the reverse diode current drops from I_{rmax} to $0.2 \cdot I_{rmax}$). The ratio of durations of these stages, K (the softness coefficient, $K = t_f/t_s$) mainly characterizes the degree of softness or sharpness of the blocking properties recovery for a diode when switching-off.

Devices with a soft character of the blocking properties of recovery at switching-off are the most desirable for applying in electric power converters, since they contribute significantly less electromagnetic interference in the operation of other parts and devices of power electronic systems. Besides, the soft character of recovery of the diode blocking capability prevents the appearance of sharp splashes of voltage on a parasitic inductivity of the power circuit of electronic power converters [5].

2. Sample fabrication

Fabrication of p-i-n GaAs diode structures was carried out in two stages with using the liquid phase epitaxy (LPE). The epitaxial growth of high-voltage lightly doped gradual GaAs p^0-i-n^0 junction layers was carried out from a confined Ga-As melt in the hydrogen atmosphere in quartz boats from 900 °C to room temperature. This procedure for GaAs p^0-i-n^0 structure growth is presented in detail in [2,4]. In this mode of LPE growth of p^0-i-n^0 structures, it is possible to obtain thick i-type layers of lightly doped GaAs with a free carrier concentration of $\sim 10^{13}$ cm^{-3}, which makes it possible to reach blocking voltages U_b of up to 2000 V [3]. The heavily doped (with Te) GaAs or AlGaAs n^+-emitter layers of diode structures were grown in graphite piston cassette. The typical free carrier distribution across the thickness of diode structures is shown in figure 1.

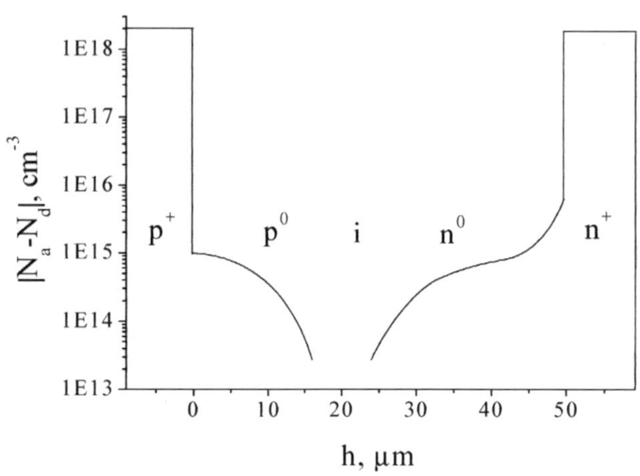

Figure 1. Typical free carrier distribution across the thickness of GaAs high voltage p^0-i-n^0 layers grown on p^+-GaAs substrates with a later-grown GaAs or AlGaAs n^+-emitter.

To study the effect of presence of a potential barrier on a heterointerface between a lightly doped base layer and a heavily doped heteroepitaxial n^+-emitter on the diode switching-off behaviour, we have fabricated test GaAs p-i-n diodes with same structures of the p^+-emitter (substrate) and lightly doped p^0-i-n^0 base layers but with different n-type emitters (see figure 1). The fabricated devices of the first group were fully homo-epitaxial and had a GaAs n^+-emitter. Devices of the second group had a hetero-epitaxial AlGaAs n^+-emitter with the AlAs content of up to 15% ensuring the presence on its boundary a potential barrier for carriers $\Delta E_g \sim 0.19$ eV. The devices had the same base thicknesses of 55μm, the emitter area of 1mm² and the breakdown voltage $U_b = 700$ V.

3. Turn-off switching of high voltage power GaAs p-i-n diodes with different n^+-emitters

Figure 2 shows oscillograms of turn-off switching for the diodes of the first and second groups from the conducting state with the current $I_f = 1.5$ A into the blocking one by applying the reverse voltage $U_r = 300$ V at a rate of current change dI/dt of about 150 A/μs.

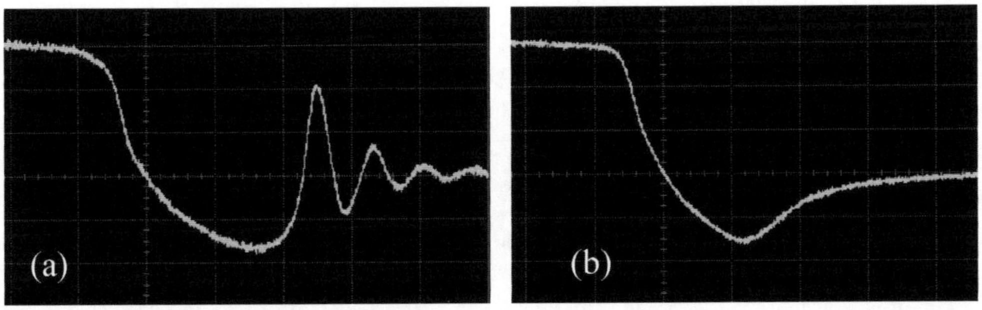

Figure 2. Waveforms of turn-off switching for GaAs p-i-n diodes with the same high voltage base layers and different n^+-emitters: a homostructure GaAs n^+-emitter (a) and a heterostructure AlGaAs one (b). Vertical scanning is 0.5 A/div and horizontal one is 20 ns/div; dI/dt is about 150 A/μs.

It is clear from figure 2 that application of the wideband heterostructure AlGaAs n^+- emitter allows changing the character of the process of GaAs p-i-n diode blocking properties recovery from a sharp switching-off mode (the softness coefficient K is about 0.15) to a soft one (the softness coefficient K is about 2). The reverse recovery of full charge for diodes of the first and second groups had similar values. The difference between K values for the first and second groups of diodes can be explained by the presence of a wideband AlGaAs n^+-emitter in the second group of diodes. On the one hand, the formation of a hetero-junction results in a higher concentration of injected carriers plasma near the diode AlGaAs n^+-emitter in the conducting state, but, on the other hand, it prevents its rapid resorption by the reverse current at switching the diode off owing to the presence of a potential barrier on the hetero-junction. In such a device, the phase of charge resorption in the n^0-base away from the p-n junction lasts much longer, since the process of space-charge layer widening in the p-n junction is hampered by the presence of residual electron-hole plasma in the region near the AlGaAs n^+-emitter, and the recovery character becomes softer.

4. Summary

Application of wideband AlGaAs n^+-emitters in the design of high voltage power GaAs p-i-n diodes allows controlling the degree of softness of the switching-off process and optimizing the dynamic characteristics of devices for operating in different modes. The carried out investigations have allowed fabricating diodes with blocking voltage of up to 700 V with a soft character of turn-off switching (with the softness coefficient K value of about 2).

Acknowledgments
The work has been supported by the Russian Science Foundation, agreement No. 14-29-00178.

5. References

[1] Danilchenko V G, Korolkov V I and Soldatenkov F Yu 2009 *Semiconductors* **43** 1055
[2] Soldatenkov F Yu, Danilchenko V G and Korolkov V I 2007 *Semiconductors* **41** 211
[3] Danilchenko V G, Korolkov V I, Ponomarev S I and Soldatenkov F Yu 2011 *Semiconductors* **45** 515
[4] Kozlov V A, Soldatenkov F Yu, Danilchenko V G, Korolkov V I and Shulpina I L 2014 Defect engineering for carrier lifetime control in high voltage GaAs power diodes *Proc. 25th Advanced Semiconductor Manufacturing Conf. (ASMC-2014) (Saratoga Springs, USA, 19-21 May 2014)* 139-144
[5] Grekhov I V, Rozhkov A V, Kostina L S, Konovalov A V and Fomenko Y L 2011 *Technical Physics* **56** 1429

Effect of bus-bar material and configuration on the efficiency of GaInP/GaAs/Ge solar cells

A V Malevskaya[1], D A Malevski and N D Il'inskaya

Ioffe Institute, Polytechnicheskaya 26, St.Petersburg 194021, Russian Federation

[1]E-mail: amalevskaya@mail.ioffe.ru

Abstract. The effect of bus-bar material and configuration was investigated for multi-junction solar cells (SC) based on GaInP/GaAs/Ge structure. Different configurations of contact bus-bars were obtained by the electro-chemical deposition of Ag and Au materials. The resistivity of contact bus-bars of gold and silver were measured. It is established that the bus-bars configuration in a form of truncated pyramids of 3-10 µm wide and of 4-8 µm thick located with a spacing less than 100 µm allows to reduce optical and ohmic losses and to increase the efficiency of SCs.

1. Introduction

The efficiency increase in conversion of the sunlight into electric power is the main task for developing solar power engineering, which motivates the interest to multijunction SCs. The key ways to raise the multi-junction SC efficiency is the increase in the number of photoactive *p-n* junctions, the reduction of the contact ohmic resistance, the increase in bus-bars conductivity, the decrease in the coefficient of sunlight reflection from a photosensitive SC surface, the reduction of leakage currents on the chip rear surface. In the present work, investigations and developments in the field of reducing ohmic and optical losses in converting the sunlight for multijunction GaInP/GaAs/Ge SCs have been carried out. Studies of the effect of contact materials and different configurations of contact bus-bars on SC parameters were conducted.

The increase in the efficiency of multijunction SCs based on the GaInP/GaAs/Ge nanoheterostructure is achieved by using sunlight concentrators [1]. Application of sunlight concentrators allows reducing sizes of SCs down to 2 mm², which results in a substantial saving of the materials being consumed. However, essential difficulties in carrying out post-growth technological operations for creating SC chips arise. The contact bus-bar width is 3-6 µm with a spacing of 50-100 µm. Minimization of the bus-bar width and the increase in the spacing between bus-bars allow reducing losses caused by shadowing of the photosensitive SC surface. It should be noted that multijunction SCs are intended for operation at ratios of the sunlight concentration greater than 500. The photocurrent density in this case exceeds 10 A/cm², which leads to raising ohmic losses and decreasing SC efficiency.

A known procedure for reducing ohmic losses at high operating currents is decreasing the spacing between contact bus-bars. However, this technical solution results in raising optical losses due to increasing shadowing of a photosensitive SC surface by contact bus-bars, which, in turn, leads to the SC efficiency decrease.

A known way for lowering down the mentioned above optical losses is decreasing the contact bus-bar width. But this technical solution results in the rise of the contact resistance, which, in turn, also decreases the SC efficiency.

In consequence of these reasons, the SC efficiency in the known designs drops at sunlight concentration ratios greater than 500. The main purpose of our research is the study of contact bus-bar resistance, the comparison of contact configurations and materials and their influence on the efficiency of SCs.

2. Experimental

The research was carried out on the GaInP/GaAs/Ge structure. The schema of SC based on this structure is presented in figure 1. It consists of Ge, GaAs and GaInP subcells, frontal and back contacts and an antireflection layer.

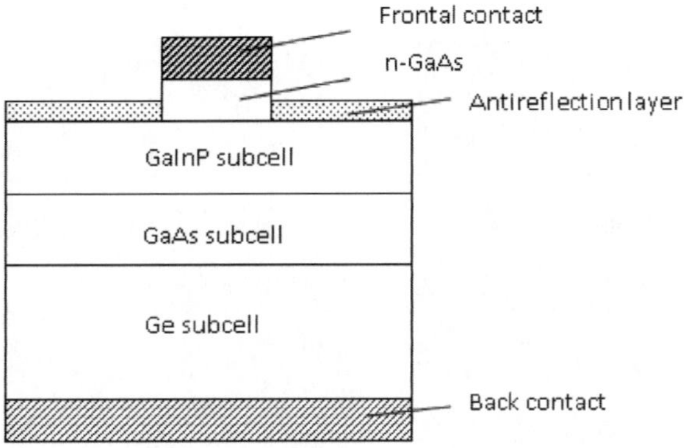

Figure 1. Schema of multi-junction solar cell based on GaInP/GaAs/Ge structure.

In fabricating the concentrator SCs with a frontal n-GaAs layer and a side part of p-Ge substrate, the use of multilayer contact systems is feasible [2]. In making a frontal ohmic contact, the successive deposition of layers is performed: the Au:Ge one based on an electric alloy (with the weight ratio of 88:12), the layers of Ni and Au. The layer thicknesses of the face contact to n-GaAs are: 30 - 150 nm for Au-Ge; 0-100 nm for Ni; 100-200 nm for Au.

Application of the AuGe-Ni-Au contact system allows obtaining a low contact resistance of the order of $1\cdot10^{-6} - 5\cdot10^{-6}$ Ohm·cm^2 to the n-GaAs layer [3]. The Au:Ge eutectic alloy possesses good adhesion to the structure surface. The Ni layer performs the function of a barrier layer between the AuGe and Au ones, which is required for the following brazing the ohmic contact – Ni hinders Au diffusion into the structure. The top Au layer is deposited for the following operation of electrochemical deposition of a metallic contact and thickening up to 2 - 8 µm.

To create contact bus-bars of 2 - 8 µm thick, the electrochemical deposition of a material with low resistivity is conducted. The most widespread materials for performing electrochemical thickening of contacts are Au and Ag. The Au resistivity is 0.023 Ohm·mm^2/m. That of Ag is 0.015 - 0.016 Ohm·mm^2/m.

The cyanide electrolyte is widely used for Au electrochemical deposition [4]. Main merits of the given electrolyte are an elaborate technology for preparing electrolyte and for the deposition process and also long lifetime. These aspects define expediency of its application in indoor research and in production.

The gold electrochemical deposition is performed through a photoresist mask. It has been elucidated during experiment that, in carrying out deposition on the semiconductor plate surface, growth of a metal under the photoresist layer takes place due to the destruction of photoresist by cyanide ions on the interface of the photoresist and semiconductive surface (figure 2(a)).

Study of the gold electrochemical deposition through a photoresist mask in the pulse current mode has been carried out. The advantage of the pulsed technique of deposition compared to deposition at direct current is the increase in density of the material being deposited and improvement of its adhesion.

During developing electrochemical deposition of gold in the pulsed current mode, a result depicted in figure 2(a) has been obtained. In depositing gold on the semiconductor structure surface, we could not avoid its growth under the photoresist. The pulsed electrochemical deposition of a gold layer was performed at the current density of 0.02-0.2 mA/mm^2, the duty cycle of 0.25 and the frequency of 30 - 40 Hz.

An optimum way to decrease the rise of the Au layer is the additional passivation of the semiconductor structure from the frontal surface by the evaporation of the dielectric layer, for example Si_3N_4 after which the surface conduction decreases. Figure 2(b) presents SEM pictures of a deposited Au layer through the mask consisted of a Si_3N_4 layer and a photoresist layer.

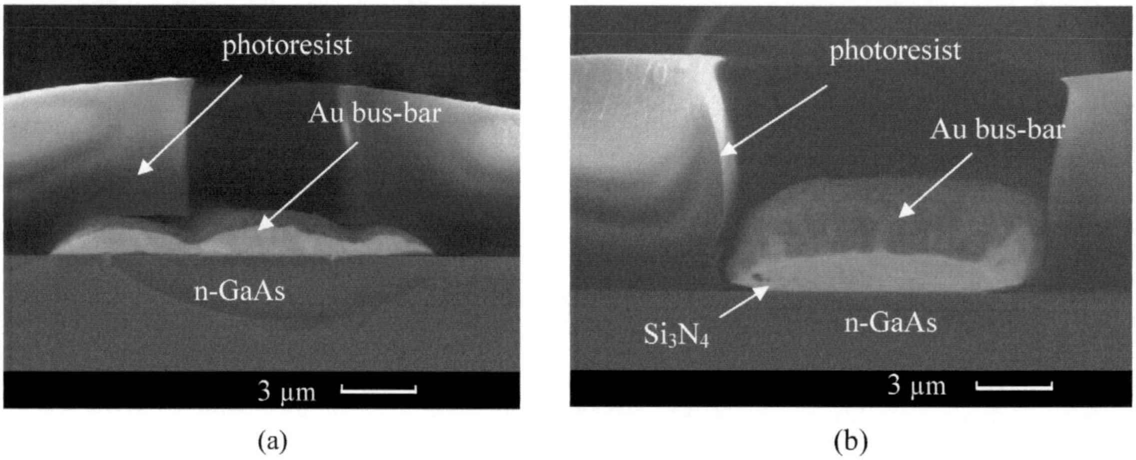

Figure 2. Scanning electron microscope (SEM) pictures of an electrochemically deposited Au bus-bar on the semiconductor structure surface through the mask of photoresist (a) and through the mask of Si_3N_4 and photoresist (b).

In fabricating a concentrator multijunction SCs, an important task is to reduce the semiconductor structure series resistance. With increasing solar radiation density, the losses on the device internal resistance rise, the essential part of which is the contact resistance. Reduction of the contact resistance is necessary for raising the current collecting from the devices and decreasing the heating associated with flow of a large density current.

A version of using silver for electrochemical thickening of the contact has been considered. The use of silver is determined by a high plasticity of the given material, which allows depositing thicker layers without contact growth up. At electrochemical growth of silver through a photoresist mask with a preset rear wall profile, silver repeats exactly the photoresist mask profile according to its high plasticity. Thus, the rear walls of contact bus-bars have smooth specular surface.

The electrochemical deposition of silver is performed at 18 - 25 ^{0}C from electrolyte with a mass content of silver of 20 - 30 g/l and pH of 7.5 - 8.5 [5]. Figure 3 presents SEM pictures of the silver layer after electrochemical deposition, which repeats exactly the profile of the photoresist mask side wall. In creating photoresist masks with different side profiles, one can obtain contact bus-bars of different configurations [6]. To protect the silver contacts from the effect of the environment, a gold layer of 0.1 - 0.2 μm thick is deposited electrochemically.

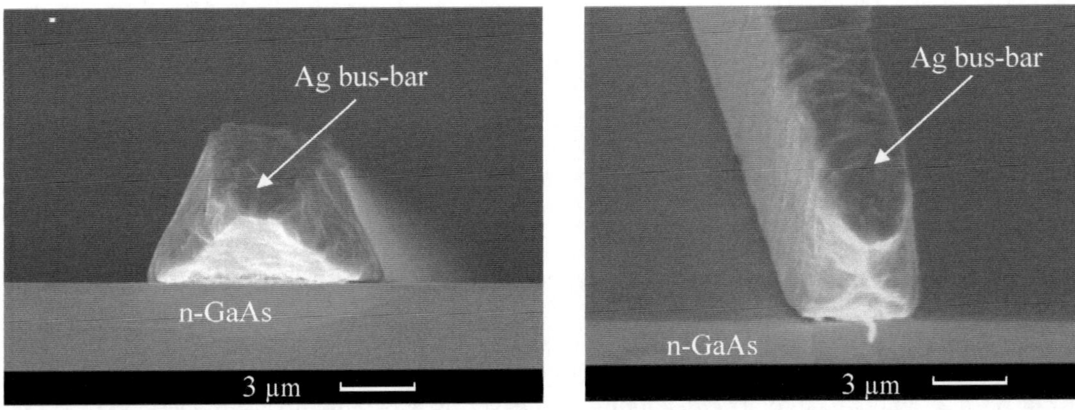

Figure 3. SEM pictures of a Ag bus-bar deposited electrochemically in a form of a truncated pyramid with the wide low base when using different photoresist mask configurations.

Results and discussion

The investigation of the contact material effect on SC parameters has been carried out. The aim of the experiment was to create contact bus-bars on a nonconducting silicon substrate by electrochemical deposition of gold and silver and to measure the resistance for bus-bars of different configurations by the 4-probe method. For the calculations, the measured values of bus-bar length, width and cross section area were used. The contact bus-bar resistance was measured on a laboratory multimeter.

The resistivity was calculated by the following expression:

$$\rho = \frac{RS}{L} \tag{1}$$

where ρ is resistivity, R is the measured resistance, S is the bus-bar cross section area, L is the bus-bar length.

The theoretical values of Ag and Au resistivity are the following: ρ(Ag)=0.015÷0.016 Ohm·mm²/m; ρ(Au)=0.023 Ohm·mm²/m. Deposition of silver and gold was conducted in the DC and AC modes. Data of measurements and calculations are presented in Table 1.

Table 1. Contact parameters

Sample No	Bus-bar material	Deposition mode	Bus-bar No	L, mm	S, mm²	R, Ohm	ρ, Ohm·mm²/m
1	Ag	DC	1	19.7	4.0×10^{-5}	12.1	0.025
			2	19.2		12.2	0.026
			3	19.2		12.1	0.025
2	Au	DC	1	15.1	4.0×10^{-5}	8.5	0.028
			2	15.1		8.9	0.024
			3	15.1		9.1	0.024
3	Au	AC	1	20.8	3.6×10^{-5}	13.2	0.023
			2	19.2		13.2	0.025
			3	18.7		11.5	0.024

It has been determined according to the carried out experiment that the resistivity of contact bus-bars of gold and silver are close by values. The golden bus-bar resistivity corresponds to the theoretical one. That of silver bus-bar is greater than the theoretical one, which is due to the structure of obtained bus-bars. In our experiment the cross section areas of compared bus-bars are close by values. However, the advantages of using silver for contact bus-bars fabrication are as follows: the cross section areas of bus-bars can be increased due to the termination of contact growth up on the conducting surface. Moreover, a form of a truncated pyramid of silver bus-bar allows increasing the contact thickness. Contact bus-bars of 3-6 μm wide can be fabricated of 5 - 8 μm thick, the similar contact bus-bars of gold can be only 2 - 3 μm thick. So, the cross section area of silver bus-bars will be 2 - 3 times greater than that of gold bas-burs. Thus, if we use silver as a contact material, the electric resistance, according to the expression (1), can be decreased by 2 - 3 times with increasing contact thickness.

The comparison of the efficiencies of GaInP/GaAs/Ge SCs fabricated with the use of silver and gold as a contact material is presented in figure 4. It is necessary to mention that the experiments were carried out on the same structure. The maximum efficiency of SCs with silver bus-bars is 35.3% at sunlight concentration (Kc) of 1000 suns, and not less than 35.0% in the range of 500 – 1500 suns. The maximum efficiency of SCs with gold bus-bars is 33.7% at sunlight concentration of 1000 suns.

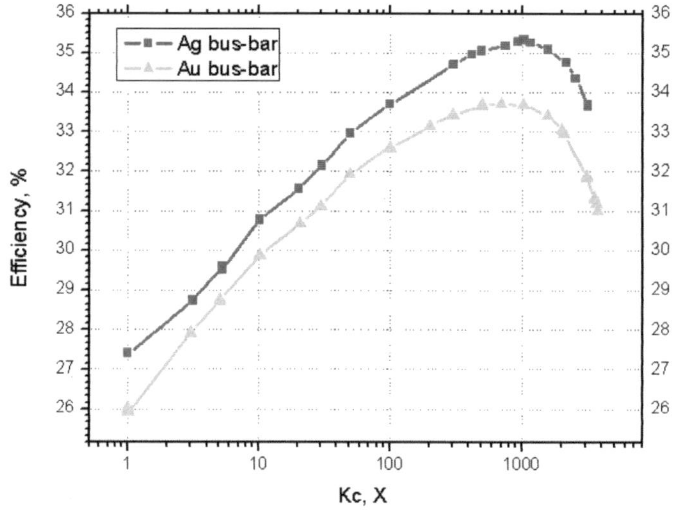

Figure 4. Efficiency of GaInP/GaAs/Ge SCs with contact bus-bars made of silver and gold.

4. Conclusion

The main advantage of gold material for contact creation is its high chemical tolerance to the environment. The advantage of using silver for creating contact bus-bars is a feasibility to make them of different configurations, which allows reducing losses at shadowing the SC photosensitive region.

Creation of silver contact bus-bars in a form of truncated pyramids of 3 - 10 μm wide and of 4 - 8 μm thick located with a spacing less than 100 μm allows reducing losses at shadowing the SC photosensitive surface, increasing absorption of the sunlight due to sunlight reflection from the side walls of contact bus-bars and decreasing ohmic losses due to increasing thickness of the current collecting bus-bar. As a result of carried out experiment, the efficiency of SCs was increased by 1.5% at the sunlight concentration of 1000 suns and the range of sunlight concentration was increased up to 1500 suns without a considerable reduction of SC efficiency.

Acknowledgements
The authors are grateful to N.A. Kalyuzhnyy and S.A. Mintairov for GaInP/GaAs/Ge structure development, to M.Z. Shvarts and N.H. Timoshina for the measurements of solar cells, to

Journal of Physics: Conference Series **690** (2016) 012039 doi:10.1088/1742-6596/690/1/012039

S.I. Troshkov for SEM pictures investigation, to F.Y. Soldatenkov for the investigation of contact materials.

References

[1] Andreev V M, Davidyuk N Y, Malevski D A, Pan`chak A N, Rumyantsev V D, Sadchikov N A, Chekalin A V and Luque A 2014 *Tech. Phys.* **59** 1650-57
[2] Mikhail E. Levinshtein 1997 *A Wiley-Interscience publication USA*
[3] Andreev V M, Il'inskaya N D, Soldatenkov F U and Usikova A A 2009 Patent № 2391741
[4] Andreev V M, Il'inskaya N D, Kalyuzhnyy N A, Lantratov V M, Malevskaya A V and Mintairov S A 2008 Patent № 2391744
[5] Andreev V M, Il'inskaya N D, Kalyuzhnyy N A, Lantratov V M, Malevskaya A V and Mintairov S A 2010 Patent № 2436194
[6] Andreev V M, Il'inskaya N D, Kalyuzhnyy N A, Lantratov V M, Malevskaya A V and Mintairov S A 2009 Patent № 2391745.

Analysis of a multi-junction solar cell with shunted subcells using SPICE

E D Filimonov, S A Kozhukhovskaia, M A Mintairov and M Z Shvarts

Ioffe Institute, 26 Politekhnicheskaya str., St Petersburg, 194021, Russia

E-mail: dexterfree1@gmail.com

Abstract. In the paper a generalized equivalent circuit based on the SPICE model of a multijunction solar cell allowing forming and studying current-voltage characteristics is presented. The negative effect of the low shunting resistance on determining true values of a subcell current at different mismatches of *p-n* junction photocurrents in measuring a spectral dependence of photosensitivity was investigated.

1. Introduction

In recent years a variety of analytical and numerical methods have been used for simulating multi-junction solar cells (MJ SCs) with applying certain commercial software. Among standard software packages, the most widely used by developers and researchers of MJ SCs is the SPICE type program (Simulation Program with Integrated Circuit Emphasis). Such programs allow forming equivalent circuits of different complexity, investigating interaction of circuit elements, analyzing the processes taking place with regard to different operation conditions and accounting for peculiarities of semiconductor structures being studied [1-3]. In the present work, formation of a MJ SC equivalent circuit was carried out on the base of the QUCS (Quite Universal Circuit Simulator) software instrument with using the SPICE model optimized for describing a triple-junction SC. The GaInP (1.9 eV)/GaAs (1.42 eV)/Ge (0.66 eV) structure has been chosen as an object for simulation.

2. The model

To describe SC I-V characteristics, a generalized model of a MJ SC allowing one to present a semiconductor PV converter as an electric circuit was used (figure 1(a)). Such a circuit consists of three distributed elements, each of which reproduces a characteristic of a definite *p-n* junction (subcell) and includes the following components:

- current source setting the value of the subcell photocurrent J_{ph};
- diode for forming the forward branch of I-V characteristic;
- diode for forming the reverse branch of I-V characteristic;
- resistance R_{rev} connected in series to the "reverse branch" diode for controlling its curve slope;
- shunt resistance R_{sh} simulating different current leakages and internal ohmic losses.

A series resistance R_S responsible for ohmic losses on contacts is also included into the circuit. Electric parameters of the circuit components have been chosen in correspondence with standard characteristics of state-of-art semiconductor SCs [4]. In the work, considered were specimens of $1 cm^2$ in area.

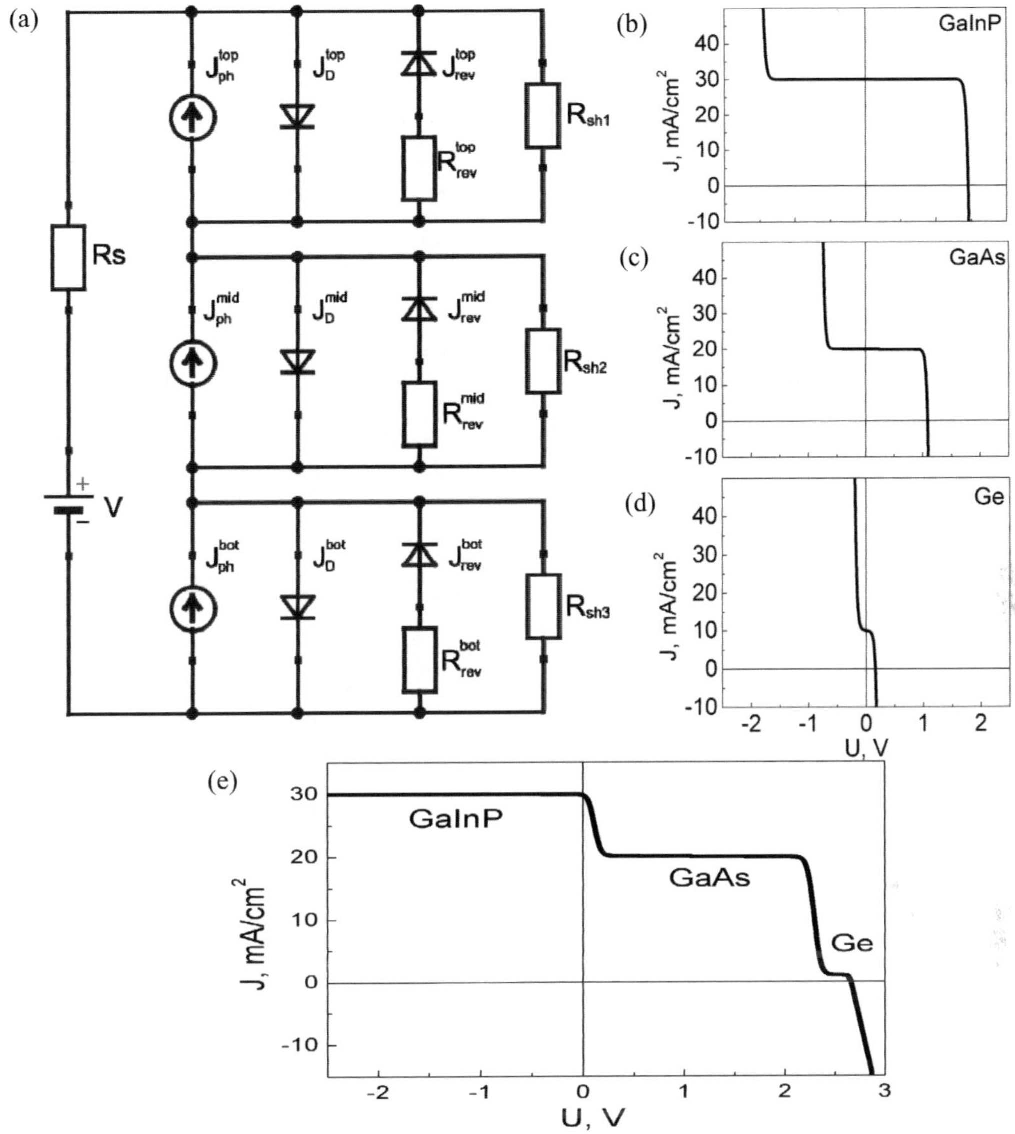

Figure 1. (a) –equivalent circuit based on the SPICE model of a MJ SC (parameters of the circuit components: J_D – *p-n* junction saturation current; J_{rev} – reverse saturation current; V – circuit voltage).(b), (c), (d) – simulated load I-V characteristics of the top, middle, bottom subcells, respectively. (e) – MJ SC general curve with accounting preset current mismatches between subcells.

In developing the model, peculiarities of the MJ SC design and methods for its photovoltaic characteristics investigating have been taken into account [5], which has allowed constructing a detailed SPICE model of a MJ SC ensuring a feasibility to form dark and load I-V characteristics of both the whole SC (figure 1(e)) and each junction separately (figures 1(b), 1(c), 1(d)). Depending on the set task, the proposed model allows one to:

- estimate the effect of the shunt and series resistances on the values of a MJ SC short circuit current or of separate subcells photocurrents being registered at a preset bias voltage;
- study the effect of diode characteristics of individual subcells (reverse saturation current, dark current) on the total I-V characteristic of a MJ SC depending on the photocurrent mismatch level;

- control the level of the mentioned photocurrent mismatch creating conditions typical for operation of a MJ SC and its subcells in both real conditions of daylight and in carrying out special investigations (first of all spectral dependencies of the external quantum yield of a photoresponse of separate subcells in a MJ SC).

It should be noted that just a feasibility to form operation states of a MJ SC determined by a strong photocurrent mismatch between *p-n* junctions is a key point at simulating the conditions for correct recording and analyzing spectral dependencies of the external quantum yield of a photoresponse of separate subcells in a MJ SC.

3. Results and discussion

Let us consider a MJ SC I-V characteristic in conditions typical for obtaining spectral dependence of a photosensitivity. For this, we created a light shift, at which the photocurrent J^{bot}_{ph1} of the Ge subcell is substantially smaller than photocurrents of other wideband subcells, and limitation by GaAs subcell current takes place within the GaInP-GaAs pair (see figure 1(e)). Registration of the J^{bot}_{ph1} value is performed at the certain bias voltage, which value depends on the SC's level of illumination: $U_{BIAS}>U'_{BIAS}$ (figure 2, for simplicity the I-V characteristic of the top GaInP subcell is not shown).

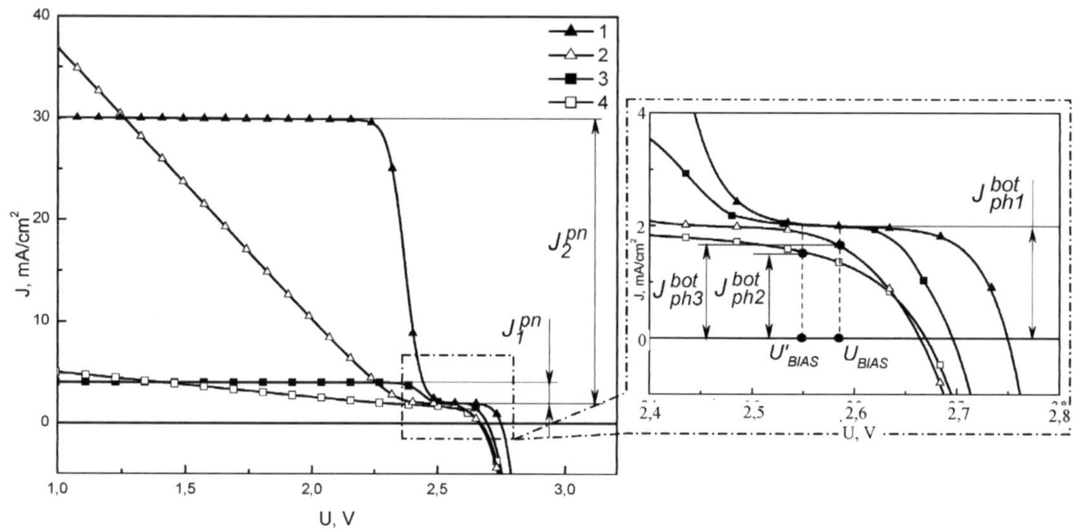

Figure 2. I-V characteristic of a MJ SC in conditions of a photocurrent limitation by a bottom narrow bandgap subcell: without (1, 3) and with (2, 4) current leakages in the top wide bandgap subcell at various current difference between the GaAs and Ge subcells J_1^{pn}, J_2^{pn}.

If there is a small shunt resistance of a *p-n* junction or a pronounced current leakage in the wideband subcell, a slope on its I-V curve appears in the vicinity of the point of short circuit current. At a certain value of shunt resistance in the GaAs subcell the influence arises from this "shunt" on to the recorded Ge subcell photocurrent, which is expressed in its decrease $J_{ph2}{}^{bot}$, $J_{ph3}{}^{bot} < J_{ph1}{}^{bot}$ (see figure 2). In case of a weak current difference J_1^{pn} between the GaAs and Ge subcells such an influence starts to play a considerable role already at an insignificant GaAs subcell I-V characteristic slope, i.e. at a rather high shunt resistance (figure 3).

The obtained uncontrollable decrease of the Ge subcell photocurrent can also emerge in measuring the spectral photoresponse, when illumination of the Ge subcell is performed by frequency modulated monochromatic radiation [6] resulting in inflated values of the external quantum yield. It is obvious that, to decrease the effect of the "shunted" *p-n* junction on the tested subcell photocurrent, it is necessary to ensure a significant exceeding by that junction photocurrent. However, at small values of the shunt resistance, the negative effect mentioned above can't be avoided.

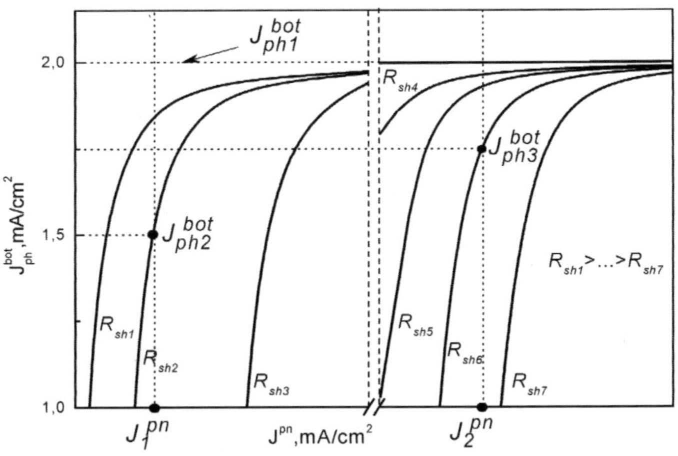

Figure 3. Dependence of a photocurrent density of the narrow bandgap subcell J_{ph}^{bot} on the current difference J^{pn} (i.e. photocurrent through the wide bandgap *p-n* junction) at different shunt resistances $R_{sh1}>R_{sh2}>...>R_{sh7}$ (example from figure 2).

Thus, it has been shown that, in investigating the spectral dependencies of the external quantum yield of the photoresponse of subcells with pronounced current leakages in the photoactive *p-n* junctions, the greatest possible difference between their photocurrents should be sought.

4. Conclusion

In the work, a distributed equivalent circuit based on the SPICE model of a MJ SC is proposed. Its feasibilities in forming I-V characteristics of semiconductor structures with different physical properties have been demonstrated. The main aim of the research was in considering the negative effect of current leakages in *p-n* junctions on the value of the tested subcell photoresponse being registered. Understanding of the processes of mutual effect of the subcells will allow determining correct conditions for recording spectral dependencies of the photoresponse. The results obtained with the help of the SPICE model agree well with the theoretical and experimental data on practical monolithic MJ SCs [7].

Acknowledgments

Support for this work comes from the Russian Foundation on Basic Research (Grant 14-08-01222-a)

References

[1] Steiner M, Guter W, Peharz G, Philipps S P, Dimroth F and Bett A W 2012 *Prog. in PV: Research and Applications* **20** 274–283

[2] Foss S E, Olaisen B R, Marstein E S and Holt A 2007 A new 2.5D distributed SPICE model of solar cell *Section for Renewable Energy, Institute for Energy Technology*

[3] Ben Or A and Appelbaum J 2012 *Prog. in PV: Research and Applications* **21** 290–297

[4] Emelyanov V M, Kalyuzhnyy N A, Mintairov M A, Mintairov S A, Shvarts M Z and Lantratov V M 2010 *Proc. of the 5th World Conf. on PV Energy Conversion (Valencia)* 406–411

[5] Gideon S, Gur M, Abraham K 2012 *Solar Energy Materials & Solar Cells* **98** 57–65

[6] Kozhukhovskaia S A, Filimonov E D and Shvarts M Z 2015 *IOP Journal of Physics: Conference Series* **586** 012014

[7] Mintairov M A, Evstropov V V, Mintairov S A, Shvarts M Z, Timoshina N Kh, Kalyuzhnyy N A 2015 *Semiconductors* **49(5)** 668-673

Simulation of photovoltaic efficiency of a tandem solar cell on Si with GaN nanowires as an emitter layer

A M Mozharov[1], A D Bolshakov[1], G E Cirlin[1,2,3,4], D A Kudryashov[1], A S Gudovskikh[1,5], I S Mukhin[1,6] and Zh I Alferov[1]

[1] St. Petersburg Academic University, 194021 St. Petersburg, Russia
[2] Institute for Analytical Instrumentation RAS, 190103 St. Petersburg, Russia
[3] St. Petersburg State University, 199034 St. Petersburg, Russia
[4] St. Petersburg State Polytechnical University, 195251 St. Petersburg, Russia
[5] St. Petersburg Electrotechnical University "LETI", 197376 St. Petersburg, Russia
[6] ITMO University, 197101 St. Petersburg, Russia

E-mail: alex000090@gmail.com

Abstract. Modeling of photovoltaic properties of multi-junction solar cells integrated on a Si substrate with an array of GaN nanowires as a top emitter has been carried out. Very good antireflection properties of the structure are demonstrated theoretically: the calculated reflectance is lower than 3% for both GaN nanowires/Si and GaN nanowires on GaPN/Si cells. According to our simulations, basic GaN nanowires on Si cell with optimised parameters (doping and NWs morphology) provides energy conversion efficiency of 20%. The approach demonstrates high potential of the GaPN(As)-based multi-junction lattice matched with Si solar cells using an array of GaN nanowires as a top emitter.

1. Introduction

The great interest confined nowadays to the studies in the field of photovoltaics is mainly justified by the inexhaustibility of the solar energy source. The other important reasons for searching the new ways of electrical energy generation are the growing energy consumption firstly caused by the very fast global computing power increase and the irreversible negative impact of hydrocarbons burning on global ecology and their upcoming depletion. That is why researches in the field of sun light energy conversion aimed at development of efficient and green solar cells are considered very important in modern science.

Our letter is dedicated to the integration of lattice-matched heterostructures on a Si substrate with GaN nanowires (NWs) in order to obtain a multi-junction solar cell (SC). We demonstrate theoretically that such a structure possesses several benefits. Firstly, among them we highlight the high energy conversion efficiency. Synthesis of NWs provides very good antireflection and carrier transport properties and allows formation of the wide band gap window. As a side effect, during growth of the NWs the top surface of the SC is being passivated.

It is well known that III-V semiconductor heterostructures provide great optical and electronic properties [1]. That is why integration of Si technology with epitaxial III-V structures is being pursued for the long time. The bottleneck of this problem is a large lattice mismatch between Si and conventional optical III-V materials [2]. NWs seem to be one of the possibilities on this way because they can be grown epitaxially on mismatched substrates due to a developed surface area [3]. Therefore, quasi one-dimensional semiconductor NWs are one of the most promising building blocks for future nano- and optoelectronic devices [4]. It is worth mentioning that NW morphology and high crystal quality should greatly affect recombination losses as compared to the polycrystalline materials [5]. In our study we have chosen GaN as the material for NWs because of its wide band gap, a high carrier mobility and a large band gap difference with Si [1,2,6].

Recently it's been reported that GaPN(As) materials allow variation of the band gap with simultaneous variation of the lattice constant [7-8] in a wide range and can be lattice-matched with Si. Using the latter materials we can produce a tandem SC with multiple junctions grown epitaxially on a Si substrate with a top emitter formed with an array of GaN NWs. A very important for our investigation property of GaN NWs as compared to other III-V materials (like GaAs, GaP) is the possibility of n-type doping of these structures up to high doping levels. This property together with a relatively high electron affinity (~ 4.1 eV) permits formation of high quality ohmic contacts with metals or transparent conductive coatings as well as good carrier transport. On the other hand, a big gap in the valence band occurs at the interface between GaN and conventional SC materials of Si, GaAs and GaP. Formation of the gap provides good restriction for positive charge carriers' transport.

The method of GaN NWs growth in a molecular-beam epitaxy machine includes deposition of a mask AlN layer on top of the SC base. This layer implements passivation of the SC surface and restricts electrical transport for carriers of both signs along the area free of NWs.

2. Modelling

Simulations of photovoltaic properties of a basic type of the multi-junction SC containing GaN NWs synthesized on a Si substrate have been carried out. In our modeling we considered different morphologies of NWs and doping levels of both semiconductors. In figure 1 you can find the schematic image of the investigated SC and its band diagram. A thin dielectric layer (e.g., SiO_2, polymethyl methacrylate (PMMA) or SU-8) is assumed to cover an unwanted 2D GaN layer (formation of this layer is a side effect of NW growth). We proposed use of conductive transparent indium tin oxide (ITO) or Al-doped zinc oxide (AZO) coating atop NWs in order to obtain good electrical contact. Ohmic losses in the top contact have been neglected.

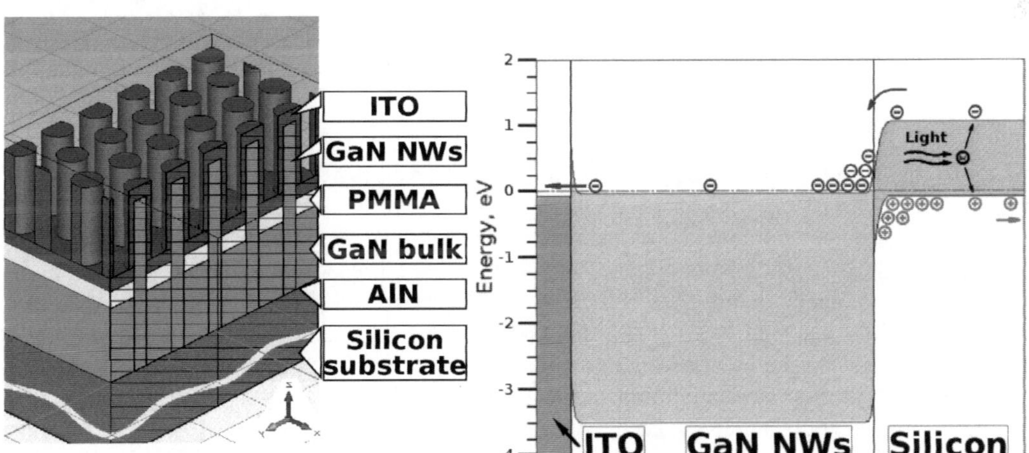

Figure 1. The schematic image of GaN/Si SC and its band diagram

In the first part of our modeling, we studied reflection properties of the structure. We have carried out the variation of nanowire diameters, lengths and surface density in a wide range in order to find the optimal configuration corresponding to minimum reflectance losses. For this purpose, we developed a 3D model of the cell in the "CST Microwave" software package. The model of a NWs array was introduced with cylindrical nanostructures having the same dimensions with a regular hexagonal position on the substrate surface. The investigated SC model includes a p-type Si substrate, a bulk GaN layer, n-type GaN nanowires, a dielectric layer of PMMA and a top contact in form of ITO layer. The bulk GaN layer thickness was considered half of the GaN NWs length. The dielectric and ITO layer thicknesses were taken as 50 nm and 100 nm, respectively.

We have calculated the model SC integral reflectance over AM1.5D spectrum over a wide range of geometrical parameters. The following steps were taken during the simulations: first, we took arbitrary values of the NWs radius and length and for these values we obtained the optimal distance between NWs (or NWs surface density) corresponding to the lowest value of the structure reflectance; second, we took the same arbitrary value of the NWs length and the optimal wire to wire distance and obtained the optimal NW radius; on the third step by using the optimal values of NW surface density and radius calculated previously, we obtained the optimal length of NWs. We repeated the described sequence, using the estimated parameters until their values converge. By using this calculation technique, we obtained the following geometrical parameters corresponding to the lowest reflectance: wire to wire distance is 540 nm, radius is 120 nm, and length is 1300 nm. According to our simulations, such an array will provide the integral reflectance value of only 3%, which is an order of magnitude less than the reflectance of pure Si. In figure 2 one can see the dependence of antireflection properties on the distance (or period) between NWs at different values of the NWs radius for the fixed optimal NWs length of 1300 nm.

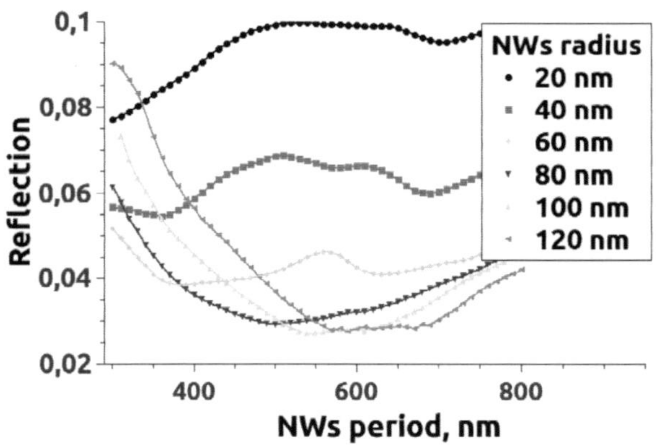

Figure 2. Dependence of antireflection properties on the period and NWs radius for the fixed NWs length

In the second part of our modeling, we studied semiconductor properties of the modeled SC. The optimal doping levels of both a Si substrate and GaN NWs corresponding to the maximum solar energy conversion efficiency were obtained. For this purpose, the Newton method with Fermi statistics was used. We assumed NWs to be defect-free and neglected the influence of the surface states at the sidewalls. The latter allowed us to simplify the problem solution to one-dimensional calculations. In our modeling we substituted the NWs array with a GaN layer which thickness has a value of a minimum travel distance for photogenerated carriers. This is an effective spacing between a Si substrate and an ITO layer which equals the GaN 3D layer thickness plus the dielectric thicknesses multiplied by the ratio of the total SC area to the area covered with NWs. We used the following latter values: the GaN layer and PMMA total thickness equals 700 nm and the area ratio is 8.

In figure 3 (a) one can find the curves representing the calculated dependencies of cell efficiency on the substrate doping level at different levels of NW doping. These graphs possess maxima at the Si acceptor concentration of around $6x10^{17}$ cm^{-3}. Further increase of the substrate doping level leads to a decrease of both the depletion region and carrier mobility in Si which causes a fall in SC efficiency. The maximum estimated efficiency reaches 20%.

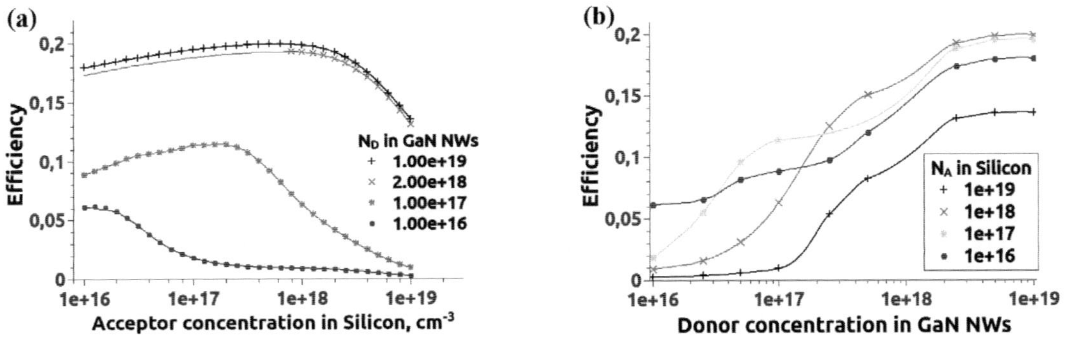

Figure 3. The calculated dependencies of cell efficiency on the substrate (a) and NWs (b) doping levels

In figure 3 (b) one can find the curves representing the calculated dependencies of cell efficiency on the NWs doping level at different levels of the substrate doping. Analysis of the graphs shows that at the donor concentration of around $5x10^{18}$ cm^{-3} the SC efficiency tends to saturation and further increase of NWs doping would not result in a significant rise in efficiency. At the optimal doping level, VOC is 0.73 V, ISC is 34 mA/cm^2 and fill factor (FF) is 85%.

3. Conclusion

Simulations of the solar cell based on the GaN NWs/Si heterostructure have been carried out. It has been demonstrated theoretically that if the doping levels of Si and GaN NWs have the values of $6x10^{17}$ and $5x10^{18}$ cm^{-3}, respectively, and the distance between NWs, their diameter and length are 540 nm, 100 nm and 1300 nm, respectively, then reflectance of the solar cell is only 3% while it's energy conversion efficiency reaches 20%.

The proposed design of a solar cell represents the basic type of a tandem cell that includes a NW array as an emitter of the top junction. The use of GaN NWs allows to improve light absorption and electrical transport properties of the SC. This approach opens new perspectives for the studies of multi-junction solar cells based on GaPN(As) optical materials lattice-matched to Si substrates.

Acknowledgements
This work was carried out in framework of the joint Russian-French project LAMSOL supported by Ministry of Science and Education of Russian Federation (project # 14.616.21.0040) and by PHC KOLMOGOROV Programme (project # 35522TL)

References

[1] Alferov Zh I 2001 Nobel Lecture (2000) *Rev Modern Phys* **73** 767
[2] Sadao Adachi 2009 *Properties of Semiconductor Alloys: Group-IV, III–V and II–VI Semiconductors* (UK: John Wiley & Sons Ltd)
[3] Chuang, L C, Moewe M, Chase C, Kobayashi N P, Chang-Hasnain C, Crankshaw S 2007 *Applied Physics Letters* **90** 043115
[4] Agarwal R and Lieber C M 2004 *Applied Physics A* **85-3** 209
[5] Harmon E S, Melloch M R, and Woodall J M 1993 *Applied Physics Letters* **63-16** 2248
[6] Mozharov A, Bolshakov A, Cirlin G and Mukhin I 2015 *Physica Status Solidi RRL* **1–4**
[7] Geisz J F, Friedman D J and Kurtz S R. 2002 *Proceedings 29th IEEE PVSC* 864
[8] Henini M 2005 *Dilute nitride semiconductors* (UK: Elsevier)

Technology development for GaN based power microwave DHFET

A N Alexeev[1], V P Chaly[2], D M Krasovitsky[2], V V Mamaev[1,3], S I Petrov[1], V G Sidorov[3] and N S Arcebashev[1,3]

[1]SemiTEq JSC, Saint-Petersburg, Russia

[2]Svetlana-Rost JSC, Saint-Petersburg, Russia

[3]Saint Petersburg State Polytechnical University, Saint-Petersburg, Russia

Toodayl88@gmail.com

Abstract. Key issues of epitaxial growth of III-nitrides on mismatched substrates are analysed. Thick AlN "templates" grown at extremely high temperatures and transition layers allow to reduce the dislocation density down to the order of 10^8 cm^{-2}.

1. Introduction

III-Nitrides group attracted much attention for their excellent prospects for military and civil microwave electronics. Lot of microwave applications have been already developed, but the problem of a lack of native bulk material suitable for substrates is still relevant. Device passivation issues also play a significant role especially in devices degradation. In this paper we will show how some of these issues are solved at the stage of heterostructure design and growth, in particular, giving excellent electron confinement in double heterostructure field effect transistors (DHFET).

2. Experimental

III-nitride heterostructures were grown in a STE3N (SemiTeq) molecular beam epitaxy system. The main advantages of the system used are extremely high substrate temperatures (up to 1200°C) and V/III ratios (up to 1000), combined with the conventional MBE opportunities such as precise composition and thickness control etc.

The topical problem of nitrides heteroepitaxy on lattice mismatched substrates leads to a high dislocation density. There are two widespread growth methods for III-N device heterostructures: long, metal-organic chemical vapor deposition (MOCVD) and molecular beam epitaxy (MBE). The typical dislocation density of MOCVD-grown GaN layers is 1-1.5 orders of magnitude less than the one of the MBE-grown films. Moreover, this difference is due to a higher substrate temperature in MOCVD, postulated as a necessary condition for lateral coalescence of the nucleation grains at the initial growth stage.

Ammonia MBE is characterized by the tendency to increase the V/III ratio and growth temperature. However, an increase of the substrate temperature above the one corresponding to the noticeable thermal decomposition of GaN (> 950°C), leads to surface roughening and to a decrease of the electron mobility [1, 2]. At the same time, the thermal stability of AlN is sufficiently higher compared to GaN. In turn, this allows increasing the growth temperature to high temperatures, at which the thermal decomposition of AlN is still negligible. In this context, it was suggested to grow a relatively thick AlN layer at the initial growth stage under higher temperatures that would allow to improve a coalescence of the nucleation blocks and to quickly realize the transition to the two-dimensional growth mode.

Importantly, a similar tendency is correct for the plasma assisted MBE - the method known by its «classic» approach with moderate growth temperatures and a V/III ratios close to 1. It should be noted that the increase in the temperature in ammonia MBE requires much more efforts than in PAMBE. Thus, the high temperature ammonia MBE opens the way for the growth of high quality structures for micro and optoelectronics with an active region grown further by ammonia or plasma-assisted MBE in one growth run. In particular, this approach is promising for growth of heterostructures for UV devices and microwave transistors in which high quality layers of AlN and AlGaN with high aluminum content are required.

The structural quality of this layer is extremely important for the properties of the entire subsequent structure. While growing the AlN layer, we used extremely high substrate temperatures in 1050°C-1200°C range. The increase in the crystalline quality of the multilayer heterostructure has been confirmed by STEM.

We demonstrate growth of thick AlN templates grown in the same run prior to the growth of DHFET structure. However, when growing AlGaN or GaN layers on such "thick" AlN layers, it is necessary to decrease the effect of strain, caused by the lattice mismatch. It was found that growing of $Al_xGa_{1-x}N$ ($0<x<0.5$) over "thick" AlN layers followed by a strained superlattice (SL) $Al_xGa_{1-x}N/Al_yGa_{1-y}N$ leads to the improvement of the morphology of the surface and structural properties, compared to growth on "thin" nucleating layers. The overall quality of GaN-based materials grown using this complex technique is competitive to MOCVD. The electron mobility in 1.5 mkm-thick Si-doped GaN layers reaches 600 - 650 $cm^2/V{\cdot}s$ at the electron concentration of $(3 - 5)\times10^{16}$ cm^{-3} (4). The dislocation density estimated by a transmission electron microscope is less than $(9 - 10)\times10^8$ cm^{-2} that is one of the best values ever reported for films directly grown by MBE on sapphire.

This heterostructure with a cap layer $Al_xGa_{1-x}N$ grown on SiC substrates allows to vary the electron sheet density and mobility in a two dimensional electron gas (2DEG) in the range of $(1.0\text{-}1.8)\times10^{13}$ cm^{-2} and 1300 – 1700 $cm^2/V{\cdot}s$, respectively, providing control of a 2DEG channel sheet resistance in the range of 230–400 Ω/sq. The barrier layer consists of 1 nm AlN and 24 nm AlGaN doped with Si. In case of using the AlN/GaN superlattice as a barrier layer, the thickness is 10 nm. The electron sheet concentration, mobility and sheet resistance for different barriers are shown in table 1.

Table 1. Electron sheet concentration, mobility and sheet resistance of heterostructures in comparison with groups

Barrier layer	$n{\cdot}10^{13}$, cm^{-2}	μ, $cm^2/V{\cdot}c$	Rs
$Al_{0.25}Ga_{0.75}N$	1.1-1.2	1300-1400	400-420
$Al_{0.3}Ga_{0.7}N$	1.5-1.6	1300-1400	290-310
$Al_{0.4}Ga_{0.6}N$	1.7-1.8	1300-1400	250-270
Superlattice	1.8-1.9	1500-1700	230-250
$In_{0.17}Al_{0.83}N$	2.3-2.4	1200-1300	210-220
$In_{0.17}Al_{0.83}N$[3]	~1.6	1600	244
$In_{0.17}Al_{0.83}N$[4]	2.27	1684	163

Distribution of sheet resistance in DHFET with a AlGaN barrier grown on a 3-inch substrate is shown in figure 1, the uniformity is about 1-2%. High quality heterostructures were grown using ammonia as well as plasma-assisted MBE on high temperature ammonia buffer layers.

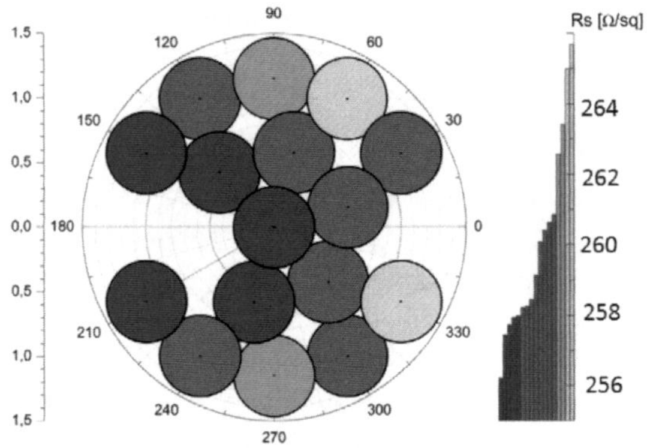

Figure 1. Sheet resistance uniformity in a heterostructure.

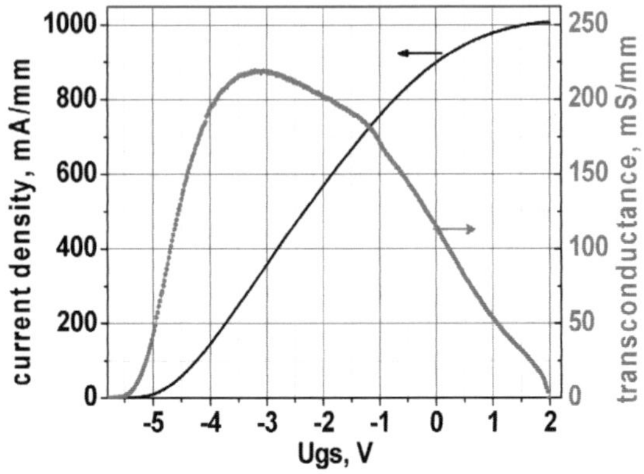

Figure 2. PCM transfer chars measured on typical AlN/AlGaN/GaN/AlGaN HS grown on a 3" SiC substrate.

An STE EB71 electron beam system and an STE RTA100 rapid thermal annealing system were used for Ti/Al/Ni/Au ohmic contacts metallization and annealing. The typical value of ohmic contact resistance corresponds to 0.4 - 0.5 Ω×mm. The STE ICPe68 and STE ICPd81 systems were used for mesa isolation etching and SiN_x passivation, correspondingly. PCM DC-parameters include current density of 1 A/mm, maximum transconductance of 200 mS/mm, and breakdown voltages over 100V after SiN_x passivation step (figure 2).

Application of this technology and DH design for growing on SiC substrates enabled one to manufacture a DHFET with a gate length of 0.5 μm for 0.03–4.0 GHz extra-broadband power amplifiers having P_{out} = 2.5 W, gain 17 – 25 dB and efficiency 30%. During aging tests all the devices have shown reliable operation for more than 3500 hours under 85 C. These device parameters confirm the high quality of the heterostructure and the chosen technological and design approaches.

Performance and reliability of nitride DHFET devices can be improved by replacing the barrier AlGaN layer by a InAlN one with In content of 17% lattice matched to GaN. At this composition stress and piezopolarisation are not present, potentially improving the stability of the heterostructure. At the same, the time electron density is even more because of higher difference in spontaneous polarization.

In the case of PAMBE, Me-rich growth conditions were used, fluxes of Al and In were equal to 0.3 and 0.1 um/hour correspondingly, and N flux corresponded to the growth rate of 0.37 um/hour. Alternatively, in the case of ammonia MBE N-rich growth conditions were used at extremely high ammonia flux of 1000 sccm. The main parameter for optimization in both cases is the substrate temperature. Increasing of substrate temperature leads to decreasing In content due to InN thermal decomposition. On the other hand, decreasing substrate temperature leads to a droplet formation and a transition to the three-dimension mode. It has been found experimentally that the optimal temperature is about 490°C for PAMBE and 620°C for ammonia MBE. Besides, a thin AlN layer (of several angstroms) between GaN and InAlN is very important for high quality DHFET growth. After the design and growth condition optimization, high quality GaN/InAlN heterostructures (with electron sheet density and mobility being in the range $(2.2–2.4)\times10^{13}$ cm^{-2} and 1200–1300 $cm^2/V\times s$, respectively) were grown using both PA-MBE and NH3-MBE at extremely high ammonia flux.

3.Summary

Extreme ammonia MBE growth regimes for as-grown AlN templates give improved quality of GaN. GaN grown on sapphire by both ammonia and PA MBE is comparable to good quality MOCVD and several times better than in conventional MBE. High quality DHFET for extra-broadband power amplifiers were fabricated. The device parameters confirm the high quality of the heterostructure and chosen technological and design approaches. High quality GaN/InAlN heterostructures were grown using PA-MBE as well as NH3-MBE at extremely high flux of ammonia.

Acknowledgements
The reported study was supported by RFBR, research project No 14-41-08048

References

[1] Webb J B, Tang H, Bardwell J A, Moisa S, Peters C and MacElwee T 2001 *J. Cryst. Growth,* **230** 584-589.
[2] Semond F, Lorenzini P, Grandjean N and Massies J 2001 *Appl. Phys. Lett* **78** 335-337.
[3] Wong M H, Wu F, Hurni C A, Choi S, Speck J S and Mishra U K 2012 *Appl. Phys. Lett.* **100** 072
[4] Ahmadi E et.al. 2014 *Semiconductor Science and Technology* **29** 045011

Cu/synthetic and impact-diamond composite heat-conducting substrates

E N Galashov, A A Yusuf and E M Mandrik

Department of Applied Physics, Scientific Research Department Novosibirsk State University, 2 Pirogov str., Novosibirsk 630090, Russia

E-mail: chessplanet86@yandex.ru

Abstract. Composite material with high thermal conductivity was obtained by the method of thermal sintering of diamonds (synthetic and impact) and copper powder and by further hot isostatic pressing. The content of diamond (the particle size is 20 – 250 μm) in the diamond copper composite was 30 – 45 weight percent. The coefficient of thermal conductivity of copper diamond composite materials based on synthetic diamonds was measured to be 700 – 750 $W \cdot m^{-1} \cdot K^{-1}$. The coefficient of thermal conductivity of copper diamond composite materials based on impact diamonds was measured to be 850 – 900 $W \cdot m^{-1} \cdot K^{-1}$. The coefficient of thermal expansion (CTE) was measured to be 5.5 – 6.5 $\cdot 10^{-6}$/°C for synthetic diamonds and 6.1 – 6.5 $\cdot 10^{-6}$/°C for impact diamonds. The obtained copper diamond composite materials are promising to be used as crystal holders for semiconductor crystals in THz and microwave devices.

1. Introduction

The technical capabilities of modern devices (locating systems, military and space applications, etc. [1-3]) lately are increasingly demanding higher and higher frequencies. However, the generation of high frequencies is accompanied by the release of an enormous amount of heat, which can adversely affect the performance of terahertz and ultra-high-frequency electronics. These electronics can be damaged by overheating. The power and the life of chips used in the terahertz and ultra-high-frequency electronics can be increased by removing heat from the crystal by raising the thermal conductivity of the crystal holder. Creating a material that can cope with this problem is a very urgent task. It should be noted that in Russia highly conducting crystal holders with thermal conductivity higher than 250 $W \cdot m^{-1} \cdot K^{-1}$ and having a coefficient of thermal expansion corresponding to CTE of expansion crystals (GaAs, ZnSe, GaN) are not commercially available.

The most heat-conducting solid material is diamond. The thermal conductivity of diamonds reaches 2300 $W \cdot m^{-1} \cdot K^{-1}$ [4]. However, the main disadvantage of diamonds is low CTE $1 \cdot 10^{-6}$/°C, whereas the base semiconductor crystals have a CTE of between 4 and $8 \cdot 10^{-6}$/°C.

We described earlier the major problems associated with the synthesis of highly conducting metal-diamond composites in our work [5]. The thermal conductivity of the substrates based on synthetic diamonds coated with tungsten carbide by the method of vapor deposition of tungsten carbonyl was in a region of 450 – 650 $W \cdot m^{-1} \cdot K^{-1}$. The coefficient of thermal expansion CTE was in the range of 5.5 – 7.5 10^{-6}/ °C.

In this paper, we propose a three-step process for the synthesis of the heat-conducting metal-diamond composites based on synthetic diamonds and the impact diamonds by the methods of thermal sintering and hot-isostatic pressing.

2. Fabrication

The synthesis of conductive crystal holders (the substrates) is carried out in 3 stages: metallization of diamonds, thermal sintering of substrates and hot isostatic pressing of substrates. To avoid the formation of pores at synthesis of metal-diamond substrates (diamonds are difficult to wet by molten metals) [6] and to avoid processes of transformation of diamond into graphite at the high-temperature synthesis it is needed to cover diamond by metals such as tungsten, molybdenum, chromium, etc. These metals have a high melting point and can prevent the process of turning diamond into graphite at the formation of diamond/heat-conducting metal substrates.

2.1. Experimental setup for coated of diamonds

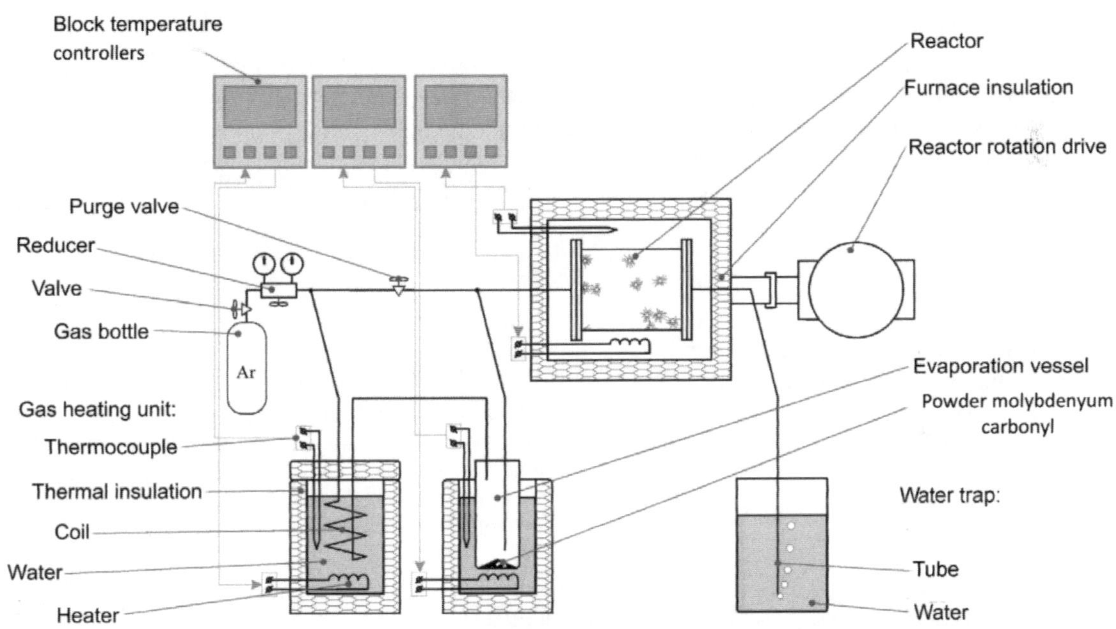

Figure 1. Diagram of the apparatus for applying coatings of molybdenum.

The diamonds sized 20 – 230 µm for synthetic ones and 50 – 100 µm for the impact ones are covered by molybdenum coating by the method of vapor deposition of molybdenum carbonyl (figure 1). The coating composition may vary within Mo-Mo_2C-MoC. The diamond coating process is described as follows: argon passes through a tube to a coil heated by a water heater. Heated argon enters an evaporation vessel, where gas and vapor of molybdenum carbonyl are mixed. Then, argon with vapor of molybdenum carbonyl moves into a reactor where diamonds are sited. The surface of diamonds is cleaned from dirt due to pre-heating. The reactor with diamonds is rotated by a reactor rotation drive, and diamonds are coated by the composition of Mo-Mo_2C-MoC. Then, unnecessary gas residues are removed into a water trap.

The use was made of initial synthetic diamonds (production of Guangdong Shuangling Diamong Industry Pty. Ltd.) and impact diamonds (Deposit Popigai impact structure, the Krasnoyarsk Territory, Yakutia).

Figure 2. Covered synthetic diamonds.

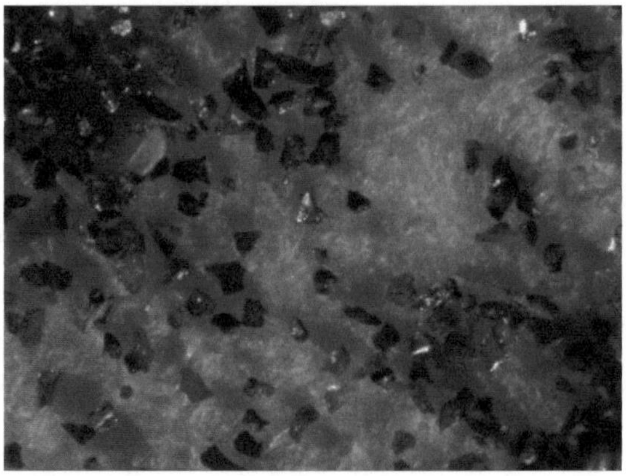

Figure 3. Covered impact diamonds.

2.2. Thermal sintering of substrates

Copper powder and diamonds are mixed in an agate mortar. The mixture of diamonds and copper is pressed under the pressure of 50 tons in a steel mold (the inner diameter is 25 mm) by a Metallkraft WPP 100 MBK hydraulic press (OPTIMUM, Germany). Then the pressed substrate is placed in a graphite mold with punches. The mold is placed in a vacuum furnace. Sintering of copper diamond substrates is occurred at 900 °C and pressure of 10^{-3} Pa for 1 hour.

The substrates based on synthetic and impact diamonds were obtained by thermal sintering of substrates (figure 4). The content of diamonds in the copper diamond composite was 30 – 45 weight percent.

2.3. Hot isostatic pressing of substrates

The sintering copper diamond composite substrate is placed in a graphite mold with punches (figure 5). This mold is placed in a hot isostatic press AIP6 – 30H. The substrates are melted at the temperature of 1050 °C and under the pressure of 200 MPa for 1 hour. The pressure was created by argon. The thickness of the melted copper diamond substrates is in the range of 0.5 – 1 mm. The surface of the substrates is smooth.

Figure 4. Thermal synthesized Cu/Diamond substrate.

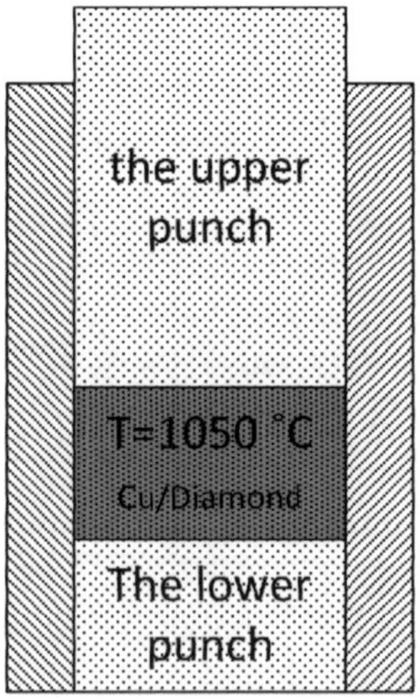

Figure 5. Press mold.

Journal of Physics: Conference Series **690** (2016) 012043 doi:10.1088/1742-6596/690/1/012043

3. Results

The X-ray diffraction patterns (XRD) were recorded using a Thermo Scientific ARL X'TRA diffractometer with an angle range of $2\theta = 15 - 80°$ at a step size of 0.05° and a time of 5 s.

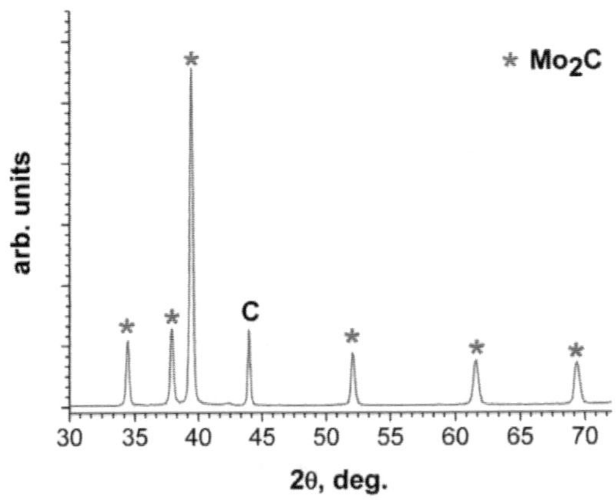

Figure 6. XRD of covered synthetic diamonds.

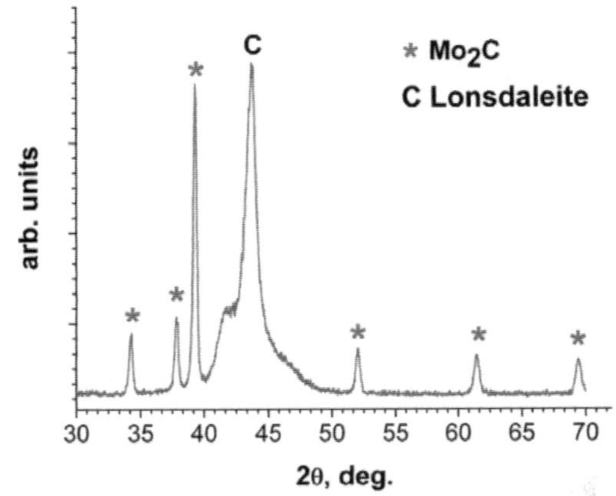

Figure 7. XRD of covered impact diamonds.

The X-ray diffraction analysis of synthetic (figure 6) and impact (figure 7) coated diamonds showed that synthetic coated diamonds have a diamond phase (cubic lattice Fd3m, *a = 3.559 Å*) and a phase of coating Mo_2C (a hexagonal lattice P63/mmc, *a = b = 3.006 Å c = 4.733 Å*). The impact coated diamonds also have a phase of coating Mo_2C, but the phase of diamonds is a lonsdaleite phase (a hexagonal lattice P63/mmc).

The electron microscopy analysis by a TM-1000 electron microscope (Hitachi Science Systems Ltd.) determined that the covered synthetic diamonds (figure 8) have the developed morphology. The thickness of the coating of synthetic diamonds was estimated. It is located in the range of 0.1 – 0.3 μm (figure 8 left). The surface of impact diamonds has a porous structure, which is the reason of strong adhesion between the coating and the surface of diamonds.

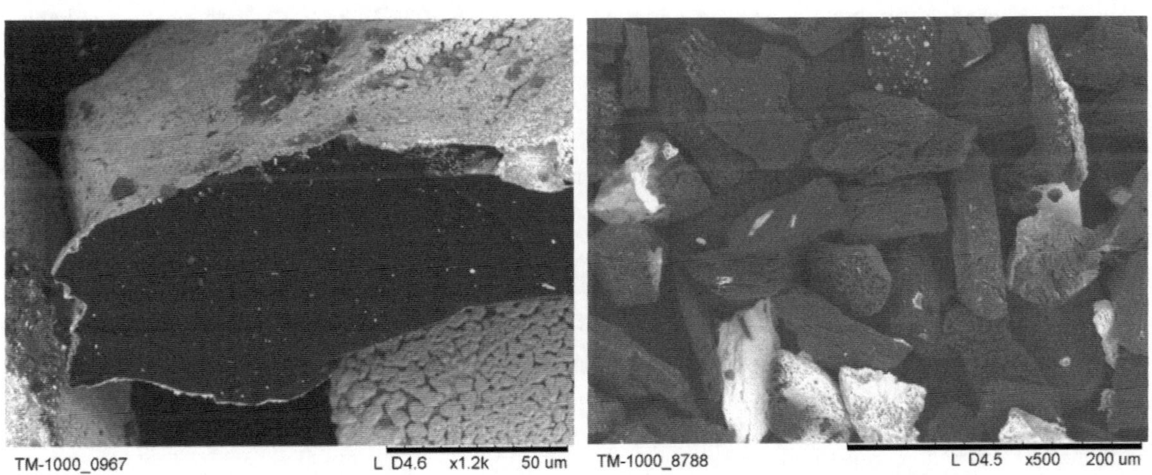

Figure 8. Electron microscopy of coated synthetic (left) and impact (right) diamonds.

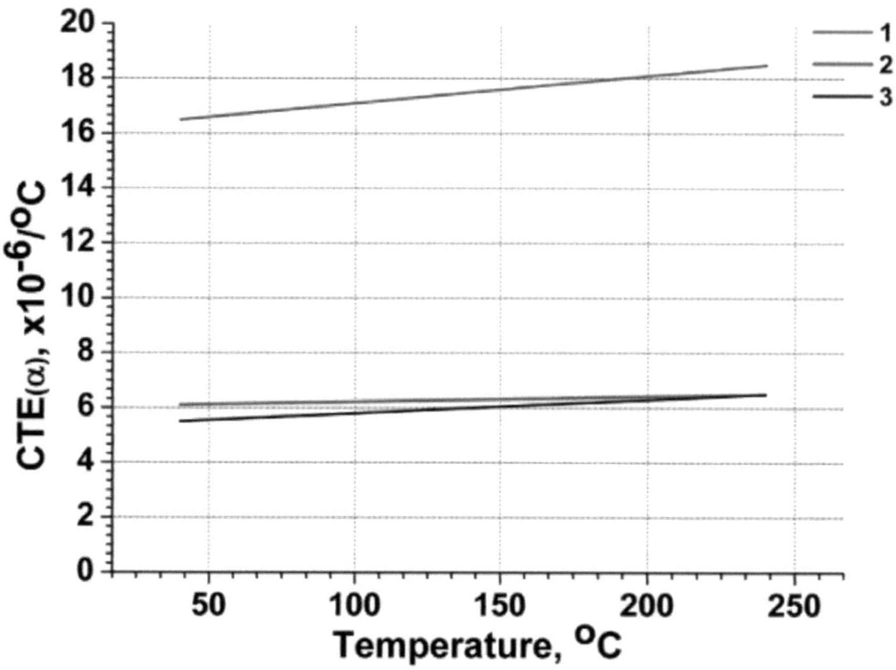

0

Figure 9. CTE of the samples. 1 – copper, 2 – copper with impact diamonds, 3 – copper with synthetic diamonds.

The thermal conductivity was measured by a Netzsch LFA 447 laser flash apparatus (Germany) using the laser beam pulse heating method. The thermal conductivity of copper diamond composites based on synthetic diamonds is in the range of 700 – 750 $W \cdot m^{-1} \cdot K^{-1}$. The coefficient of thermal expansion (figure 9) was measured by a Netzsch TMA 402 F3 Hyperion thermomechanical analyzer in the temperature range of 293 – 600 K. The CTE parameter of the copper diamond composites based on synthetic diamonds is in the range of $6.4 - 6.5 \cdot 10^{-6}$/ °C. The thermal conductivity of copper substrate was measured to be 400 $W \cdot m^{-1} \cdot K^{-1}$. The coefficient of thermal expansion of copper substrate was measured to be in range of 16.5 – 18.3 10^{-6}/°C. The thermal conductivity of copper diamond composites based on synthetic diamonds is in the range of 850 – 900 $W \cdot m^{-1} \cdot K^{-1}$. Perhaps, this is due to the fact that the thermal conductivity of impact diamonds is much higher than that of synthetic diamonds. Also, the structure of the impact diamond lattice and the Mo_2C coating are identical (a hexagonal lattice P63/mmc), whereas the synthetic diamond structure (a cubic lattice Fd3m) and the coating structure (a hexagonal lattice P63/mmc) are different. The CTE parameter of the copper diamond composites based on impact diamonds is in the range of $6.1 - 6.5 \cdot 10^{-6}$/ °C.

4. Conclusions

The technological route for the fabrication of copper diamonds composites with a smooth surface is designed and demonstrated in the present study. The smooth surface of substrates has no significant irregularities and allows creating a compact contact between the substrate surface and the semiconductor material. The substrates can provide the efficient heat dissipation from electronic devices to prevent silicon and other semiconducting materials from overheating. The high thermal conductivity is the specific feature of copper diamond composite based on impact diamonds compared with copper diamond composite based on synthetic diamonds.

5. Acknowledgments

This work was supported in part by the Government of the Russian Federation (contract no. 14.Z50.31.0019).

Journal of Physics: Conference Series **690** (2016) 012043 doi:10.1088/1742-6596/690/1/012043

References

[1] Maiskaya M 2013 Towards submillimeter wavelengths *Electronics* **6** 44–59

[2] Maiskaya M 2011 Development of the terahertz gap. Semiconductors invade the submillimeter range *Electronics* **6** 74–87

[3] Chekrygin I, Chekrygin A and Chekrygin V 2010 Terahertz technology in medicine *20th Int. Crimean Conference "Microwave & Telecommunication Technology"* 1115–1116

[4] Goela J S and Shinde S L 2006 High Thermal Conductivity Materials *Springer* 271

[5] Galashov E N, Yusuf A A and Mandrik E M 2015 Cu/Diamond composite heat-conducting shims *BOOK of ABSTRACTS 2nd International School and Conference on Optoelectronics, Photonics, Engineering and Nanostructures «Saint-Petersburg OPEN 2015»* 363

[6] Mikhailov A, Kozlov M, Kagarmanov N and Balandin P 1974 Study to improve the wear resistance of the matrix of diamond drilling tools *Collection N.T.* **39** 211

Institute of Physics
Dirac House, Temple Back
Bristol BS1 6BE UK

ISSN: 1742-6588
ISBN 978-1-5108-2235-1

9 781510 822351

CURRAN ASSOCIATES INC.
proceedings.com

27th Micromechanics and Microsystems Europe Workshop 2016

Journal of Physics: Conference Series Volume 757

Cork, Ireland
28 – 30 August 2016

IOP
Institute of Physics